Principles and Practice of Analytical Chemistry

Fourth Edition

F. W. FIFIELD and D. KEALEY
Kingston University

BLACKIE ACADEMIC & PROFESSIONAL
An Imprint of Chapman & Hall

London · Glasgow · Weinheim · New York · Tokyo · Melbourne · Madras

Published by
Blackie Academic & Professional, an imprint of Chapman & Hall
Wester Cleddens Road, Bishopbriggs, Glasgow G64 2NZ

Chapman & Hall, 2–6 Boundary Row, London SE1 8HN, UK

Blackie Academic & Professional, Wester Cleddens Road, Bishopbriggs, Glasgow G64 2NZ, UK

Chapman & Hall GmbH, Pappelallee 3, 69469 Weinheim, Germany

Chapman & Hall USA, 115 Fifth Avenue, Fourth Floor, New York NY 10003, USA

Chapman & Hall Japan, ITP-Japan, Kyowa Building, 3F, 2-2-1 Hirakawacho, Chiyoda-ku, Tokyo 102, Japan

DA Book (Aust.) Pty Ltd, 648 Whitehorse Road, Mitcham 3132, Victoria, Australia

Chapman & Hall India, R. Seshadri, 32 Second Main Road, CIT East, Madras 600 035, India

First edition 1975
Second edition 1983
Third edition 1990
This edition 1995

© 1995 Chapman & Hall

Typeset in 10/12 pt Times by AFS Image Setters Ltd, Glasgow
Printed in Great Britain by TJ Press (Padstow) Ltd., Padstow, Cornwall

ISBN 0 7514 0226 5

A catalogue record for this book is available from the British Library
Library of Congress Catalog Card Number: 94-74246

∞ Printed on acid-free text paper, manufactured in accordance with ANSI/NISO Z39.48-1992 (Permanence of Paper).

Contents

Preface   ix

Acknowledgements   xi

CHAPTER 1   INTRODUCTION   1

The scope of analytical chemistry. The function of analytical chemistry.
Analytical problems and their solution. The nature of analytical
methods. Trends in analytical methods and procedures. Glossary of
terms.

CHAPTER 2   THE ASSESSMENT OF ANALYTICAL DATA   14

2.1   Definitions and basic concepts   14

2.2   The nature and origin of errors   16

2.3   The evaluation of results and methods   18

The reliability of measurements. The analysis of data. The application
of statistical tests. Limits of detection. Quality control charts.
Standardization of analytical methods. Chemometrics.

CHAPTER 3   pH, COMPLEXATION AND SOLUBILITY
EQUILIBRIA   37

3.1   Chemical reactions in solution   38

Equilibrium constants. Kinetic factors in equilibria.

3.2   Solvents in analytical chemistry   42

Ionizing solvents. Non-ionizing solvents.

3.3   Acid-base equilibria   43

Weak acid and weak base equilibria. Buffers and pH control. The pH
of salt solutions.

3.4   Complexation equilibria   50

The formation of complexes in solution. The chelate effect.

3.5   Solubility equilibria   53

Solubility products.

CHAPTER 4   SEPARATION TECHNIQUES   55

4.1   Solvent extraction   56

Efficiency of extraction. Selectivity of extraction. Extraction systems.
Extraction of uncharged metal chelates. Methods of extraction.
Applications of solvent extraction.

4.2   Chromatography   75

4.2.1 Gas chromatography. 4.2.2 High performance liquid chromatography. 4.2.3 Supercritical fluid chromatography. 4.2.4 Thin-layer
chromatography. 4.2.5 Ion-exchange chromatography. 4.2.6 Gel-
permeation chromatography.

4.3   Electrophoresis   164

Factors affecting ionic migration. Effect of temperature, pH and ionic strength. Electroosmosis. Supporting medium. Detection of separated components. Applications of tradional zone electrophoresis. High performance capillary electrophoresis.

CHAPTER 5   TITRIMETRY AND GRAVIMETRY   184

5.1   Titrimetry   184

Definitions. Titrimetric reactions. Acid–base titrations. Applications of acid-base titrations. Redox titrations. Applications of redox titrations. Complexometric titrations. EDTA. Applications of EDTA titrations. Titrations with complexing agents other than EDTA. Precipitation titrations.

5.2   Gravimetry   211

Precipitation reactions. Practical gravimetric procedures. Applications of gravimetry.

CHAPTER 6   ELECTROCHEMICAL TECHNIQUES   223

6.1   Potentiometry   227

Electrode systems. Direct potentiometric measurements. Potentiometric titrations. Null-point potentiometry. Applications of potentiometry.

6.2   Polarography, stripping voltammetry and amperometric techniques   243

Diffusion currents. Half-wave potentials. Characteristics of the DME. Quantitative analysis. Modes of operation used in polarography. The dissolved oxygen electrode and biochemical enzyme sensors. Amperometric titrations. Applications of polarography and amperometric titrations.

6.3   Electrogravimetry and coulometry   257

Coulometry. Coulometry at constant potential. Coulometric titrations. Applications of coulometric titrations.

6.4   Conductometric titrations   261

Ionic conductances.

CHAPTER 7   AN INTRODUCTION TO ANALYTICAL SPECTROMETRY   267

Electromagnetic radiation. Atomic and molecular energy. The absorption and emission of electromagnetic radiation. The complexity of spectra and the intensity of spectral lines. Analytical spectrometry. Instrumentation.

CHAPTER 8   ATOMIC SPECTROMETRY   282

8.1   Arc/spark atomic (optical) emission spectrometry   287

Instrumentation. Sample preparation. Qualitative and quantitative analysis. Interferences and errors associated with the excitation process. Applications of arc/spark emission spectrometry.

8.2   Glow discharge atomic emission spectrometry   294

Instrumentation. Applications.

8.3 Plasma emission spectrometry 296

Instrumentation. Sample introduction for plasma sources. Analytical measurements. Applications of plasma emission spectrometry.

8.4 Inductively coupled plasma–mass spectrometry (ICP–MS) 304

Principles. Instrumentation. Applications.

8.5 Flame emission spectrometry 308

Instrumentation. Flame characteristics. Flame processes. Emission spectra. Quantitative measurements and interferences. Applications of flame photometry and flame atomic emission spectrometry.

8.6 Atomic absorption spectrometry 317

Absorption of characteristic radiation. Instrumentation. Sample vaporization. Quantitative measurements and interferences. Atomic fluorescence spectrometry.

8.7 X-ray emission spectrometry 331

X-ray processes. Instrumentation. Applications of X-ray emission spectrometry.

CHAPTER 9 MOLECULAR SPECTROMETRY 351

9.1 Visible and ultraviolet spectrometry 361

Polyatomic organic molecules. Metal complexes. Qualitative analysis – the identification of structural features. Quantitative analysis – absorptiometry. Choice of colorimetric and spectrophotometric procedures. Fluorimetry. Applications of UV/visible spectrometry and fluorimetry.

9.2 Infrared spectrometry 377

Diatomic molecules. Polyatomic molecules. Characteristic vibration frequencies. Factors affecting group frequencies. Qualitative analysis – the identification of structural features. Quantitative analysis. Sampling procedures. Applications of infrared spectrometry.

9.3 Nuclear magnetic resonance spectrometry (nmr) 393

Instrumentation. The nmr process. Chemical shift. Spin–spin coupling. Carbon-13 nmr. Pulsed Fourier transform nmr (ft-nmr). Quantitative analysis – the identification of structural features. Quantitative analysis. Applications of nmr spectrometry.

9.4 Mass spectrometry 425

Instrumentation. Principle of mass spectrometry. Characteristics and interpretation of molecular mass spectra. Applications of mass spectrometry.

9.5 Spectrometric identification of organic compounds 436

CHAPTER 10 RADIOCHEMICAL METHODS IN ANALYSIS 447

10.1 Nuclear structure and nuclear reactions 448

Decay reactions. The kinetics of decay reactions. Bombardment reactions and the growth of radioactivity.

10.2 Instrumentation and measurement of radioactivity 455

Radiation detectors. Some important electronic circuits. The statistics of radioactive measurements.

10.3   Analytical uses of radionucleides   465

Chemical pathway studies. Radioisotope dilution methods. Radioimmunoassay. Radioactivation analysis. Environmental monitoring.

CHAPTER 11   THERMAL TECHNIQUES,   475
11.1   Thermogravimetry   477

Instrumentation. Applications of TG.

11.2   Differential thermal analysis (DTA)   482

Instrumentation. Applications of DTA

11.3   Differential scanning calorimetry (DSC)   489

Instrumentation. Applications of DSC. DTA and DSC.

11.4   Thermomechanical analysis (TMA) and dynamic
          mechanical analysis (DMA)   493

Instrumentation. Applications of TMA. Dynamic mechanical analysis.

11.5   Pyrolysis – gas chromatography   497

Instrumentation.

**CHAPTER 12   OVERALL ANALYTICAL PROCEDURES AND
               THEIR AUTOMATION**   504
12.1   Sampling and sample pretreatment   504

Representative samples and sample storage. Sample concentration and
clean-up: solid phase extraction.

12.2   Examples of analytical problems and procedures   509

1: Evaluation of methods for the determination of fluoride in water
samples. 2: Analysis of a competitive product. 3: The assessment of the
heavy metal pollution in a river estuary. 4: The analysis of
hydrocarbon products in a catalytic reforming study.

12.3   The automation of analytical procedures.   519

The automation of repetitive analysis. Constant monitoring and on
line analysis. Laboratory robotics.

**CHAPTER 13   THE ROLE OF COMPUTERS AND MICRO-
               PROCESSORS IN ANALYTICAL CHEMISTRY**   528
13.1   Introduction   528

Instrument optimization. Data recording and storage. Data processing
and data analysis (chemometrics). Laboratory management. Expert
systems.

13.2   Computers and microprocessors   534

Mini- and microcomputers. Microprocessors.

13.3   Instrument-computer interfaces   537
13.4   The scope of microprocessor control and computers in
          analytical laboratories   542

1. A microprocessor-controlled potentiometric titrator.
2. An infrared spectrometer interfaced to a dedicated microcomputer.
3. A computing integrator for chromatographic analysis.
4. A microprocessor-based X-ray or γ-ray spectrometer.

Index   551

Preface to the fourth edition

There have been significant advances in both analytical instrumentation and computerised data handling during the five years since the third edition was published in 1990.

Windows-based computer software is now widely available for instrument control and real-time data processing and the use of laboratory information and management systems (LIMS) has become commonplace. Whilst most analytical techniques have undergone steady improvements in instrument design, high-performance capillary electrophoresis (HPCE or CE) and two-dimensional nuclear magnetic resonance spectrometry (2D-NMR) have developed into major forces in separation science and structural analysis respectively. The powerful and versatile separation technique of CE promises to rival high-performance liquid chromatography, particularly in the separation of low levels of substances of biological interest. The spectral information provided by various modes of 2D-NMR is enabling far more complex molecules to be studied than hitherto. The electrophoresis section of chapter 3 and the NMR section of chapter 9 have therefore been considerably expanded in the fourth edition along with a revision of aspects of atomic spectrometry (chapter 8). New material has been included on fluorescence spectrometry (chapter 9), the use of Kováts Retention Indices in gas chromatography (chapter 3) and solid phase extraction for sample cleanup and concentration (chapter 12). Additions to high performance liquid chromatography (chapter 3) reflect the growing importance of chiral stationary phases, solvent optimization and pH control, continuous regeneration cartridges for ion chromatography and HPLC-MS. Throughout the book there have been numerous other changes and additions to enhance clarity and presentation including a number of new or improved diagrams and some additional worked examples on the statistical assessment of analytical data (chapter 2).

The earlier editions have been widely used by both undergraduate and postgraduate students of analytical chemistry, and the fourth edition should continue to provide a sound basis for this readership. Industrial trainees and those in related disciplines who require a knowledge of analytical chemistry will find this a suitable text for reading and reference purposes.

We continue to benefit from discussions with many of our colleagues at Kingston University, and particularly with Mr. P. J. Haines whose knowledge of thermal techniques has proved invaluable. The Publisher's reviewers and users of the book continue to be a source of helpful and much appreciated comments.

DK
FWF

Acknowledgements

The following figures are reproduced with permission of the publishers:

Figure 7.8 from Christian and O'Reilly, *Instrumental Analysis*, 2nd edn., (1986) by permission of Allyn and Bacon, U.K.

Figure 10.17 from Cyclic GMP RIA Kit, Product Information 1976, by permission of Amersham International, U.K.

Figures 8.14 and 8.15 from Date and Gray, *Applications of Inductively Coupled Plasma Mass Spectrometry* (1989); figure 2.7 from Kealey, *Experiments in Modern Analytical Chemistry* (1986); by permission of Blackie, U.K.

Figure 8.24 from Manahan, *Quantitative Chemical Analysis* (1986) by permission of Brookes Cole, U.K.

Figures 8.27 and 8.28(a) and (b) from Allmand and Jagger, *Electron Beam X-ray Microanalysis Systems,* by permission of Cambridge Instruments Ltd., U.K.

Figures 4.20, 4.24(a) and (c) and 4.25 from Braithwaite and Smith, *Chromatographic Methods* (1985); figures 11.2, 11.3, 11.4, 11.10 and 11.17 from Brown, *Introduction to Thermal Analysis* (1988); by permission of Chapman and Hall.

Figures 11.23, 11.25 and 11.26 reprinted from Irwin, *Analytical Pyrolysis* (1982) by courtesy of Marcel Dekker Inc. NY.

Figure 4.26(b) from Euston and Glatz, *A new Hplc Solvent Delivery System*, Techn. Note 88–2 (1988) by permission of Hewlett-Packard, Waldbronn, Germany.

Figures 4.10, 4.16, 6.4, 6.11(a) and (b), 6.12(a) and (b), 9.1, 9.4 and 9.50(a) and (b) from *Principles of Instrumental Analysis*, 2nd edn., by Douglas Skoog and Donald West, Copyright © 1980 by Saunders College/Holt, Rinehart and Winston, Copyright © 1971 by Holt, Rinehart and Winston. Reprinted by permission of Holt, Rinehart and Winston, CBS College Publishing;

figures 9.36, 9.37, 9.38, 9.39 and problems 9.6, 9.7 and 9.8 from *Introduction to Spectroscopy* by Donald L. Pavia *et al.*, Copyright © 1979 by W. B. Saunders Company. Reprinted by permission of W. B. Saunders Company, CBS College Publishing.

Figure 8.39 from *X-ray Microanalysis of Elements in Biological Tissue*, by permission of Link Systems, U.K.

Figure 4.24(b) from Williams and Howe, *Principles of Organic Mass Spectrometry* (1972), by permission of McGraw-Hill Book Co. Ltd., U.K.

Figure 9.2(b) from *50XC/55XC FTIR Spectrometer Brochure*, by permission of Nicolet Analytical Instruments, Madison, Wisconsin, U.S.A.

Figure 8.38 from Walinga, *Advantages and Limitations of Energy Dispersive X-ray Analysis*, Phillips Bulletin (1972) by permission of NV Philips Gloeilampenfabrieken, Netherlands.

Figure 8.25 from Brown and Dymott, *The use of platform atomisation and matrix modification as methods of interference control in graphite furnace analysis*, by permission of Philips Scientific and Analytical Equipment.

Figures 11.21 and 11.24 from Frearson and Haskins, *Chromatography and Analysis*, Issue 7, (1989) by permission of RGC Publications.

Figures 4.14, 4.27, 9.2(a), 11.11, 11.20, 12.1 and 12.6(b) from *Instrumental Methods of Analysis*, 7th edn., H. H. Willard, L. L. Merritt, J. A. Dean and F. A. Settle, © 1988 Wadsworth, Inc. Reprinted by permission of the publisher.

Figures 4.26(c), 4.31 and 13.3 from Snyder and Kirkland, *Introduction to Modern Liquid Chromatography*, 2nd edn., (1979); 9.40(a), (b) and (c) from Cooper, *Spectroscopic Techniques for Organic Chemists* (1980); 9.45 from Millard, *Quantitative Mass Spectrometry* (1978); 4.13, 4.14, 4.26(a), 4.28, 4.29(a), 4.32, 4.33, 4.36 and 4.38 from Smith, *Gas and Liquid Chromatography in Analytical Chemistry* (1988); figures 4.35 and 13.2 from **Berridge, *Techniques for the Automated Optimisation of Hplc Separations*** (1985) reproduced by permission of John Wiley and Sons Limited; 11.1, 11.5, 11.6, 11.12, 11.13, 11.14, 11.18 and 11.19 from Wendlandt, *Thermal Analysis*, 3rd edn., (1986); reprinted by permission of John Wiley and Sons Inc., all rights reserved.

Figure 10.16 from Chapman, *Chemistry in Britain* **15** (1979) 9, by permission of the Royal Society of Chemistry.

Figure 6.4 is reprinted courtesy of Orion Research Incorporated, Cambridge, Mass., U.S.A. 'ORION' is a registered trademark of Orion Research Incorporated.

Chapter 1

Introduction

Is there any iron in moon dust? How much aspirin is there in a headache tablet? What trace metals are there in a tin of tuna fish? What is the purity and chemical structure of a newly prepared compound? These and a host of other questions concerning the composition and structure of matter fall within the realms of analytical chemistry. The answers may be given by simple chemical tests or by the use of costly and complex instrumentation. The techniques and methods employed and the problems encountered are so varied as to cut right across the traditional divisions of inorganic, organic and physical chemistry as well as embracing aspects of such areas as biochemistry, physics, engineering and economics. Analytical chemistry is therefore a subject which is broad in its scope whilst requiring a specialist and disciplined approach. An enquiring and critical mind, a keen sense of observation and the ability to pay scrupulous attention to detail are desirable characteristics in anyone seeking to become proficient in the subject. However, it is becoming increasingly recognized that the role of the analytical chemist is not to be tied to a bench using a burette and balance, but to become involved in the broader aspects of the analytical problems which are encountered. Thus, discussions with scientific and commercial colleagues, customers and other interested parties, together with on-site visits can greatly assist in the choice of method and the interpretation of analytical data thereby minimizing the expenditure of time, effort and money.

The purpose of this book is to provide a basic understanding of the principles, instrumentation and applications of chemical analysis as it is currently practised. The amount of space devoted to each technique is based upon its application in industry as determined in a national survey of analytical laboratories. Some little used techniques have been omitted altogether. The presentation is designed to aid rapid assimilation by emphasizing unifying themes common to groups of techniques and by including short summaries at the beginning of each section.

1

THE SCOPE OF ANALYTICAL CHEMISTRY

Analytical chemistry has bounds which are amongst the widest of any technological discipline. An analyst must be able to design, carry out, and interpret his measurements within the context of the fundamental technological problem with which he is presented. The selection and utilization of suitable chemical procedures requires a wide knowledge of chemistry, whilst familiarity with and the ability to operate a varied range of instruments is essential. Finally, an analyst must have a sound knowledge of the statistical treatment of experimental data to enable him to gauge the meaning and reliability of the results that he obtains.

When an examination is restricted to the identification of one or more constituents of a sample, it is known as *qualitative analysis*, while an examination to determine how much of a particular species is present constitutes a *quantitative analysis*. Sometimes information concerning the spatial arrangement of atoms in a molecule or crystalline compound is required or confirmation of the presence or position of certain organic functional groups is sought. Such examinations are described as *structural analysis* and they may be considered as more detailed forms of analysis. Any species that are the subjects of either qualitative or quantitative analysis are known as *analytes*.

There is much in common between the techniques and methods used in qualitative and quantitative analysis. In both cases, a sample is prepared for analysis by physical and chemical 'conditioning', and then a measurement of some property related to the analyte is made. It is in the degree of control over the relation between a measurement and the amount of analyte present that the major difference lies. For a qualitative analysis it is sufficient to be able to apply a test which has a known sensitivity limit so that negative and positive results may be seen in the right perspective. Where a quantitative analysis is made, however, the relation between measurement and analyte must obey a strict and measurable proportionality; only then can the amount of analyte in the sample be derived from the measurement. To maintain this proportionality it is generally essential that all reactions used in the preparation of a sample for measurement are controlled and reproducible and that the conditions of measurement remain constant for all similar measurements. A premium is also placed upon careful calibration of the methods used in a quantitative analysis. These aspects of chemical analysis are a major preoccupation of the analyst.

THE FUNCTION OF ANALYTICAL CHEMISTRY

Chemical analysis is an indispensable servant of modern technology whilst it partly depends on that modern technology for its operation. The two have in fact developed hand in hand. From the earliest days of quantitative chemistry in the latter part of the eighteenth century, chemical analysis has provided an important basis for chemical development. For example, the combustion studies of La Voisier and the atomic theory proposed by Dalton had their bases in quantitative analytical evidence. The transistor provides a more recent example of an invention which would have been almost impossible to develop without sensitive and accurate chemical analysis. This example is particularly interesting as it illustrates the synergic development that is so frequently observed in differing fields. Having underpinned the development of the transistor, analytical instrumentation now makes extremely wide use of it. In modern technology, it is impossible to over-estimate the importance of analysis. Some of the major areas of application are listed below.

(a) *Fundamental Research*
The first steps in unravelling the details of an unknown system frequently involve the identification of its constituents by qualitative chemical analysis. Follow up investigations usually require structural information and quantita-tive measurements. This pattern appears in such diverse areas as the formula-tion of new drugs, the examination of meteorites, and studies on the results of heavy ion bombardment by nuclear physicists.

(b) *Product Development*
The design and development of a new product will often depend upon establishing a link between its chemical composition and its physical properties or performance. Typical examples are the development of alloys and of polymer composites.

(c) *Product Quality Control*
Most manufacturing industries require a uniform product quality. To ensure that this requirement is met, both raw materials and finished products are subjected to extensive chemical analysis. On the one hand, the necessary constituents must be kept at the optimum levels, while on the other impurities such as poisons in foodstuffs must be kept below the maximum allowed by law.

(d) *Monitoring and Control of Pollutants*
Residual heavy metals and organo-chlorine pesticides represent two well known pollution problems. Sensitive and accurate analysis is required to enable the distribution and level of a pollutant in the environment to be assessed and routine chemical analysis is important in the control of industrial effluents.

(e) *Assay*
In commercial dealings with raw materials such as ores, the value of the ore is set by its metal content. Large amounts of material are often involved, so that taken overall small differences in concentration can be of considerable commercial significance. Accurate and reliable chemical analysis is thus essential.

(f) *Medical and Clinical Studies*
The level of various elements and compounds in body fluids are important indicators of physiological disorders. A high sugar content in urine indicating a diabetic condition and lead in blood are probably the most well-known examples.

ANALYTICAL PROBLEMS AND THEIR SOLUTION

The solutions of all analytical problems, both qualitative and quantitative, follow the same basic pattern. This may be described under seven general headings.

(1) *Choice of Method*
The selection of the method of analysis is a vital step in the solution of an analytical problem. A choice cannot be made until the overall problem is defined, and where possible a decision should be taken by the client and the analyst in consultation. Inevitably, in the method selected, a compromise has to be reached between the sensitivity, precision and accuracy desired of the results and the costs involved. For example, X-ray fluorescence spectrometry may provide rapid but rather imprecise quantitative results in a trace element problem. Atomic absorption spectrophotometry, on the other hand, will supply more precise data, but at the expense of more time consuming chemical manipulations.

(2) *Sampling*
Correct sampling is the cornerstone of reliable analysis. The analyst must decide in conjunction with his technological colleagues how, where, and when a sample should be taken so as to be truly representative of the parameter that is to be measured.

(3) *Preliminary Sample Treatment*
For quantitative analysis, the amount of sample taken is usually measured by mass or volume. Where a homogeneous sample already exists, it may be subdivided without further treatment. With many solids such as ores, however, crushing and mixing are a prior requirement. The sample often needs additional preparation for analysis, such as drying, ignition and dissolution.

(4) *Separations*

A large proportion of analytical measurements is subject to interference from other constituents of the sample. Newer methods increasingly employ instrumental techniques to distinguish between analyte and interference signals. However, such distinction is not always possible and sometimes a selective chemical reaction can be used to mask the interference. If this approach fails, the separation of the analyte from the interfering component will become necessary. Where quantitative measurements are to be made, separations must also be quantitative or give a known recovery of the analyte.

(5) *Final Measurement*

This step is often the quickest and easiest of the seven but can only be as reliable as the preceding stages. The fundamental necessity is a known proportionality between the magnitude of the measurement and the amount of analyte present. A wide variety of parameters may be measured (table 1.1).

(6) *Method Validation*

It is pointless carrying out the analysis unless the results obtained are known to be meaningful. This can only be ensured by proper validation of the method before use and subsequent monitoring of its performance. The analysis of validated standards is the most satisfactory approach. Validated standards have been extensively analysed by a variety of methods, and an accepted value for the appropriate analyte obtained. A standard should be selected with a matrix similar to that of the sample. In order to ensure continued accurate analysis, standards must be re-analysed at regular intervals.

(7) *The Assessment of Results*

Results obtained from an analysis must be assessed by the appropriate statistical methods and their meaning considered in the light of the original problem.

THE NATURE OF ANALYTICAL METHODS

It is common to find analytical methods classified as *classical* or *instrumental*, the former comprising 'wet chemical' methods such as gravimetry and titrimetry. Such a classification is historically derived and largely artificial as there is no fundamental difference between the methods in the two groups. All involve the correlation of a physical measurement with the analyte concentration. Indeed, very few analytical methods are entirely instrumental, and most involve chemical manipulations prior to the instrumental measurement.

Table 1.1 A general classification of important analytical techniques

GROUP	PROPERTY MEASURED
gravimetric	weight of pure analyte or of a stoichiometric compound containing it
volumetric	volume of standard reagent solution reacting with the analyte
spectrometric	intensity of electromagnetic radiation emitted or absorbed by the analyte
electrochemical	electrical properties of analyte solutions
radiochemical	intensity of nuclear radiations emitted by the analyte
mass spectrometric	abundance of molecular fragments derived from the analyte
chromatographic	physico-chemical properties of individual analytes after separation
thermal	physico-chemical properties of the sample as it is heated and cooled

A more satisfactory general classification is achieved in terms of the physical parameter that is measured (table 1.1).

TRENDS IN ANALYTICAL METHODS AND PROCEDURES

There is constant development and change in the techniques and methods of analytical chemistry. Better instrument design and a fuller understanding of the mechanics of analytical processes enable steady improvements to be made in sensitivity, precision, and accuracy. These same changes contribute to more economic analysis as they frequently lead to the elimination of time-consuming separation steps. The ultimate development in this direction is a non-destructive method, which not only saves time but leaves the sample unchanged for further examination or processing.

The automation of analysis, sometimes with the aid of laboratory robots, has become increasingly important. For example, it enables a series of bench analyses to be carried out more rapidly and efficiently, and with better precision, while in other cases continuous monitoring of an analyte in a production process is possible. Two of the most important developments in recent years have been the incorporation of microprocessor control into analytical instruments and their interfacing with micro- and minicomputers. The microprocessor has brought improved instrument control, performance and, through the ability to monitor the condition of component parts, easier routine maintenance. Operation by relatively inexperienced personnel can be facilitated by simple interactive keypad dialogues including the storage and re-call of standard methods, report generation and diagnostic testing of the system. Microcomputers with sophisticated data handling and graphics software packages have likewise made a considerable impact on the collection, storage, processing, enhancement and interpretation of analytical data. *Laboratory Information and Management Systems* (LIMS), for the automatic logging of large numbers of samples, *Chemometrics*, which involve computerized and often sophisticated statistical analysis of data, and *Expert Systems*, which provide interactive computerized guidance and assessments in the solving of analytical problems, have all become important in optimizing chemical analysis and maximizing the information it provides.

Analytical problems continue to arise in new forms. Demands for analysis at 'long range' by instrument packages steadily increase. Space probes, 'borehole logging' and deep sea studies exemplify these requirements. In other fields, such as environmental and clinical studies, there is increasing recognition of the importance of the exact chemical form of an element in a sample rather than the mere level of its presence. Two well-known examples are the much greater toxicity of organo-lead and organo-mercury compounds compared with their inorganic counterparts. An identification and determination of the element in a specific chemical form presents the analyst with some of his more difficult problems.

GLOSSARY OF TERMS

The following list of definitions, though by no means exhaustive, will help both in the study and practice of analytical chemistry.

Accuracy
   The closeness of an experimental measurement or result to the true or accepted value (p. 14).

Analyte
   Constituent of the sample which is to be studied by quantitative measurements or identified qualitatively.

Assay
   A highly accurate determination, usually of a valuable constituent in a material of large bulk, e.g. minerals and ores. Also used in the assessment of the purity of a material, e.g. the physiologically active constituent of a pharmaceutical product.

Background

That proportion of a measurement which arises from sources other than the analyte itself. Individual contributions from instrumental sources, added reagents and the matrix can, if desired, be evaluated separately.

Blank

A measurement or observation in which the sample is replaced by a simulated matrix, the conditions otherwise being identical to those under which a sample would be analysed. Thus, the blank can be used to correct for background effects and to take account of analyte other than that present in the sample which may be introduced during the analysis, e.g. from reagents.

Calibration

1. A procedure which enables the response of an instrument to be related to the mass, volume or concentration of an analyte in a sample by first measuring the response from a sample of known composition or from a known amount of the analyte, i.e. a *standard*. Often, a series of standards is used to prepare a *calibration curve* in which instrument response is plotted as a function of mass, volume or concentration of the analyte over a given range. If the plot is linear, a *calibration factor* (related to the slope of the curve) may be calculated. This facilitates the rapid computation of results without reference to the original curve.

2. Determination of the accuracy of graduation marks on volumetric apparatus by weighing measured volumes of water, or determinations of the accuracy of weights by comparison with weights whose value is known with a high degree of accuracy.

Concentration

The amount of a substance present in a given mass or volume of another substance. The abbreviations w/w, w/v and v/v are sometimes used to indicate whether the concentration quoted is based on the weights or volumes of the two substances. Concentration may be expressed in several ways. These are shown in table 1.2.

Constituent

A component of a sample; it may be further classified as:

major	$>10\%$
minor	$0.01–10\%$
trace	1–100 ppm $(0.000\ 1\% – 0.01\%)$
ultra-trace	<1 ppm

Table 1.2 Alternative methods of expressing concentration*

UNITS	NAME AND SYMBOL
moles of solute per dm^3	$mol\,dm^{-3}$, M
equivalents of solute per dm^3	normal, N
milli-equivalents of solute per dm^3	$meq\,dm^{-3}$
grams of solute per dm^3	$g\,dm^{-3}$
parts per million	ppm (γ)
milligrams of component per kg	$mg\,kg^{-1}$
milligrams of solute per dm^3	$mg\,dm^{-3}$
parts per billion	ppb
nanograms of component per kg	$ng\,kg^{-1}$
nanograms of solute per dm^3	$ng\,dm^{-3}$
parts per trillion	ppt
picograms of component per kg	$pg\,kg^{-1}$
picograms of solute per dm^3	$pg\,dm^{-3}$
parts per hundred	% (w/w, w/v, v/v)
millimoles of solute per $100\,cm^3$	mM %
grams of solute per $100\,cm^3$	g %
milligrams of solute per $100\,cm^3$	mg %
micrograms of solute per $100\,cm^3$	μg %
nanograms of solute per $100\,cm^3$	ng %
micrograms of solute per cm^3	$\mu g\,cm^{-3}$ } $\equiv$ ppm
micrograms per gram	$\mu g\,g^{-1}$
nanograms of solute per cm^3	$ng\,cm^{-3}$ } $\equiv$ ppb
nanograms per gram	$ng\,g^{-1}$
picograms of solute per cm^3	$pg\,cm^{-3}$ } $\equiv$ ppt
picograms per gram	$pg\,g^{-1}$

*The table includes most of the methods of expressing concentration that are in current use, although some are not consistent with S.I.

Detection Limit
The smallest amount or concentration of an analyte that can be detected by a given procedure and with a given degree of confidence (p. 27).

Determination
A quantitative measure of an analyte with an accuracy of considerably better than 10% of the amount present.

Equivalent
That amount of a substance which, in a specified chemical reaction,

produces, reacts with or can be indirectly equated with one mole (6.023×10^{23}) of hydrogen ions. This confusing term is *obsolete* but its use is still to be found in some analytical laboratories.

Estimation

A semi-quantitative measure of the amount of an analyte present in a sample, i.e. an approximate measurement having an accuracy no better than about 10% of the amount present.

Interference

An effect which alters or obscures the behaviour of an analyte in an analytical procedure. It may arise from the sample itself, from contaminants or reagents introduced during the procedure or from the instrumentation used for the measurements.

Internal Standard

A compound or element added to *all* calibration standards and samples in a constant known amount. Sometimes a major constituent of the samples to be analysed can be used for this purpose. Instead of preparing a conventional calibration curve of instrument response as a function of analyte mass, volume or concentration, a *response ratio* is computed for each calibration standard and sample, i.e. the instrument response for the analyte is divided by the corresponding response for the fixed amount of added internal standard. Ideally, the latter will be the same for each pair of measurements but variations in experimental conditions may alter the responses of both analyte and internal standard. However, their *ratio* should be unaffected and should therefore be a more reliable function of the mass, volume or concentration of the analyte than its response alone. The analyte in a sample is determined from its response ratio using the calibration graph and should be independent of sample size.

Masking

Treatment of a sample with a reagent to prevent interference with the response of the analyte by other constituents of the sample (p. 41).

Matrix

The remainder of the sample of which the analyte forms a part.

Method

The overall description of the instructions for a particular analysis.

Precision

The random or indeterminate error associated with a measurement or result. Sometimes called the variability, it can be represented statistically by the standard deviation or relative standard deviation (coefficient of variation) (p. 15).

Primary Standard

A substance whose purity and stability are particularly well-established and with which other standards may be compared.

Procedure

A description of the practical steps involved in an analysis.

Reagent

A chemical used to produce a specified reaction in relation to an analytical procedure.

Sample

A substance or portion of a substance about which analytical information is required.

Sensitivity

1. The change in the response from an analyte relative to a small variation in the amount being determined. The sensitivity is equal to the slope of the calibration curve, being constant if the curve is linear.
2. The ability of a method to facilitate the detection or determination of an analyte.

Validation of Methods

In order to ensure that results yielded by a method are as accurate as possible, it is essential to validate the method by analysing standards which have an accepted analyte content, and a matrix similar to that of the sample. The accepted values for these validated standards are obtained by extensive analysis, using a range of different methods. Internationally accepted standards are available.

Standard

1. A pure substance which reacts in a quantitative and known stoichiometric manner with the analyte or a reagent.
2. The pure analyte or a substance containing an accurately known amount of it which is used to calibrate an instrument or to standardize a reagent solution.

Standard Addition

A method of quantitative analysis whereby the response from an analyte is measured before and after adding a known amount of that analyte to the sample. The amount of analyte originally in the sample is determined from a calibration curve or by simple proportion if the curve is linear. The main advantage of the method is that all measurements of the analyte are made in the same matrix which eliminates interference effects arising

Table 1.3   Physical quantities and units including S.I. and C.G.S.

PHYSICAL QUANTITY	S.I.		C.G.S.	
	UNIT	SYMBOL	UNIT	SYMBOL
length, l	metre	m	centimetre	cm
mass, m	kilogram	kg	gram	g
time, t	second	s	second	s
energy, E	joule	J	erg	—
			electron volt	eV
			calorie	cal
thermodynamic temperature, T	kelvin	K	kelvin	K
amount of substance, n	mole	mol	mole	mol
force, F	newton	N	dyne	—
volume, V	cubic metre	m^3	cubic centimetre	cm^3 (ml)
	cubic decimetre	dm^3	litre	l
electric current, I or i	ampere	A	ampere	A
electric potential difference, E	volt	V	volt	V
electric resistance, R	ohm	Ω	ohm	Ω
electric conductance, G	siemens	S	mho	Ω^{-1}
quantity of electricity, Q	coulomb	C	coulomb	C
electric capacitance, C	farad	F	farad	F
frequency, ν	hertz	Hz	cycles per second	cps
wavenumber, σ ($\bar{\nu}$)			reciprocal centimetre	cm^{-1}
wavelength, λ	metre	m	centimetre	cm
	millimetre	mm	millimetre	mm
	micrometre	μm	micron	μ
	nanometre	nm	millimicron	$m\mu$
			Angstrom	Å
magnetic flux density, B	tesla	T	gauss	G
disintegration rate	curie	Ci	curie	Ci
	becquerel	Bq		
nuclear cross-sectional area	barn	b	barn	b

from differences in the overall composition of sample and standards (p. 32, 108).

Standardization

Determination of the concentration of an analyte or reagent solution from its reaction with a standard or primary standard.

Technique

The principle upon which a group of methods is based.

Physical quantities relevant to analytical measurements and the units and symbols used to express them are given in table 1.3. Both S.I. and C.G.S. units have been included because of current widespread use of the latter and for ease of comparison with older literature. However, only the S.I. nomenclature is now officially recognized and the use of the C.G.S. system should be progressively discouraged.

Further Reading
Skoog, D. A. and West, D. M., *Fundamentals of Analytical Chemistry*, 4th Ed., CBS College Publishing, New York, 1982.

Chapter 2

The Assessment of Analytical Data

A critical attitude towards the results obtained in analysis is necessary in order to appreciate their meaning and limitations. Precision is dependent on the practical method and beyond a certain degree cannot be improved. Inevitably there must be a compromise between the reliability of the results obtained and the use of the analyst's time. To reach this compromise requires an assessment of the nature and origins of errors in measurements; relevant statistical tests may be applied in the appraisal of the results. With the development of microcomputers and their ready availability, access to complex statistical methods has been provided. These complex methods of data handling and analysis have become known collectively as *chemometrics*.

2.1 Definitions and Basic Concepts

True result. The 'correct' value for a measurement which remains unknown except when a standard sample is being analysed. It can be estimated from the results with varying degrees of precision depending on the experimental method.
Accuracy. The nearness of a measurement or result to the true value. Expressed in terms of *error*.
Error. The difference between the true result and the measured value. It is conveniently expressed as an *absolute error*, defined as the actual difference between the true result and the experimental value in the same units. Alternatively, the *relative error* may be computed, i.e. the error expressed as a percentage of the measured value or in 'parts per thousand'.
Mean. The arithmetic average of a replicate set of results.
Median. The middle value of a replicate set of results.
Degree of freedom. An independent variable. The number of degrees of freedom possessed by a replicate set of results equals the total number of

14

results in the set. When another quantity such as the mean is derived from the set, the degrees of freedom are reduced by one, and by one again for each subsequent derivation made.

Precision. The variability of a measurement. As in the case of error, above, it may be expressed as an absolute or relative quantity. Standard deviations are the most valuable precision indicators (*vide infra*).

Spread. The numerical difference between the highest and lowest results in a set. It is a measure of precision.

Deviation (e.g. from the mean or median). The numerical difference, with respect to sign, between an individual result and the mean or median of the set. It is expressed as a relative or absolute value.

Standard deviation σ. A valuable parameter derived from the normal error curve (p. 17) and expressed by:

$$\sigma = \left[\frac{\sum_{i=1}^{i=N} (x_i - \mu)^2}{N} \right]^{\frac{1}{2}} \tag{2.1}$$

where x_i is a measured result, μ is the true mean and N is the number of results in the set. Unfortunately, μ is never known and $\bar{x}$ the mean derived from the set of results has to be used. In these circumstances the degrees of freedom are reduced by one and an estimate of the true standard deviation is calculated from:

$$s = \left[\frac{\sum_{i=1}^{i=N} (x_i - \bar{x})^2}{N - 1} \right]^{\frac{1}{2}} \tag{2.2}$$

A better estimate of the standard deviation may often be obtained by the pooling of results from more than one set. Thus, s may be calculated from K sets of data.

$$s = \left[\frac{\sum_{i=1}^{N_1} (x_i - \bar{x}_1)^2 + \sum_{i=1}^{N_2} (x_i - \bar{x}_2)^2 + \cdots \sum_{i=1}^{N_K} (x_i - \bar{x}_K)^2}{M - K} \right]^{\frac{1}{2}} \tag{2.3}$$

where $M = N_1 + N_2 + \cdots N_K$. One degree of freedom is lost with each set pooled. A common requirement is the computation of the pooled value for two sets of data only. In this case the simplified equation (2.4) may conveniently be used:

$$s^2 = \{(n_1 - 1)s_1^2 + (n_2 - 1)s_2^2\}/(n_1 + n_2 - 2) \tag{2.4}$$

Table 2.1   Standard deviations from arithmetically combined data

ARITHMETIC QUANTITY	STANDARD DEVIATION
x	σ_x
y	σ_y
$1/x$	σ_x/x^2
$x \pm y$	$(\sigma_x^2 + \sigma_y^2)^{\frac{1}{2}}$
x^2	$2\sigma_x x$
$x^{\frac{1}{2}}$	$\sigma_x/2x^{\frac{1}{2}}$
xy	$(\sigma_x^2 y^2 + \sigma_y^2 x^2)^{\frac{1}{2}}$
x/y	$x/y[(\sigma_x^2/y^2 + \sigma_y^2/x^2)]^{\frac{1}{2}}$

Standard deviations for results obtained by the arithmetic combination of data will be related to the individual standard deviations of the data being combined. The exact relation will be determined by the nature of the arithmetic operation (table 2.1).

The *relative standard deviation* or *coefficient of variation* ($s \cdot 100/\bar{x}$) is often used in comparing precisions.

Variance. The square of the standard deviation (σ^2 or s^2). This is often of practical use as the values are additive, e.g. $s_{x+y+z}^2 \ldots = s_x^2 + s_y^2 + s_z^2 \ldots$

2.2   The Nature and Origin of Errors

On the basis of their origin, errors may usually be classified as *determinate* or *indeterminate*. The first are those having a value which is (in principle at least) measurable and for which a correction may be made. The second fluctuate in a random manner and do not have a definite measurable value.

Determinate errors may be *constant* or *proportional*. The former have a fixed value and the latter increase with the magnitude of the measurement. Thus their overall effects on the result will differ. These effects are summarized in figure 2.1. The errors usually originate from one of three major sources: operator error; instrument error; method error. They may be detected by blank determinations, the analysis of standard samples, and independent analyses by alternative and dissimilar methods. Proportional variation in error will be revealed by the analysis of samples of varying sizes. Proper training should ensure that operator errors are eliminated. However, it may not always be possible to eliminate instrument and method errors entirely and in these circumstances the error must be assessed and a correction applied.

Indeterminate errors arise from the unpredictable minor inaccuracies of

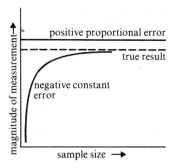

Figure 2.1. The effects of constant and proportional errors on a measurement of concentration

the individual manipulations in a procedure. A degree of uncertainty is introduced into the result which can be assessed only by statistical tests. The deviations of a number of measurements from the mean of the measurements should show a symmetrical or *Gaussian distribution* about that mean. Figure 2.2 represents this graphically and is known as a *normal error curve*. The general equation for such a curve is

$$y = \frac{\exp\left[-(x - \mu)^2/2\sigma^2\right]}{\sigma(2\pi)^{\frac{1}{2}}} \tag{2.5}$$

where μ is the mean and σ is the standard deviation. The width of the curve

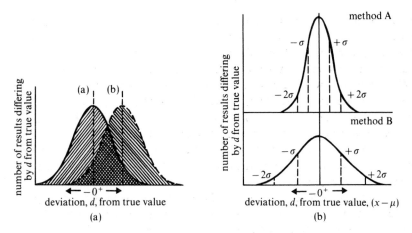

Figure 2.2. Normal error curves. (a) Curve (a) shows a normal distribution about the true value. Curve (b) shows the effect of a determinate error on the normal distribution. (b) Curves showing the results of the analysis of a sample by two methods of differing precision. Method A is the more precise, or reliable

is determined by σ, which is a useful measure of the spread or precision of a set of results, and is unique for that set of data. An interval of $\mu \pm \sigma$ will contain 68.3% of the statistical sample, whilst the intervals $\mu \pm 2\sigma$ and $\mu \pm 3\sigma$ will contain 95.5% and 99.7% respectively.

2.3   The Evaluation of Results and Methods

A set of replicate results should number at least twenty-five if it is to be a truly representative 'statistical sample'. The analyst will rarely consider it economic to make this number of determinations and therefore will need statistical methods to enable him to base his assessment on fewer data, or data that have been accumulated from the analysis of similar samples. Any analytical problem should be examined at the outset with respect to the precision, accuracy and reliability required of the results. Analysis of the results obtained will then be conveniently resolved into two stages—an examination of the reliability of the results themselves and an assessment of the meaning of the results.

THE RELIABILITY OF MEASUREMENTS

When considering the reliability of the results, any determination which deviates rather widely from the mean should be first investigated for gross experimental or arithmetic error. Except in cases where such errors are revealed, questionable data should only be rejected when a proper statistical test has been applied. This process of *data rejection* presents the analyst with an apparent paradox. If the limits for acceptance are set too narrowly, results which are rightly part of a statistical sample may be rejected and narrow limits may therefore only be applied with a low *confidence* of containing all statistically relevant determinations. Conversely wide limits may be used with a high confidence of including all relevant data, but at a risk of including some that have been subject to gross error. A practical compromise is to set limits at a *confidence level* of 90% or 95%.

There are two criteria which are commonly used to gauge the rejection of results. Of these, the most convenient to use is based on the $\bar{x} \pm 2\sigma$ interval which contains 95.5% of the relevant measurements. Some workers believe this limit is too wide, and regard the *Q-test* at a 90% confidence level to be a more acceptable alternative. A *rejection quotient Q* is defined as

$$Q = (x_n - x_{n-1})/(x_n - x_1) \tag{2.6}$$

where x_n is the questionable result in a set $x_1, x_2, x_3, \ldots, x_n$. Q is calculated for the questionable data and compared with a table of critical values (table 2.2). The result is rejected if Q(experimental) exceeds Q(critical).

Table 2.2   Critical values of Q at the 90% confidence
level

NUMBER OF RESULTS	Q_{crit}(90% confidence)
2	—
3	0.94
4	0.76
5	0.64
6	0.56
7	0.51
8	0.47
9	0.44
10	0.41

THE ANALYSIS OF DATA

Once the reliability of a replicate set of measurements has been established
the mean of the set may be computed as a measure of the true mean. Unless
an infinite number of measurements is made this true mean will always remain
unknown. However, the *t-factor* may be used to calculate a *confidence
interval* about the experimental mean, within which there is a known (90%)
confidence of finding the true mean. The limits of this confidence interval

Table 2.3   Values of t for various levels of probability

DEGREES OF FREEDOM	CONFIDENCE LEVEL (%)				
	80	90	95	99	99.9
1	3.08	6.31	12.7	63.7	637
2	1.89	2.92	4.30	9.92	31.6
3	1.64	2.35	3.18	5.84	12.9
4	1.53	2.13	2.78	4.60	8.60
5	1.48	2.02	2.57	4.03	6.86
6	1.44	1.94	2.45	3.71	5.96
7	1.42	1.90	2.36	3.50	5.40
8	1.40	1.86	2.31	3.36	5.04
9	1.38	1.83	2.26	3.25	4.78
10	1.37	1.81	2.23	3.17	4.59
11	1.36	1.80	2.20	3.11	4.44
12	1.36	1.78	2.18	3.06	4.32
13	1.35	1.77	2.16	3.01	4.22
14	1.34	1.76	2.14	2.98	4.14
∞	1.29	1.64	1.96	2.58	3.29

are given by:

$$\bar{x} \pm ts/N^{\frac{1}{2}} \tag{2.7}$$

where $\bar{x}$ is the experimental mean, t is a statistical factor derived from the normal error curve (values in table 2.3), s is the estimated standard deviation and N is the number of results.

Example 2.1

If the analysis of a sample for iron content yields a mean result of 35.40% with a standard deviation of 0.30%, the size of the confidence interval will vary inversely with the number of measurements made. For two measurements, the confidence interval (90%) is

$$35.40 \pm 6.31 \times 0.30 \times 2^{-\frac{1}{2}} = 35.40\% \pm 1.34\%$$

and for five measurements

$$35.40 \pm 2.13 \times 0.30 \times 5^{-\frac{1}{2}} = 35.40\% \pm 0.29\%$$

(N.B. s has been derived from the set of data and $N - 1$ degrees of freedom are used in evaluating t.) The essential conclusion, here, is that five analyses at most are required to get a reasonable estimate of the true mean.

When a comparison of two separate replicate sets of data is required, the first stage is normally to compare their respective precisions by means of the *F-test*. This test uses the ratio of the variances of the two sets to establish any statistically significant difference in precision. F is calculated from

$$F = s_x^2/s_y^2 \tag{2.8}$$

(By convention the larger variance is always taken as numerator.) The value of F thus obtained is compared with critical values computed on the assumption that they will be exceeded purely on a probability basis in only 5% of

Table 2.4   Critical values for F at the 5% level

DEGREES OF FREEDOM (DENOMINATOR)	DEGREES OF FREEDOM (NUMERATOR)						
	3	4	5	6	12	20	∞
3	9.28	9.12	9.01	8.94	8.74	8.64	8.53
4	6.59	6.39	6.26	6.16	5.91	5.80	5.63
5	5.41	5.19	5.05	4.95	4.68	4.56	4.36
6	4.76	4.53	4.39	4.28	4.00	3.87	3.67
12	3.49	3.26	3.11	3.00	2.69	2.54	2.30
20	3.10	2.87	2.71	2.60	2.28	2.12	1.84
∞	2.60	2.37	2.21	2.10	1.75	1.57	1.00

cases (table 2.4). When the experimental value of F exceeds the critical value then the difference in variance or precision is deemed to be statistically significant.

Having established that the standard deviations of two sets of data agree at a reasonable confidence level it is possible to proceed to a comparison of the mean results derived from the two sets, using the t-test in one of its forms. As in the previous case, the factor is calculated from the experimental set of results and compared with the table of critical values (table 2.3). If t_{exp} exceeds the critical value for the appropriate number of degrees of freedom, the difference between the means is said to be significant. When there is an accepted value for the result based on extensive previous analysis t is computed from equation (2.9)

$$t = [(\bar{x} - \mu)/s]N^{\frac{1}{2}} \qquad (2.9)$$

where $\bar{x}$ is the mean of the experimental set, μ the accepted value, s the experimental standard deviation and N the number of results.

If there is no accepted value and two experimental means are to be compared, t can be obtained from equation (2.10) with $(M + N - 2)$ degrees of freedom.

$$t = [(\bar{x} - \bar{y})/s][MN/(M + N)]^{\frac{1}{2}} \qquad (2.10)$$

where $\bar{x}$ is the mean of M determinations, $\bar{y}$ the mean of N determinations and s the pooled standard deviation (equation (2.3)).

THE APPLICATION OF STATISTICAL TESTS

Table 2.5, together with the subsequent worked examples, illustrates the application of the statistical tests to real laboratory situations. Equation (2.10) is a simplified expression derived on the assumption that the precisions of the two sets of data are not significantly different. Thus the application of the F-test (equation 2.8)) is a prerequisite for its use. The evaluation of t in more general circumstances is of course possible, but from a much more complex expression requiring tedious calculations. Recent and rapid developments in programmable desk calculators are removing the tedium and making use of the general expression more acceptable. The references at the end of the chapter will serve to amplify this point.

Example 2.2
In a series of replicate analyses of a sample the following data (%) were obtained:

4.20 7.01 7.31 7.54 7.55 7.58 7.59

Table 2.5 Some practical problems with relevant statistical tests

PRACTICAL PROBLEMS	RELEVANT TESTS
One result in a replicate set differs rather widely from the rest. Is it a significant result?	Examine for gross error. Apply Q-test (equation (2.6))
Two operators analysing the same sample by the same method obtain results with different spreads. Is there a significant difference in precision between the results?	Examine data for unreliable results. Apply F-test (equation (2.8))
A new method of analysis is being tested by the analysis of a standard sample with an accurately known composition. Is the difference between the experimental value and the accepted value significant?	Examine data for unreliable results. Apply t-test (equation (2.9))
Two independent methods of analysis have been used to analyse a sample of unknown composition. Is the difference between the two results significant and thus indicative of an error in one method?	Examine data for unreliable results. Establish that both sets have similar precisions by F-test. Apply T-test (equation (2.10))
With what confidence can the mean of a set of experimental results be quoted as a measure of the true mean?	Calculate the confidence interval (equation (2.7))
If the standard deviation for a method is known, how many results must be obtained to provide a reasonable estimate of the true mean?	Use the confidence interval method (equation (2.7))
Is a determinate error fixed or proportional?	Graphical plot of results against sample weight (figure 2.1)

In the assessment of these data for reliability the first step of arranging them in rank order has been carried out.

On inspection 4.20 is rejected as having a gross error, and 7.01 is seen as questionable and requiring to be tested by the Q-test.

$$Q_{exp} = \frac{7.31 - 7.01}{7.59 - 7.01} = 0.52$$

Comparison with critical values in table 2.2 shows Q_{crit} for 6 results to be 0.56. Thus $Q_{exp} < Q_{crit}$ and 7.01 is retained.

Example 2.3
The accepted value for the chloride content of a standard sample obtained from extensive previous analysis is 54.20%. Five analyses of the same sample are carried out by a new instrumental procedure, 54.01, 54.24, 54.05, 54.27, 54.11% being the results obtained. Is the new method giving results consistent with the accepted value?

(a) A preliminary examination shows no unreliable results.

(b) The mean and the standard deviations are then calculated (equation (2.2)).

x	$(x - \bar{x})$	$(x - \bar{x})^2$	
54.01	-0.13	0.016 9	$s = (0.052\ 8/4)^{\frac{1}{2}} = 0.115$
54.24	$+0.10$	0.010 0	
54.05	-0.09	0.008 9	
54.27	$+0.13$	0.016 9	
54.11	-0.03	0.000 9	
$\Sigma x = 270.68$		$\Sigma(x - \bar{x})^2$	
$\bar{x} = 54.14$		$= 0.052\ 8$	

(c) Equation (2.9) is then applied to compare the means

$$t = [(54.20 - 54.14)/0.115]5^{\frac{1}{2}}$$
$$= 0.06 \times 2.236/0.115 = 1.17$$

From t-criteria the confidence level for this difference to be significant is less than 80%.
Conclusion: The new instrumental method is giving the same results as the accepted method.

Example 2.4
The technique of a trainee operator is being assessed by comparing his results with those obtained by an experienced operator. Do the results obtained indicate a significant difference between the skill of the two operators?

The trainee operator carried out six determinations yielding a mean of 35.25% with a standard deviation of 0.34%. The experienced operator obtained a mean of 35.35% and a standard deviation of 0.25% from five determinations.

(a) The F-test is used to compare the standard deviations (equation (2.8))

$$F_{exp} = 0.34^2/0.25^2 = 1.85$$

F_{crit} from table 2.4 is 6.26 and there is no significant difference in the standard deviations (at the 95 % level).

(b) Equation (2.10) now enables the two means to be compared. If s is first computed from the pooled data (equation (2.3)) and found to be 0.29 % then

$$t = [(35.35 - 35.25)/0.29](30/11)^{\frac{1}{2}}$$
$$= 0.57$$

Conclusion: The probability of the difference in means being significant is very low, and there is no difference between the skill of the two operators.

Example 2.5
In an investigation of a determinate error a series of replicate measurements were made using a range of sample weights. The results obtained are tabulated below.

Sample weight/g	Analyte/%
0.113	9.67
0.351	9.96
0.483	10.04
0.501	10.03
0.711	10.09
0.867	10.12
0.904	10.13

In order to decide whether the error is constant or proportional plot a graph of results against sample weight. Take care in selecting the scale to ensure that the trends are not obscured. The graph (figure 2.3) shows clearly that the error is negative and constant (see figure 2.1).

The Estimation of the Overall Precision of a Method from its Unit Operations
A frequent problem in analysis is the estimation of the overall precision of a method before it has been used or when insufficient data are available to carry out a statistical analysis. In these circumstances the known precision limits for the unit operations of the method (volume measurement, weighing, etc.) may be used to indicate its precision. Table 2.6, gives the normal precision limits for 'Grade A' volumetric equipment.

If the absolute standard deviations for a set of unit operations are a, b, c, ..., then s, the overall standard deviation for the method is given by:

$$s = (a^2 + b^2 + c^2 + \cdots)^{\frac{1}{2}} \qquad (2.11)$$

when the individual measurements are combined as a sum or difference

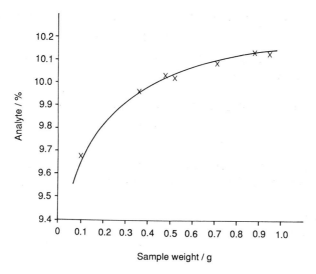

Figure 2.3. Plot of analyte content against sample weight to investigate a determinate error

(table 2.1). Conveniently, the relative standard deviations may be used to express the overall standard deviation when the individual measurements are combined as a product or quotient.

$$s_r = (a_r^2 + b_r^2 + c_r^2 + \cdots)^{\frac{1}{2}} \qquad (2.12)$$

Table 2.6 Precision limits for laboratory equipment used in volumetric analysis

burettes	± 0.02 cm^3 for each reading
pipettes (to deliver)	(5 ± 0.01) cm^3 (10 ± 0.02) cm^3 (25 ± 0.03) cm^3 (50 ± 0.05) cm^3 (100 ± 0.08) cm^3
volumetric flasks (to contain)	(25 ± 0.03) cm^3 (50 ± 0.05) cm^3 (100 ± 0.08) cm^3 (250 ± 0.12) cm^3 (500 ± 0.15) cm^3 $(1\ 000 \pm 0.30)$ cm^3 $(2\ 000 \pm 0.50)$ cm^3
analytical balances (to weigh to)	$\pm 0.000\ 1$ g

Example 2.6

Consider as an example the standardization of a solution of hydrochloric acid by titration against a weighed amount of sodium carbonate. The strength of the hydrochloric acid will be computed from

$$\text{concentration of HCl} = \frac{\text{weight of Na}_2\text{CO}_3}{\text{vol of HCl used}} \cdot \frac{2M_r(\text{HCl})}{M_r(\text{Na}_2\text{CO}_3)}$$

where M_r represents the relative molecular mass of the compound. The results of an analysis are as follows:

weight of bottle + Na$_2$CO$_3$ (1)	16.254 1 $\pm$ 0.000 1 g
weight of bottle + Na$_2$CO$_3$ (2)	16.041 9 $\pm$ 0.000 1 g
weight of Na$_2$CO$_3$ used	0.212 2 g
final burette reading	45.21 $\pm$ 0.02 cm^3
initial burette reading	0.52 $\pm$ 0.02 cm^3
volume of HCl used	44.69 cm^3

The overall precision of the weighing is now computed using equation (2.11).

$$s = [(0.000\ 1)^2 + (0.000\ 1)^2]^{\frac{1}{2}} = 0.000\ 14\ \text{g}$$

The relative standard deviation for the weighing is then

$$0.000\ 14/0.212\ 2 = 0.000\ 66 \qquad \text{i.e. about } 0.07\%$$

Similarly, the overall precision of the volume measurement is obtained. An allowance for the uncertainty in the colour change observation at the end point must be included (e.g. ± 0.03 cm^3). Thus

$$s = [(0.02)^2 + (0.02)^2 + (0.03)^2]^{\frac{1}{2}} = 0.041\ \text{cm}^3$$

whence the relative standard deviation is

$$0.041/44.69 = 0.000\ 92 \qquad \text{i.e. about } 0.09\%$$

Finally, the estimated precision for the determination of the concentration of hydrochloric acid is obtained using equation (2.12).

$$s_r = [(0.07)^2 + (0.09)^2]^{\frac{1}{2}} = 0.11\%$$

One important point to remember is that the absolute standard deviations for the unit processes are constant, but the relative standard deviations will decrease with the magnitude of the sample and the titre. In other words, within limits the larger the sample taken the better the precision of the results.

Significant Figures

Results are normally given to a certain number of *significant figures*. All the digits in a number that are known with certainty plus the first that is uncertain, constitute the significant figures of the number. In the case of a zero it is taken as significant when it is part of the number but not where it merely indicates the magnitude. Thus a weight of 1.042 1 g which is known within the limits of ± 0.000 1 g has five significant figures whilst one of 0.042 1 g which is known within the same absolute limits has only three. When a derived result is obtained from addition or subtraction of two numbers, its significant figures are determined from the *absolute uncertainties*. Consider the numbers 155.5 ± 0.1 and 0.085 ± 0.001 which are added together to give 155.585. Uncertainty appears at the fourth digit, whence the result should be rounded off to 155.6. If the derived result is a product or quotient of the two quantities, the *relative uncertainty* of the least certain quantity dictates the significant figures. 0.085 has the greatest relative uncertainty at 12 parts per thousand. The product $155.5 \times 0.085 = 13.327$ 5 has an absolute deviation of 13.327 $5 \times 0.012 = 0.16$. Uncertainty thus appears in the third digit and the result is rounded off to 13.3.

LIMITS OF DETECTION

It is important in analysis at trace levels to establish the smallest concentration or absolute amount of an analyte that can be detected. The problem is one of discerning a difference between the response given by a 'blank' and that given by the sample, i.e. detecting a weak signal in the presence of background noise. All measurements are subject to random errors, the distribution of which should produce a normal error curve. The spread of replicate measurements from the blank and from the sample will therefore overlap as the two signals approach each other in magnitude. It follows that the chances of mistakenly identifying the analyte as present when it is not or vice versa eventually reach an unacceptable level. The detection limit must therefore be defined in statistical terms and be related to the probability of making a wrong decision.

Figure 2.4(a) shows normal error curves (B and S) with true means μ_B and μ_S for blank and sample measurements respectively. It is assumed that for measurements made close to the limit of detection, the standard deviations of the blank and sample are the same, i.e. $\sigma_B = \sigma_S = \sigma$. In most cases, a 95% confidence level is a realistic basis for deciding if a given response arises from the presence of the analyte or not, i.e. there is a 5% risk in reporting the analyte 'detected' when it is not present and vice versa. Thus, point L on curve B represents an upper limit above which only 5% of blank measurements with true mean μ_B will lie whilst point L on curve S represents a lower limit below which only 5% of sample measurements with true mean μ_S will lie. If μ_S now approaches μ_B until points L on each curve coincide (figure

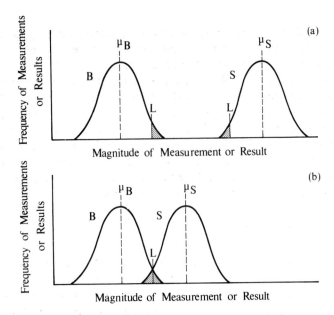

Figure 2.4.   Normal error curves for blank B and sample S measurements

2.4(b)), the point of coincidence represents the *practical detection limit* for a single measurement, i.e. if a measurement falls at or below L, it has a 95% probability of arising from background sources or random noise only, whilst if it falls above L it has a 95% probability of arising from the presence of the analyte. Furthermore, it follows that μ_S must now represent the *theoretical detection limit* because a true mean lying below μ_S would have a normal distribution with more than 5% of values below L. Because the chances of making an incorrect decision were chosen to be equal in this case (5% probability), then μ_S is given by

$$\mu_S = 2L \qquad (2.13)$$

Individual practical results falling between L and μ_S must be regarded as 'detected' but should be reported as 'less than μ_S'.

The value of L and hence of μ_S is related to σ and is given by

$$L = \mu_B + 1.64\sigma \qquad (2.14)$$

where a large number of blanks (preferably more than 20) have been measured or

$$L = \mu_B + 2.33\sigma \qquad (2.15)$$

where a smaller number of paired measurements of sample and blank have been made.

Example 2.7

The determination of iron at trace levels can be accomplished by the spectro-photometric measurement of its complex with the reagent *o*-phenanthroline. The sensitivity of a particular method is 53 ppm per unit absorbance and the standard deviation of the blank estimated from 25 measurements is 0.002 of an absorbance unit. The practical detection limit is therefore $1.64\sigma_B$ or 0.003 3 of an absorbance unit which corresponds to 0.17 ppm, and the theoretical detection limit is $3.28\sigma_B$ or 0.006 6 of an absorbance unit which corresponds to 0.35 ppm. (The value of μ_B in this case is assumed to be zero.) Hence if a result is less than 0.17 ppm, the conclusion is that iron is 'not detected'. If the value lies between 0.17 ppm and 0.35 ppm, the iron content should be reported as 'less than 0.35 ppm'.

QUALITY CONTROL CHARTS

Chemical analysis finds important applications in the quality control of in-dustrial processes. In an ideal situation a continuous analysis of the process stream is made and some aspects of this are discussed in Chapter 12. How-ever, such continuous analysis is by no means always possible, and it is common to find a process being monitored by the analysis of separate samples taken at regular intervals. The analytical data thus obtained need to be capable of quick and simple interpretation, so that rapid warning is available if a process is going out of control and effective corrective action can be taken. One method of data presentation which is in widespread use is the *control chart*. A number of types of chart are used but where chemical data are concerned the most common types used are *Shewhart charts* and *cusum charts*. Only these types are discussed here. The charts can also be used to monitor the performance of analytical methods in analytical laboratories.

Shewhart Charts

In an explanation of the construction and operation of control charts it is helpful to consider a simple example such as the mixing of two materials in a process stream. It is important to recognize that there are two aspects of the composition of the process stream which need to be controlled. Firstly, the overall composition of the mixture may be assessed by averaging the results on a run of about five samples and plotting the results on an *averages chart*. Secondly, short term variations in composition are reflected by the range of results in a run plotted on a *ranges chart*. Shewhart charts are thus used in pairs (figure 2.5a and 2.5b). It is usual to mark the charts with *warning limits* (*inner control limits*) and *action limits* (*outer control limits*). These represent the values at which there is respectively 95 % (1.96σ) or 99.9 % (3.09σ) confidence that the process is going out of control. The use of the averages chart is thus

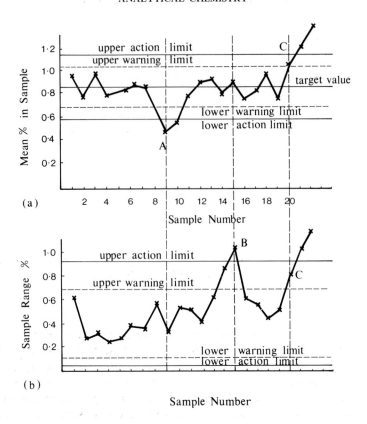

Figure 2.5. A typical pair of Shewhart charts. (a) Averages chart and (b) ranges chart. Point A shows a lack of control of averages only, point B of ranges only and point C of both together

based on an estimate of the variability and an assumption that this remains constant. It is hence invalidated if a significant change in variability occurs. Therefore the ranges chart fulfils a second important function in providing a check on such changes in variability.

Figure 2.5 shows a pair of Shewhart charts. Point A illustrates the pattern which occurs when the averages are going out of control but the ranges remain in control. This would suggest perhaps that the supply of one of the feedstocks for the process is being interrupted. Point B shows the ranges going out of control with the averages remaining under control. One explanation of this would be that the mixing had become inefficient. Finally, point C shows both averages and ranges going out of control which implies a serious fault in the process.

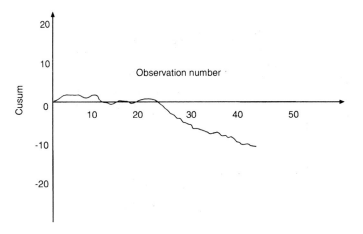

Figure 2.6. A typical cusum plot showing a lack of control with a negative deviation developing after 20 observations

Cusum Charts

An alternative method of monitoring the behaviour of a process is the *cumulative sum technique* (*cusum*). The cusum plot provides a simple method of diagnosing when a process begins to move out of control. A sequence of analyses is made on the process stream at regular intervals. The values thus obtained are compared with the target value for that particular component and the deviations aggregated to provide the cumulative sum. For a process in control, the deviations will have small positive or negative values and the cusum will remain close to zero, with the plot having a roughly horizontal pattern. If the process goes out of control, the cusum will become increasingly positive or negative with the corresponding change in slope signifying the point at which loss of control ensues. One major advantage of such charts is that the computations required are minimal and the plotting very simple. Thus charts may be constructed and used by operators of limited technical knowledge. Figure 2.6 shows a cusum plot for a process going out of control.

STANDARDIZATION OF ANALYTICAL METHODS

Quantitative analysis demands that an analytical measurement can be accurately and reliably related to the composition of the sample in a strict proportionality (p. 2). The complexity of relationships, especially for instrumental techniques, means that the proportionalities need to be practically established in *calibration procedures*. For a typical simple calibration, a range of standards is prepared containing varying amounts of

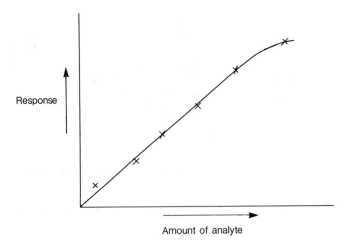

Figure 2.7.   A typical calibration curve

the analyte. These are then analysed by the standard method and a *calibration curve* of signal vs. amount of analyte is plotted. Results for unknowns are then interpolated from this graph (figure 2.7).

Many examples of calibration curves will be found in later pages of this text. A good calibration curve will be linear over a good range of analyte quantities. Ultimately curvature must be anticipated at the higher ranges and uncertainty and deteriorating precision at the low ones.

A particular issue that must be considered for all calibration procedures is the possibility of *matrix effects* on the analyte signal. If such effects are present they may be allowed for in many cases by *matrix matching* of the standard to the sample. This of course requires an accurate knowledge of the sample matrix. Where this is not available the method of *standard addition* is often effective. This involves 'spiking' at least three equal aliquots of the sample with different amounts of the analyte and then measuring the response of both spiked and unspiked aliquots. A plot of response vs analyte extrapolated back will give abscissae intercepts from which the amount of analyte in the sample may be deduced (figure 2.8).

Even this method, however, does not guarantee to overcome all matrix effects, and is often limited where the matrix is extremely complex and remains largely unknown (as in naturally occurring materials such as rocks and biological specimens). The use of *recognized standards* has proved invaluable for the standardization and assessment of methods in these circumstances. Pioneered in geochemistry, this involves the collection and thorough mixing of suitable standard materials. These are then distributed for analysis by recognized laboratories using as many different techniques as

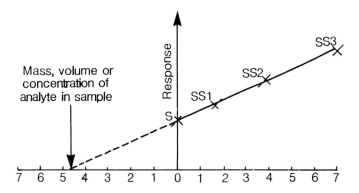

Figure 2.8.   Calibration curve for standard addition; S = unspiked sample; SS1, SS2, SS3 = spiked samples

possible. These ensuing results ultimately converge to give an accepted composition for the standard which may subsequently be used for the assessment and calibration of various analytical methods. Standards are exemplified by a range of rock types for geochemistry and bovine liver and kale leaves for biological analysis.

CHEMOMETRICS

Advanced mathematical and statistical techniques used in analytical chemistry are referred to under the umbrella term of chemometrics. This is a loose definition and chemometrics is not readily distinguished from the more rudimentary techniques discussed in the earlier parts of this chapter, except in terms of sophistication. The techniques are applied to the development and assessment of analytical methods as well as to the assessment and interpretation of results. Once the province of the mathematician, the computational powers of the microcomputer now make such techniques routinely accessible to analysts. Hence, although it would be inappropriate to consider the detail of the methods in a book at this level, it is nevertheless important to introduce some of the salient features to give an indication of their value. Two important applications in analytical chemistry are in method optimization, and in pattern recognition of results.

Analytical methods often contain many different variables which need to be optimized to attain best performance. The different variables are not always independent. For example, pH and polarity in a solution may be interdependent. Optimization by changing one variable at a time while keeping the others nominally constant is neither efficient nor necessarily successful. This point is well illustrated in chapter 4 where the optimization of chromatographic separations is discussed. (p. 132).

A sample may be characterized by the determination of a number of different analytes. For example, a hydrocarbon mixture can be analysed by use of a series of uv absorption peaks. Alternatively, in a sediment sample a range of trace metals may be determined. Collectively, these data represent patterns characteristic of the samples, and similar samples will have similar patterns. Results may be compared by vectorial presentation of the variables, when the variables for similar samples will form clusters. Hence the term *cluster analysis*. Where only two variables are studied, clusters are readily recognized in a two-dimensional graphical presentation. For more complex systems with more variables, i.e. *n*, the clusters will be in *n*-dimensional space. *Principal component analysis* (*pca*) explores the interdependence of pairs of variables in order to reduce the number to certain principal components. A practical example could be drawn from the sediment analysis mentioned above. Trace metals are often attached to sediment particles by sorption on to the hydrous oxides of Al, Fe and Mn that are present. The Al content could be a principal component to which the other metal contents are related. Factor analysis is a more sophisticated form of principal component analysis.

Problems

1. Define the terms *precision* and *accuracy* as they are used in analytical chemistry. Indicate how they may be estimated quantitatively. To what extent is the estimate of accuracy dependent upon precision?

2. In an experiment the following replicate set of volume measurements (cm^3) was recorded:

$$25.35, 25.80, 25.28, 25.50, 25.45, 25.43$$

(a) Calculate the mean of the raw data.
(b) Using the rejection quotient (*Q*-test) reject any questionable results.
(c) Recalculate the mean and compare it with the value obtained in 2(a).

3. Calculate the absolute standard deviation and the relative standard deviation (%) of the following replicate set of data. Comment upon the precision of the measurement.

Weight of component/g:

$$5.346, 5.362, 5.379, 5.335, 5.341.$$

4. The three sets of replicate results below were accumulated for the analysis of the same sample. Pool these data to obtain the most efficient estimate of the mean analyte content and the standard deviation.

Lead content/ppm:

Set 1	Set 2	Set 3
9.76	9.87	9.85
9.42	9.64	9.91
9.53	9.71	9.42
	9.81	9.49

5. Calculate the confidence interval (90 %) about the mean of the data in question 3.

6. An exercise was carried out to compare the precisions of analytical measurements being made in two different laboratories. A completely homogeneous sample was supplied to both laboratories and the following analytical results (%) obtained by the same method.

Laboratory 1	Laboratory 2
34.97	35.02
34.85	34.96
34.94	34.99
34.88	35.07
	34.85

Comment upon the precisions attained in the two laboratories.

7. On the basis of extensive past analysis the iron content of an homogeneous ore sample is accepted to be 19.85 %. An analysis of seven samples gave an average of 20.06 % Fe. Within what probability levels is a determinate error indicated in the analysis if the absolute standard deviation of the results is 0.14 %?

8. The efficiency of mixing of powdered solids in a process stream is being investigated. Two samples of the mixture, each of about 1 kg, are taken from different parts of the same hopper. After grinding to a fine powder, analytical samples were obtained using the coning and quartering technique. Subsequent analysis gave the following results for one component:

Sample A—six analyses, mean 12.32 %
Sample B—four analyses, mean 12.63 %.

The precisions were found to be similar and the pooled standard deviation (absolute) computed to be 0.071 %. Comment upon the efficiency of the mixing.

9. Replicate analysis by a new method for a constituent of a standard sample gave the following results (%):

0.45, 0.48, 0.46, 0.60, 0.42, 0.50, 0.41.

The content of the standard is accepted to be 0.47 %.
Assess these data and draw conclusions concerning the suitability of the method of analysis for routine use.

10. A standard reference material with an accepted analyte content of 5.85 ppm was used to compare the performance of two alternative methods of analysis. Solutions were prepared using a range of sample weights. Each solution was analysed by the two

methods. Use the results obtained to assess any determinate errors exhibited by the methods.

Sample weight/g	Result, Method 1/ppm	Result, Method 2/ppm
0.125	6.37	5.72
0.612	5.97	5.69
0.334	6.10	5.70
0.781	5.93	5.71
0.271	6.20	5.68
0.457	6.03	5.74

Further Reading

BARNETT, V. and LEWIS, T., *Outliers in Statistical Data*, Wiley, 1978.

CALCUTT, R. and BODDY, R., *Statistics for Analytical Chemists*, Chapman and Hall, 1983.

KATEMAN, F. and PIJPERS, F. W., *Quality Control in Analytical Chemistry*, Wiley-Interscience, 1981.

KEALEY, D., *Experiments in Modern Analytical Chemistry*. Blackie, Glasgow, 1986.

LEE, J. D. and LEE, T. D., *Statistics and Computer Methods in BASIC*, Van Nostrand, 1983.

MILLER, J. C. and MILLER, J. N., *Statistics for Analytical Chemistry (3rd Ed.)*, Ellis Horwood/Wiley, 1994.

Chapter 3

pH, Complexation and Solubility Equilibria

Since so many analytical procedures involve solution chemistry an understanding of the principles is essential. Complex formation, precipitation reactions and the control of pH are three aspects with special relevance in analysis.

In the simplest situation where a solution of a weak acid is used the degree of dissociation of the acid and hence the composition of the solution will depend on the pH of the solution. If, as is often the case, a complexing agent is itself a weak acid, its complexing reactions will also be pH dependent.

The formation of a metal ion complex may be used to facilitate its precipitation or solvent extraction, whilst modification of its charge will influence its electrochemical properties.

DEFINITIONS

Acid. A species which donates protons in a reaction (Lowry-Brønsted) or accepts electrons (Lewis).

Base. A species which accepts protons in a reaction (Lowry-Brønsted) or donates electrons (Lewis).

Complex. A compound which is formed in a *complexation reaction* between two or more species which are capable of independent existence. Most complexes of analytical importance involve dative bonds between Lewis bases and metal ions.

Ligand. A species with at least one Lewis basic site which can participate in complex formation.

Strong electrolyte. A compound which is almost completely dissociated in solution.

Weak electrolyte. A compound which remains significantly undissociated in solution.

'p'-*Notation*—defined by

$$pX = -\log_{10}X \qquad (3.1)$$

and used for convenience in handling quantities which vary in magnitude through many powers of ten. For example, the H_3O^+ concentration of an aqueous solution may vary from 10 mol dm^{-3} to 10^{-14} mol dm^{-3}. Expressed in the 'p'-notation this becomes pH $= -1$ to pH $= 14$. Similarly the dissociation constant for acetic acid $K_a = 1.75 \times 10^{-5}$ or $10^{-4.76}$ becomes $pK_a = 4.76$.

3.1   Chemical Reactions in Solution

The rate of a chemical reaction and the extent to which it proceeds play an important role in analytical chemistry. The fundamental problem which faces the analyst arises because thermodynamic data will indicate the position of equilibrium that can be reached but not the time taken to reach that position. Similarly, a compound may be thermodynamically unstable because its decomposition will lead to a net decrease in free energy, whilst a high activation energy for the decomposition reaction restricts the rate of decomposition. In practical terms such a compound would be stable, e.g. NO. It is thus essential to consider all analytical reactions from both thermodynamic and kinetic viewpoints.

EQUILIBRIUM CONSTANTS

An equation may be written for a generalized reaction

$$aA + bB + cC + \cdots = xX + yY + zZ + \cdots \qquad (3.2)$$

where a, b, c, x, y, z are numbers indicating its stoichiometry. The position of equilibrium is expressed by the thermodynamic *equilibrium constant* K_e which is defined by

$$K_e = \frac{[X]^x[Y]^y[Z]^z \ldots}{[A]^a[B]^b[C]^c \ldots} \qquad (3.3)$$

In this expression, the square brackets refer to the activity of the component although it is more convenient to use its concentration. This approximation is generally satisfactory, except at very high concentrations, and is particularly suitable for analytical use. Where it is necessary to distinguish between the constant obtained using concentrations and the true thermodynamic equilibrium constant K_e, the former may be termed the *equilibrium quotient* and assigned the symbol Q. The exact relation between K_e and Q

has been the subject of much investigation and speculation. In this text, no distinction will be made. For all cases the values of equilibrium constants quoted are based on concentrations and represent approximations to the true values. Equilibrium constants are variously named depending upon the nature of the equilibria to which they refer, e.g. *dissociation constants, formation constants* or *solubility products*. The most convenient concentration units for analytical chemistry are $mol\,dm^{-3}$ and the dimensions of the constants will be derived from these and determined by the stoichiometry of the reaction.

e.g.

$a + b + c + \cdots$ (reactants)	$x + y + z + \cdots$ (products)	dimensions of K_e
3	3	dimensionless
3	4	$mol\,dm^{-3}$
3	2	$mol^{-1}\,dm^3$

Care must be exercised in making direct comparisons between K_e values, and due attention should be given to their dimensions.

Equilibria in Analytical Reactions
In analytical chemistry it is often necessary to *shift* the position of equilibrium in a reaction so as to obtain larger concentrations of the desired products. For example, sometimes a complete reaction between an analyte and a reagent or between an interfering ion and a masking reagent is essential. One important and widely used method of achieving a shift, involves the removal of one of the products from the system, e.g. by distillation or precipitation. In these circumstances to maintain K_e constant the reactants will be steadily converted to products until one or all are exhausted. An alternative approach that is often useful uses an excess of the reagent which in most cases causes a shift towards the products of the reaction. This is the so called *common ion effect*. Too large an excess, however, can partially reverse the shift or lead to some other undesirable effect (p. 215).

When comparing similar or parallel reactions, consideration of the changes in *Gibbs free energy* ΔG, *enthalpy* ΔH and *entropy* ΔS can be valuable. The equilibrium constant is related to these quantities by two fundamental thermodynamic expressions

$$\Delta G^{\ominus} = -RT \ln K_e \qquad (3.4)$$

and

$$\Delta G^{\ominus} = \Delta H^{\ominus} - T\Delta S^{\ominus} \qquad (3.5)$$

From these it will be seen that a more negative value of $\Delta H^{\ominus}$ or a more positive value of $\Delta S^{\ominus}$ will in each case lead to an increase in K_e. Furthermore, there is a hypothesis that if the *degree of disorder* in a system increases it will almost certainly be accompanied by an increase in entropy (i.e. ΔS becomes more positive). A reaction that results in a net increase in the number of species will in these terms be favoured. Although this hypothesis has been usefully applied in a number of cases, its use is necessarily limited by the difficulty of deciding which side of a reaction has the greater disorder.

Conditional Equilibrium Constants

The discussion of equilibrium given above will be useful in developing ideas later in the book. However, there are cases where a modified approach is needed, particularly when competing equilibria are being considered. A typical situation might be the formation of a compound between metal ion M^{n+} and a ligand L^- in an aqueous solution which also contains a weak acid AH. Equation (3.6) summarizes the equilibria.

$$
\begin{array}{l}
MOH^{(n-1)+} = OH^- + \boxed{M^{n+} + nL^- = ML_n} \\[4pt]
\;\parallel \qquad\qquad\qquad\quad + \qquad\quad + \\[2pt]
M(OH)_2^{(n-2)+} \qquad\qquad A^- \qquad\; H^+ \\[2pt]
\qquad\qquad\qquad\qquad\;\; \parallel \qquad\quad \parallel \qquad\qquad (3.6)\\[2pt]
\text{etc.} \qquad\qquad\quad MA^{(n-1)+}\; LH \\[2pt]
\qquad\qquad\qquad\qquad\;\; \parallel \\[2pt]
\qquad\qquad\qquad\quad MA_2^{(n-2)+}
\end{array}
$$

It is the reaction in the box that is of analytical importance. The position of equilibrium is conveniently expressed by a modified or *conditional equilibrium constant* K'_{ML} in which allowance has been made for the competing side reactions. The fraction of M^{n+} which has not reacted with L^- and remains present as M^{n+} is given by

$$
\alpha_M = \frac{[M]}{[M] + [MOH] + [M(OH)_2] \,----+\, [MA] + [MA_2]\,---} \qquad (3.7)
$$

and similarly for L^- not reacted with M^{n+}

$$
\alpha_L = \frac{[L]}{[L] + [LH] \,----} \qquad (3.8)
$$

(charges have been omitted for simplicity).

The conditional equilibrium constant may be defined in terms of the true thermodynamic equilibrium constant K_{ML} and the appropriate values of α.

$$
K'_{ML} = K_{ML}\alpha_L\alpha_M \qquad (3.9)
$$

Such equilibrium constants enable calculations and deductions to be made for real systems and may be used to assess the progress of a particular reaction amongst a number of competing or interfering reactions. From this consideration the possibility of *masking* interfering reactions also emerges. Suppose the solution above contains a second metal ion N^{n+} which can also react with L^-. If the amount of L^- is limited N^{n+} will be in competition with M^{n+}. Its effect, however, may be masked if A^- can be selected to react preferentially with N^{n+}, reducing α_N and hence K'_{NL} relative to K'_{ML}. These ideas will be particularly useful when handling polyprotic acid and complex forming reactions (p. 200).

Solvent Effects

Assessment of a typical analytical reaction in solution requires consideration of the solvent participation. It is likely that both reactants and products will undergo significant solvation, and the *solvation energy* will materially affect ΔG° and ΔH° for the analytical reaction. Solvation energies may be high, typically in the range -400 to -4000 kJ mol^{-1} for hydration. Solvent molecules participate in the formation of ordered species in solution with the reactants and products of the reaction, thus affecting ΔS° also. Notably the interpretation of chelating processes depends upon solvent considerations (p. 52).

Temperature Effects on Equilibrium Constants

If heat is applied to an endothermic reaction K_e will increase and, conversely, heat applied to an exothermic reaction will result in a decrease in K_e. At the same time heat will tend to increase the disorder of the system and favour the side of the reaction with the greatest potential disorder. However the degree of disorder can rarely be assessed easily and the magnitude of the latter effect is difficult to determine. Overall the variation of K_e is unlikely to exceed a few per cent per kelvin.

KINETIC FACTORS IN EQUILIBRIA

The rate at which reactions occur is of theoretical and practical importance, but it is not relevant to give a detailed account of reaction kinetics, as analytical reactions are generally selected to be as fast as possible. However, two points should be noted. Firstly, most ionic reactions in solution are so fast that they are *diffusion controlled*. Mixing or stirring may then be the rate controlling step of the reaction. Secondly, the reaction rate varies in proportion to the cube of the thermodynamic temperature, so that heat may have a dramatic effect on the rate of reaction. Heat is applied to reactions to attain the position of equilibrium quickly rather than to displace it.

3.2   Solvents in Analytical Chemistry

Solvents are conveniently divided into those which promote the ionization
of a solute (*ionizing solvents*) and those which do not (*non-ionizing solvents*).
As always such divisions are not entirely satisfactory but table 3.1 illustrates
the classification of some familiar solvents.

Table 3.1   Some typical solvents

IONIZING	INTERMEDIATE	NON-IONIZING
water	alcohols containing	chloroform
hydrochloric acid	up to four carbon	hexane
acetic acid	atoms	benzene
ammonia		ethers
sulphur dioxide		esters
bromine trifluoride		
amines		

IONIZING SOLVENTS

Some ionizing solvents are of major importance in analytical chemistry
whilst others are of peripheral interest. A useful subdivision is into *protonic
solvents* such as water and the common acids, or *non-protonic solvents* which
do not have protons available. Typical of the latter subgroup would be
sulphur dioxide and bromine trifluoride. Non-protonic ionizing solvents have
little application in chemical analysis and subsequent discussions will be
restricted to protonic solvents. Ionizing solvents have one property in
common, *self-ionization*, which reflects their ability to produce ionization of
a solute; some typical examples are given in table 3.2. Equilibrium constants
for these reactions are known as *self-ionization constants*.

It will be seen from these examples that the process of self-ionization in a
protonic solvent involves the transfer of a proton from one solvent molecule

Table 3.2   Some self-ionization equilibria

$$2H_2O \qquad = H_3O^+ + OH^-$$
$$2CH_3COOH = CH_3COOH_2^+ + CH_3COO^-$$
$$2HCl \qquad = H_2Cl^+ + Cl^-$$
$$2NH_3 \qquad = NH_4^+ + NH_2^-$$
$$2RNH_2 \qquad = RNH_3^+ + RNH_2^-$$

(R = alkyl or aryl)

to another. Thus, the solvent is acting simultaneously as a Lowry-Brønsted acid and as a base.

A final subdivision of ionizing protonic solvents can be made in terms of the behaviour of the solvent towards available protons from a solute. A basic or *protophilic* solvent such as ammonia or an amine will coordinate protons strongly, and in so doing accentuate the acidic properties of the acid.

$$RNH_2 + RCOOH = RNH_3^+ + RCOO^- \quad (3.10)$$

On the other hand an acid or *protogenic* solvent will be a poor proton acceptor, accentuating basic properties. This effect is exemplified by the solution of nitric acid in anhydrous hydrofluoric acid, which shows how nitric acid (normally regarded as a strong acid) can behave as a Lowry-Brønsted base.

$$H_2F_4 + HNO_3 = HF_4^- + H_2NO_3^+ \quad (3.11)$$

Water and other hydroxylic solvents both donate and accept protons with reasonable facility and are termed *amphiprotic*.

NON-IONIZING SOLVENTS

The major uses of non-ionizing solvents in chemical analysis are two-fold. They may be used simply to provide media for the dissolution and reaction of covalent materials, or they may play a more active part in a chemical process. For example, oxygen containing organic solvents can be used to effect the solvent extraction of metal ions from acid aqueous solutions; the lone pair of electrons possessed by the oxygen atom forming a dative bond with the proton followed by the extraction of the metal ion as an association complex.

This process is discussed more fully in chapter 4.

3.3 Acid-Base Equilibria

The Lowry-Brønsted concept provides a basis for the interpretation of reactions in protonic solvents. This concept may be summarized by considering the generalized equilibrium

$$AH + B = A^- + BH^+ \quad (3.12)$$

where AH and B represent the acid and base for the forward reaction and A^- and BH^+ the *conjugate base* and *conjugate acid* for the reverse reaction. Thus the dissociation of a typical acid in water may be represented by

$$AH + H_2O = A^- + H_3O^+ \qquad (3.13)$$

with the water acting as a base and H_3O^+ being the conjugate acid. The equilibrium constant is given by

$$K_a = \frac{[A^-][H_3O^+]}{[AH][H_2O]} \qquad (3.14)$$

and is known as the *acid dissociation constant*. In the case of a base a parallel treatment may be used in which the water acts as an acid

$$B + H_2O = BH^+ + OH^- \qquad (3.15)$$

and K_b, the *base dissociation constant*, is given by

$$K_b = \frac{[BH^+][OH^-]}{[B][H_2O]} \qquad (3.16)$$

The solvent term $[H_2O]$ varies by a negligible amount in such reactions and is incorporated in the constants K_a and K_b. If the concentrations are expressed in mol dm^{-3}, K_a and K_b have the same units. Table 3.3 lists pK_a values for some typical compounds. The behaviour of the conjugate base may be represented in line with equation (3.15).

$$A^- + H_2O = AH + OH^- \qquad (3.17)$$

whence

$$K_b = \frac{[AH][OH^-]}{[A^-]}$$

and

$$K_a K_b = \frac{[A^-][H_3O^+][AH][OH^-]}{[AH][A^-]}$$

$$= [H_3O^+][OH^-] = K_w \qquad (3.18)$$

K_w is the self ionization constant for water (table 3.2) and equation (3.18) reflects the not surprising inverse relation between K_a and K_b. It is only when K_a and K_b for a compound are of different magnitudes that it may be classified as an acid or base. An example which is difficult to classify is hypoiodous acid (HOI) where $K_a = 2.5 \times 10^{-11}$ mol dm^{-3} and $K_b = 3.2 \times 10^{-10}$ mol dm^{-3}. Although K_b has been widely used in the past, it is a quantity which is largely redundant, for K_a (or pK_a) may be used to express the strength of bases as well as acids, see table 3.3.

Table 3.3 shows that values of K_a and K_b vary over a wide range and also that there is no clear dividing line between strong acids and weak acids or strong bases and weak bases. However as a rough guide weak acids may be regarded as those having values of pK_a in the range 4 to 10, those having p$K_a = 4$ being called 'strong weak acids' and those with p$K_a = 8$ to 10 'very weak acids'. The pH of a solution of a strong acid or base may be related directly to the concentration of the acid or base. However, weak acid or base systems present a rather more complex pattern.

Table 3.3 pK_a values for some Lowry-Brønsted acids

ACID	pK_1	pK_2	pK_3	pK_4
acetic CH_3COOH	4.76			
ammonium NH_4^+	9.25			
chloracetic $CH_2ClCOOH$	2.86			
diethanolammonium $NH_2(CH_2CH_2OH)_2^+$	9.00			
ethanolammonium $NH_3(CH_2CH_2OH)^+$	9.49			
ethylenediaminetetraacetic (EDTA) $(HOOCCH_2)_2N—CH_2—CH_2—N(CH_2COOH)_2$	2.0	2.67	6.16	10.27
formic $HCOOH$	3.75			
hydrocyanic HCN	9.22			
nitric HNO_3	-1.4			
nitrilotriacetic (NTA) $N(CH_2COOH)_3$	1.66	2.95	10.28	
nitrous HNO_2	3.29			
phenol C_6H_5OH	9.98			
phosphoric H_3PO_4	2.17	7.21	12.36	
sulphuric H_2SO_4	-1.96			
sulphurous H_2SO_3	1.76	7.21		

WEAK ACID AND WEAK BASE EQUILIBRIA

Equation (3.14) may be rewritten

$$[AH]/[A^-] = [H_3O^+]/K_a \qquad (3.19)$$

and presented in a log form using the 'p' notation

$$-\log_{10}[AH]/[A^-] = pH - pK_a \qquad (3.20)$$

This is a useful equation as it gives the relation between the pH of a solution, the dissociation constant for the acid, and the composition of the solution. When the relation is represented graphically (figure 3.1) some further valuable

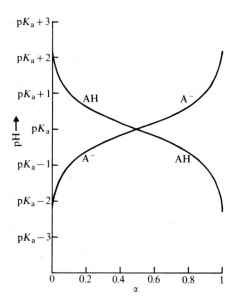

Figure 3.1.   Distribution diagram of α against pH for acid AH and its conjugate base A^-.

points emerge. As it is the difference between the pH of the solution and pK_a for the acid which is important rather than the pH alone, the pH axis is graduated in units of pH relative to pK_a and the horizontal axis in the fractional concentrations of the two species α_{AH}, and α_{A^-}, which are given by

$$\alpha_{AH} = \frac{[AH]}{[AH] + [A^-]}, \qquad \alpha_{A^-} = \frac{[A^-]}{[AH] + [A^-]} \qquad (3.21)$$

The first point to be noted is that if AH and A^- are present in equal quantities, pH $= pK_a$. Secondly, when the pH of the solution is in the range $pK_a \pm 1$, relatively large changes in α_{AH} or α_{A^-} correspond to small changes in pH. To put this another way, the solution can tolerate quite large changes in its composition whilst the pH is maintained almost constant. When this situation exists the solution is said to be *buffered* and the solution is known as a *buffer solution*.

BUFFERS AND pH CONTROL

It is often necessary to control the pH of a solution in circumstances where hydrogen ions are being generated or absorbed. The buffer solution concept developed above is a versatile and widely used method of pH control. A buffer may be prepared to meet a particular pH requirement by the use of standard pK_a values in conjunction with equation (3.21). This equation gives a good approximation provided that the solution contains at least 0.1 mol dm^{-3} of the buffer and that the required pH lies close to the value of pK_a. For example, if a solution needs to be buffered to pH $= 9$, the ammonium ion ($pK_a = 9.62$) can be used. Substitution in equation (3.20) enables the necessary solution composition to be calculated

$$-\log_{10}([NH_4^+]/[NH_3]) = 9.00 - 9.26$$

whence the ratio $[NH_4^+]/[NH_3] = 1.82$. A typical buffer solution might thus be prepared containing 1 mol dm^{-3} of NH_3 and 1.82 mol dm^{-3} of NH_4^+ ions. The pH obtained will be independent of the buffer concentration provided that the ratio of acid to conjugate base is maintained constant. However if the buffer solution is to stabilize the pH when large amounts of H^+ ions are involved it must be reasonably concentrated (i.e. greater than 0.1 mol dm^{-3}).

In practice, a buffer solution is prepared by the partial neutralization of the selected weak acid or base with a suitable strong acid or base, or by the addition of the calculated amount of the corresponding salt. The assumption is made that the salt is completely dissociated in solution, e.g. an $NH_3 + NH_4^+$ buffer may be produced by the partial neutralization of an ammonia solution with hydrochloric acid or by the addition of the appropriate quantity of ammonium chloride to an initial ammonia solution. To facilitate this practical approach, equation (3.20) may be written in a different form,

$$pH = pK_a - \log_{10}([Acid]/[Salt]) \qquad (3.22)$$

The approximate pH range over which a buffer solution remains effective can be deduced from figure 3.1. The limits of effective buffering can be seen as the points at which the ratio $[AH]/[A^-]$ becomes $10:1$ or $1:10$ whence substitution in equation (3.20) yields

$$pH = pK_a \pm 1 \qquad (3.23)$$

Table 3.4 shows some typical buffer solutions.

Table 3.4   Some typical buffer solutions

SOLUTIONS	pH RANGE
phthalic acid and potassium hydrogen phthalate	2.2–4.2
citric acid and sodium citrate	2.5–7.0
acetic acid and sodium acetate	3.8–5.8
sodium dihydrogen phosphate and disodium hydrogen phosphate	6.2–8.2
ammonia and ammonium chloride	8.2–10.2
borax and sodium hydroxide	9.2–11.2

THE pH OF SALT SOLUTIONS

When an acid in solution is exactly neutralized with a base the resulting solution corresponds to a solution of the salt of the acid–base pair. This is a situation which frequently arises in analytical procedures and the calculation of the exact pH of such a solution may be of considerable importance. The neutralization point or end point in an acid–base titration is a particular example (chapter 5). Salts may in all cases be regarded as strong electrolytes so that a salt AB derived from acid AH and base B will dissociate completely in solution. If the acid and base are strong, no further reaction is likely and the solution pH remains unaffected by the salt. However if either or both acid and base are weak a more complex situation will develop. It is convenient to consider three separate cases, (a) weak acid–strong base, (b) strong acid–weak base and (c) weak acid–weak base.

(a) *Weak Acid–Strong Base Solutions*
The conjugate base A^- will react with water and undergo *hydrolysis*,

$$A^- + H_2O = AH + OH^- \qquad (3.24)$$

producing undissociated acid and hydroxyl ions with an accompanying rise in pH. The equilibrium constant for this reaction is known as the *hydrolysis constant* K_h.

$$K_h = \frac{[AH][OH^-]}{[A^-][H_2O]} \qquad (3.25)$$

(As in equations (3.14) and (3.16) the solvent term $[H_2O]$ is by convention omitted. K_h is simply related to K_w and K_a by equation (3.26), and as such is a redundant constant whose use should be discouraged.)

$$\frac{K_w}{K_a} = \frac{[H^+][OH^-][AH]}{[H^+][A^-]} = \frac{[OH^-][AH]}{[A^-]} = K_h \qquad (3.26)$$

Equation (3.24) shows that the amounts of AH and OH^- generated in the hydrolysis are equal. Furthermore, if it is assumed that only a small amount of the salt is hydrolysed, the concentration C of the salt dissolved is approximately the same as the concentration of A^-. Then from (3.26)

$$K_w/K_a = [OH^-]^2/C \qquad \text{and} \qquad [OH^-] = (CK_w/K_a)^{\frac{1}{2}}$$

from equation (3.18), $[H^+] = K_w/[OH^-]$ whence $[H^+] = (K_w K_a/C)^{\frac{1}{2}}$ and finally

$$pH = \tfrac{1}{2}pK_w + \tfrac{1}{2}pK_a + \tfrac{1}{2}\log C \qquad (3.27)$$

The pH of the solution will be dependent upon both pK_a for the acid AH and on the concentration of the salt dissolved in the solution. For example, the pH of solutions of sodium cyanide may be calculated as follows:

$$pK_a \text{ for HCN} = 9.22$$
$$pH = 7.0 + 4.6 + \tfrac{1}{2}\log C$$
$$= 11.6 + \tfrac{1}{2}\log C$$

Thus, when

$$C = 1 \text{ mol dm}^{-3} \qquad pH = 11.6$$
$$C = 0.1 \text{ mol dm}^{-3} \qquad pH = 11.1$$
$$C = 2 \text{ mol dm}^{-3} \qquad pH = 11.8$$

(b) *Strong Acid–Weak Base*
Similar reasoning shows hydrolysis leading to the production of hydrogen ions and a drop in pH,

$$B^+ + H_2O = BOH + H^+ \qquad (3.28)$$

and enables an analogous expression for pH to be derived, i.e.

$$pH = \tfrac{1}{2}pK_w - \tfrac{1}{2}pK_b - \tfrac{1}{2}\log C \qquad (3.29)$$

(c) *Weak acid–Weak base*
Using the same approach again, both hydrolytic processes seen above will be expected to occur and the pH of the solution will depend on the relative values of pK_a and pK_b, but be independent of the concentration,

$$pH = \tfrac{1}{2}pK_w + \tfrac{1}{2}pK_a - \tfrac{1}{2}pK_b \qquad (3.30)$$

3.4 Complexation Equilibria

The thermodynamic stability of a complex ML_n formed from an acceptor metal ion M and ligand groups L may be approached in two different but related ways. (The difference between the two approaches lies in the way in which the formation reaction is presented.) Consistent with preceding sections, an equilibrium constant may be written for the formation reaction. This is the *formation constant* K_f. In a simple approach, the effects of the solvent and ionic charges may be ignored. A stepwise representation of the reaction enables a series of *stepwise formation constants* to be written (table 3.5).

Table 3.5   Stepwise formation constants
$(mol^{-1}\ dm^3)$

$M + L = ML$	$K_{f_1} = \dfrac{[ML]}{[M][L]}$
$ML + L = ML_2$	$K_{f_2} = \dfrac{[ML_2]}{[ML][L]}$
$ML_2 + L = ML_3$	$K_{f_3} = \dfrac{[ML_3]}{[ML_2][L]}$
$ML_{n-1} + L = ML_n$	$K_{f_n} = \dfrac{[ML_n]}{[ML_{n-1}][L]}$

Alternatively, a set of *overall formation constants* may be defined (table 3.6).

Table 3.6   Overall formation constants
$(mol^{-n}\ dm^{3n})$

$M + L = ML$	$\beta_1 = \dfrac{[ML]}{[M][L]}$
$M + 2L = ML_2$	$\beta_2 = \dfrac{[ML_2]}{[M][L]^2}$
$M + 3L = ML_3$	$\beta_3 = \dfrac{[ML_3]}{[M][L]^3}$
$M + nL = ML_n$	$\beta_n = \dfrac{[ML_n]}{[M][L]^n}$

The relations between values of K_f and values of β are simple and easily seen. For example, β_3 may be multiplied by $[ML]/[ML]$ and $[ML_2]/[ML_2]$ to give

$$\beta_3 = \frac{[ML_3]}{[M][L]^3} \frac{[ML]}{[ML]} \frac{[ML_2]}{[ML_2]}$$

an expression which can be rearranged in the form

$$\beta_3 = \frac{[ML]}{[M][L]} \frac{[ML_2]}{[ML][L]} \frac{[ML_3]}{[ML_2][L]}$$

or

$$\beta_3 = K_{f_1} K_{f_2} K_{f_3} \tag{3.31}$$

THE FORMATION OF COMPLEXES IN SOLUTION

It is implied above that all the constituent species from M and L to ML_n may exist together in solution. The solution composition will depend on the nature of M and L and the amounts present together with the relative values of K_f. As a general rule K_f values show a steady decrease as shown in the $Cd^{2+} + CN^-$ system. This steady decrease is brought about by three major

Table 3.7

ML_n	$K_f/mol^{-1} dm^3$
$Cd(CN)^+$	5.0×10^5
$Cd(CN)_2$	1.3×10^5
$Cd(CN)_3^-$	4.3×10^4
$Cd(CN)_4^{2-}$	3.5×10^3

factors: statistical; coulombic; steric. Once the first ligand group is attached the next stage will decrease in probability because there are less sites available. Furthermore, the positive charge characteristically present on the metal atom will be reduced, and with it the coulombic attraction for the ligand. This may even be converted to a repulsion as subsequent ligand groups are attached. The third factor concerns the bulky nature of many ligands which will place a steric restriction on the reaction. This restriction reaches its ultimate in the EDTA type of ligand (p. 199), which forms a cage around the acceptor atom and prevents the attachment of any further ligands, irrespective of the nature of the acceptor ion, figure 5.4.

The effect of this variation in values of K_f can be seen by returning to the $Cd^{2+} + CN^-$ system. Commonly a ligand is added to a solution of the acceptor ion, first forming ML then ML_2, ML_3 and finally ML_4, each becoming the predominant species in turn until ML_4 is formed to the exclusion of all others when an excess of the ligand has been added. Figure 3.2 illustrates this pattern graphically.

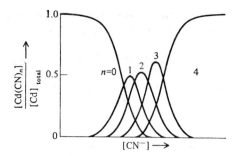

Figure 3.2.   The relation between solution composition and the amount of cyanide added in the $Cd^{2+} + CN^-$ system.

THE CHELATE EFFECT

Many ligand molecules contain more than one donating centre and are called *multidentate ligands*. If two or more dative links are formed with the same acceptor ion, a ring compound is produced. Compounds of this type are known as *chelates* and the ligands are *chelating agents*. A typical example is ethylenediamine which reacts with many metal ions to form stable 5-membered ring compounds,

Other important examples are 'oxine' and EDTA. Almost invariably, the formation of a ring system leads to a large increase in stability relative to the comparable non-ring compound. Thus Ni(II)ethylenediamine $[Nien]^{2+}$ may be compared with $[Ni(NH_3)_2]^{2+}$. For each, the number of ligand-acceptor bonds is the same, and furthermore, both involve nitrogen donor groups. It is tempting to compare the stability quantitatively by the stability constants β which are $10^{7.51}$ and $10^{5.00}$ respectively. However, such a comparison is not valid because of the different dimensions of the two constants and such stability comparisons can only be made in a loose sort of way. To rationalize the experimentally observed stability it is necessary to consider the formation reaction in a slightly more detailed way. Stage 1 of this reaction involves the breaking of $Ni—H_2O$ bonds in the $[Ni(H_2O)_6]^{2+}$ species and stage 2 the formation of $Ni—N$ bonds. Both stages are similar in chelate and non-chelate systems so it is unlikely that $\Delta H^{\ominus}$ or $\Delta G^{\ominus}$ will be significantly different.

Comparisons of $\Delta S^{\ominus}$ between the two formation/reactions however, show differences, as equations (3.32) and (3.33) illustrate.

$$[Ni(H_2O)_6]^{2+} + 2NH_3(aq) = [Ni(H_2O)_4(NH_3)_2]^{2+} + 2H_2O \quad (3.32)$$

$$[Ni(H_2O)_6]^{2+} + en = [Ni(H_2O)_4en]^{2+} + 2H_2O \quad (3.33)$$
$$\text{(en = ethylenediamine)}$$

In the first equation the disorder remains the same, whilst in the second case it increases, and is accompanied by an entropy increase, i.e. ΔS becomes more positive and the reaction is entropy favoured. Hence the chelate effect may be regarded primarily as an entropy effect with some additional stability conferred by delocalization of π electrons into the ring system. This latter effect is particularly likely if the ligands are highly conjugated.

3.5 Solubility Equilibria

The solubility of solids in liquids is an important process for the analyst, who frequently uses dissolution as a primary step in an analysis or uses precipitation as a separation procedure. The dissolution of a solid in a liquid is favoured by the entropy change as explained by the principle of maximum disorder discussed earlier. However it is necessary to supply energy in order to break up the lattice and for ionic solids this may be several hundred kilojoules per mole. Even so many of these compounds are soluble in water. After break up of the lattice the solute species are dispersed within the solvent, requiring further energy and producing some weakening of the solvent–solvent interactions. The energy needed to bring about this change can only be supplied by the solvation of the solute species utilizing van der Waals' type ion–dipole interactions and to a lesser extent dipole–dipole reactions, etc. The total energy available from this source, the *solvation energy*, may be in the range -400 to $-4\,000$ kJ mol^{-1} for aqueous systems containing ionic solids. On the other hand non-ionizing solvents will have a much lower ability to produce solvation and insufficient energy is available to break up ionic lattices and produce a solution. Solvent–solute interactions between an ionizing solvent and a covalent solid are small but may be large enough to overcome the low lattice energy for many solids although their solubility will still be very low as a result of the associated nature of the solvent. Covalent solvent–solute systems have minimal interactions and may approximate in behaviour to ideal mixtures. The dissolution equilibrium for an ionic compound AB may be summarized by

$$AB(\text{solid}) + \text{Solvent} = A^+(\text{solv}) + B^-(\text{solv}) \quad (3.34)$$

SOLUBILITY PRODUCTS

The solubility of a sparingly soluble salt is expressed by the equilibrium constant for the reaction, equation (3.34), e.g.

$$K = \frac{[A^+][B^-]}{[AB][S]^n}$$

Both [AB] and [S] are unchanged in solubility reactions of sparingly soluble salts, hence the above equation may be rewritten thus

$$K_{sp} = [A^+][B^-]$$

K_{sp}, known as the *solubility product*, is widely used as a measure of the solubility of sparingly soluble salts. It should be noted that the dimensions of this constant will change according to the stoichiometry of the reaction. Temperature effects on solubility products are readily assessed as most solubility reactions are clearly seen as endothermic and disorder increasing. Raising the temperature will thus increase K_{sp} together with the solubility of the solid.

Problems
1. Calculate the concentration of sodium acetate needed to produce a pH of 5.0 in a solution of acetic acid (0.1 M) at 25°C. pK_a for acetic acid is 4.756 at 25°C.

2. What is the pH at 25°C of a solution which is 1.5 M with respect to formic acid and 1 M with respect to sodium formate? pK_a for formic acid is 3.751 at 25°C.

3. Calculate the pH of a 2 M solution of sodium benzoate. pK_a for benzoic acid is 4.01 at 25°C, $pK_w = 14$.

4. What is the pH of a 0.025 M solution of ammonium acetate at 25°C? pK_a of acetic acid at 25°C is 4.76, pK_a of the ammonium ion at 25°C is 9.25, pK_w is 14.

Further Reading
BETTERIDGE, J. and HALLAM, H. E., *Modern Analytical Methods*, RSC Monograph, London, 1973.
BLACKBURN, T. R., *Equilibrium*, Holt, Rinehart and Winston, New York, 1969.
HARTLEY, F. R., BURGESS, C. and ALCOCK, R., *Solution Equilibrium*, Ellis Horwood, Chichester, 1980.
JEFFEREY, G. H., BASSETT, J., MENDHAM, J. and DENNEY, R. C., *Vogel's Textbook of Quantitative Chemical Analysis (5th Ed.)*, Longman, 1989.
PERRIN, D. D., *Masking and Demasking of Chemical Reactions*, Wiley, New York, 1970.
PERRIN, D. D., *Organic Complexing Reagents: Structural Behaviour and Application to Inorganic Analysis*, Knieger, 1979.

Chapter 4

Separation Techniques

If it were possible to identify or quantitatively determine any element or compound by simple measurement no matter what its concentration or the complexity of the matrix, separation techniques would be of no value to the analytical chemist. Most procedures fall short of this ideal because of interference with the required measurement by other constituents of the sample. Many techniques for separating and concentrating the species of interest have thus been devised. Such techniques are aimed at exploiting differences in physico-chemical properties between the various components of a mixture. Volatility, solubility, charge, molecular size, shape and polarity are the most useful in this respect. A change of phase, as occurs during distillation, or the formation of a new phase, as in precipitation, can provide a simple means of isolating a desired component. Usually, however, more complex separation procedures are required for multi-component samples. Most depend on the selective transfer of materials between two immiscible phases. The most

Table 4.1 Classification of separation techniques

TECHNIQUE	PHASE SYSTEM
solvent extraction	liquid–liquid
gas chromatography	gas–liquid
	gas–solid
liquid chromatography	liquid–liquid
	liquid–solid
thin-layer chromatography	liquid–solid
	liquid–liquid
ion-exchange and	liquid–solid
gel-permeation chromatography	liquid–liquid
supercritical fluid chromatography	supercritical fluid–liquid or solid
electrophoresis	liquid

widely used techniques and the phase systems associated with them are summarized in table 4.1.

All separation techniques involve one or more chemical equilibria, consequently the degree of separation achieved can vary greatly according to experimental conditions. To a large extent, attainment of optimum conditions has to be approached empirically rather than by application of rigid theory. In the following sections, which deal with *solvent extraction*, *chromatography* and *electrophoresis*, the minimum theory necessary for an understanding of the basic principles is presented.

4.1 Solvent Extraction

SUMMARY

Principles

Selective transfer of material in microgram to gram quantities between two immiscible liquid phases; separations based on solubility differences; selectivity achieved by pH control and complexation.

Apparatus and Instrumentation

Separating funnels for batch extraction; special glass apparatus for continuous extraction; automatic shakers used for discontinuous countercurrent distribution.

Applications

Very widespread use, mainly for the determination of metals as trace and minor constituents; organic materials separated or concentrated according to type. Batch methods are rapid, simple and versatile; applicable to very wide range of samples and concentrations.

Disadvantages

Sometimes requires large quantities of organic solvents; poor resolution of mixtures of organic materials except by countercurrent distribution which is slow.

Solvent extraction, sometimes called liquid–liquid extraction, involves the selective transfer of a substance from one liquid phase to another. Usually, an aqueous solution of the sample is extracted with an immiscible organic solvent. For example, if an aqueous solution of iodine and sodium chloride is shaken with carbon tetrachloride, and the liquids allowed to separate, most of the iodine will be transferred to the carbon tetrachloride layer, whilst the sodium chloride will remain in the aqueous layer. The extraction of a solute in this manner is governed by the *Nernst partition* or *distribution law* which states that at equilibrium, a given solute will always be distributed between two essentially immiscible liquids in the same proportions. Thus, for solute A distributing between an aqueous and an organic solvent,

$$[A]_o/[A]_{aq} = K_D \qquad (4.1)$$

where square brackets denote concentrations (strictly activities) and K_D is known as the equilibrium distribution or partition coefficient which is independent of total solute concentration.

It should be noted that constant temperature and pressure are assumed, and that A must exist in exactly the same form in both phases. Equilibrium is established when the chemical potentials (free energies) of the solute in the two phases are equal and is usually achieved within a few minutes by vigorous shaking. The value of K_D is a reflection of the relative solubilities of the solute in the two phases.

In many practical situations solute A may dissociate, polymerize or form complexes with some other component of the sample or interact with one of the solvents. Under these circumstances the value of K_D does not reflect the overall distribution of the solute between the two phases as it refers only to the distributing species. Analytically, the total amount of solute present in each phase at equilibrium is of prime importance, and the extraction process is therefore better discussed in terms of the distribution ratio D where

$$D = (C_A)_o/(C_A)_{aq} \qquad (4.2)$$

and (C_A) represents the total concentration of all forms of solute A. If no interactions involving A occurred in either phase, D would be equal to K_D. Considerable variation in the experimental value of D can be achieved by altering solution conditions so that solvent extraction is a very versatile technique.

EFFICIENCY OF EXTRACTION

The efficiency of an extraction depends on the magnitude of D and on the relative volumes of the liquid phases. The percentage extraction is given by

$$E = 100 \, D/[D + (V_{aq}/V_o)] \qquad (4.3)$$

where V_{aq} and V_o are the volumes of the aqueous and organic phases respectively, or

$$E = 100\,D/(D + 1) \qquad (4.4)$$

when the phases are of equal volume.

If D is large, i.e. $> 10^2$, a single extraction may effect virtually quantitative transfer of the solute, whereas with smaller values of D several extractions will be required. The amount of solute remaining in the aqueous phase is readily calculated for any number of extractions with equal volumes of organic solvent from the equation

$$(C_{aq})_n = C_{aq}[V_{aq}/(DV_o + V_{aq})]^n \qquad (4.5)$$

where $(C_{aq})_n$ is the amount of solute remaining in the aqueous phase, volume V_{aq}, after n extractions with volumes V_o of organic phase, and C_{aq} is the amount of solute originally present in the aqueous phase.

If the value of D is known, equation (4.5) is useful for determining the optimum conditions for quantitative transfer. Suppose, for example, that the complete removal of 0.1 g of iodine from 50 cm^3 of an aqueous solution of iodine and sodium chloride is required. Assuming the value of D for carbon tetrachloride/water is 85, then for a single extraction with 25 cm^3 of CCl_4,

$$(C_{aq})_1 = 0.1[50/(85 \times 25 + 50)]^1$$
$$= 0.002\,3 \text{ g in } 50 \text{ cm}^3$$

i.e. 97.7% of the I_2 is extracted.

For three extractions with 8.33 cm^3 of CCl_4,

$$(C_{aq})_3 = 0.1[50(85 \times 8.33 + 50)]^3$$
$$= 0.000\,029 \text{ g in } 50 \text{ cm}^3$$

i.e. 99.97% of the I_2 is extracted which for most purposes can be considered quantitative.

It is clear therefore that extracting several times with small volumes of organic solvent is more efficient than one extraction with a large volume. This is of particular significance when the value of D is less than 10^2.

SELECTIVITY OF EXTRACTION

Often, it is not possible to extract one solute quantitatively without partial extraction of another. The ability to separate two solutes depends on the relative magnitudes of their distribution ratios. For solutes A and B, whose distribution ratios are D_A and D_B, the separation factor β is defined as the ratio D_A/D_B where $D_A > D_B$. Table 4.2 shows the degrees of separation

Table 4.2 Separation of two solutes with one extraction, assuming equal volumes of each phase

D_A	D_B	β	% A EXTRACTED	% B EXTRACTED
	10	10	99.0	90.9
	1	10^2	99.0	50.0
10^2	10^{-1}	10^3	99.0	9.1
	10^{-2}	10^4	99.0	1.0
	10^{-3}	10^5	99.0	0.1

achievable with one extraction, assuming that $D_A = 10^2$, for different values of D_B and β. For an essentially quantitative separation β should be at least 10^5.

A separation can be made more efficient by adjustment of the proportions of organic and aqueous phases. The optimum ratio for the best separation is given by the Bush–Densen equation

$$V_o/V_{aq} = (1/D_A D_B)^{\frac{1}{2}} \tag{4.6}$$

Successive extractions, whilst increasing the efficiency of extraction of both solutes, may lead to a poorer separation. For example, if $D_A = 10^2$ and $D_B = 10^{-1}$, one extraction will remove 99.0% of A and 9.1% of B whereas two extractions will remove 99.99% of A but 17% of B. In practice, a compromise must frequently be sought between completeness of extraction and efficiency of separation. It is often possible to enhance or suppress the extraction of a particular solute by adjustment of pH or by complexation. This introduces the added complication of several interrelated chemical equilibria which makes a complete theoretical treatment more difficult. Complexation and pH control are discussed more fully in chapter 3.

EXTRACTION SYSTEMS

The basic requirement for a solute to be extractable from an aqueous solution is that it should be uncharged or can form part of an uncharged ionic aggregate. Charge neutrality reduces electrostatic interactions between the solute and water and hence lowers its aqueous solubility. Extraction into a less polar organic solvent is facilitated if the species is not hydrated, or if the coordinated water is easily displaced by hydrophobic coordinating groups such as bulky organic molecules. There are three types of chemical compound which can fulfil one or more of these requirements:

1. essentially covalent, neutral molecules
   e.g. I_2, $GeCl_4$, C_6H_5COOH

2. uncharged metal chelates
   e.g. metal complexes of acetylacetone, 8-hydroxyquinoline, dithizone, etc.
3. ion-association complexes
   e.g. $(C_6H_5CH_2)_3NH^+$, $GaCl_4^-$
   $Fe(o\text{-phenanthroline})_3^{2+}$, $2ClO_4^-$
   $[(C_2H_5)_2O]_3H^+$, $FeCl_4^-$

The partition of all three types should obey the Nernst law, but in most cases the concentrations of extractable species are affected by chemical equilibria involving them and other components of the system. These must be taken into account when calculating the optimum conditions for quantitative extraction or separation.

Extraction of covalent, neutral molecules
In the absence of competing reactions in either phase and under controlled conditions, the extraction of a simple molecule can be predicted using equations (4.3) to (4.5). However, the value of the distribution ratio D may be pH dependent or it may alter in the presence of a complexing agent. It may also be affected by association of the extracting species in either phase. These effects are considered in turn.

pH Effect
Consider the extraction of a carboxylic acid from water into ether. The partition coefficient is given by

$$K_D = \frac{[RCOOH]_{et}}{[RCOOH]_{aq}} \qquad (4.7)$$

In water, dissociation occurs

$$RCOOH = RCOO^- + H^+$$

and the acid dissociation constant is given by

$$K_a = \frac{[RCOO^-]_{aq}[H^+]_{aq}}{[RCOOH]_{aq}} \qquad (4.8)$$

The distribution ratio, which involves the total concentration of solute in each phase, is

$$D = \frac{[RCOOH]_{et}}{[RCOOH]_{aq} + [RCOO^-]_{aq}} \qquad (4.9)$$

Substituting for $[RCOO^-]_{aq}$ in (4.9) and rearranging

$$D = \frac{[RCOOH]_{et}}{[RCOOH]_{aq}(1 + K_a/[H^+]_{aq})} \tag{4.10}$$

$$\therefore D = \frac{K_D}{1 + K_a/[H^+]_{aq}} \tag{4.11}$$

At low pH, where the acid is undissociated, $D \simeq K_D$ and the acid is extracted with greatest efficiency. At high pH, where dissociation of the acid is virtually complete, D approaches zero and extraction of the acid is negligible. Graphical representation of equation (4.11) for benzoic acid shows the optimum range for extraction, figure 4.1(a). Curves of this type are useful in assessing the separability of acids of differing K_a values. A similar set of equations and extraction curve can be derived for bases, e.g. amines.

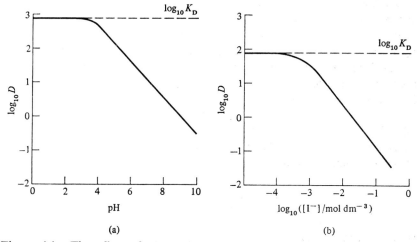

Figure 4.1. The effect of pH and complex formation on extracting species. (a) $\log_{10} D$ against pH for the extraction of benzoic acid, $K_a = 6.3 \times 10^{-5}$ mol dm^{-3}. (b) $\log_{10} D$ against $\log_{10}([I^-]/\text{mol dm}^{-3})$ for the extraction of I_2

Effect of Complex Formation

Returning to the extraction of iodine from an aqueous solution of iodine and sodium chloride, the effect of adding iodide to the system is to involve the iodine in formation of the triiodide ion

and

$$I_2 + I^- = I_3^-$$

$$K_f = \frac{[I_3^-]_{aq}}{[I_2]_{aq}[I^-]_{aq}} \tag{4.12}$$

where K_f is the formation constant of the triiodide ion.

The partition coefficient is given by

$$K_D = \frac{[I_2]_o}{[I_2]_{aq}} \tag{4.13}$$

and the distribution ratio by

$$D = \frac{[I_2]_o}{[I_2]_{aq} + [I_3^-]_{aq}} \tag{4.14}$$

Substituting for $[I_3^-]_{aq}$ in (4.14) and rearranging,

$$D = \frac{K_D}{1 + K_f[I^-]_{aq}} \tag{4.15}$$

Thus, the presence of iodide affects D in such a way that at very low concentrations $D \simeq K_D$ and iodine is extracted with greatest efficiency. At high iodide concentrations, $K_f[I^-]_{aq} \gg 1$, and D is reduced with a consequent reduction in the extraction of iodine, figure 4.1(b).

Effect of Association
The distribution ratio is increased if association occurs in the organic phase. Carboxylic acids form dimers in solvents of low polarity such as benzene and carbon tetrachloride,

$$2RCOOH = (RCOOH)_2$$

$$K_{dimer} = \frac{[(RCOOH)_2]_o}{[RCOOH]_o^2} \tag{4.16}$$

and

$$D = \frac{[RCOOH]_o + 2[(RCOOH)_2]_o}{[RCOOH]_{aq}} \tag{4.17}$$

Substituting for $[(RCOOH)_2]_o$ in (4.17) and re-arranging,

$$D = \frac{[RCOOH]_o + 2K_{dimer}[RCOOH]_o^2}{[RCOOH]_{aq}} \tag{4.18}$$

$$\therefore D = K_D(1 + 2K_{dimer}[RCOOH]_o) \tag{4.19}$$

If K_{dimer} is large D becomes larger than K_D at low pH, resulting in a more efficient extraction of the acid. Dimerization is only slight in oxygenated solvents and extraction into them is therefore less efficient than into benzene or carbon tetrachloride.

EXTRACTION OF UNCHARGED METAL CHELATES

The process of chelation was discussed in chapter 3. To form uncharged chelates which can readily be extracted into organic solvents the reagent must behave as a weak acid whose anion can participate in charge neutral-ization and contain hydrophobic groups to reduce the aqueous solubility of the complex. The formation and extraction of the neutral chelate is best considered stepwise as several equilibria are involved.

For example a monobasic reagent HR dissociates in aqueous solution (dissociation constant K_a) and is distributed between the organic and aqueous phases (distribution coefficient K_{D_R}). Thus

$$HR = H^+ + R^-$$

$$K_a = \frac{[H^+]_{aq}[R^-]_{aq}}{[HR]_{aq}} \tag{4.20}$$

$$K_{D_R} = \frac{[HR]_o}{[HR]_{aq}} \tag{4.21}$$

The hydrated metal ion $M(H_2O)_x^{n+}$ reacts with the reagent anion R^- to form the neutral chelate MR_n (formation constant K_f), i.e.

$$M(H_2O)_x^{n+} + nR^- = MR_n + xH_2O$$

$$K_f = \frac{[MR_n]_{aq}}{[M(H_2O)_x^{n+}]_{aq}[R^-]_{aq}^n} \tag{4.22}$$

The metal chelate distributes itself between the aqueous and organic phases according to the Nernst law

$$K_{D_C} = \frac{[MR_n]_o}{[MR_n]_{aq}} \tag{4.23}$$

and the corresponding distribution ratio is

$$D = \frac{[MR_n]_o}{([MR_n]_{aq} + [M(H_2O)_x^{n+}]_{aq})} \tag{4.24}$$

If several simplifying assumptions are made:

1. the concentrations of chelated species other than MR_n are negligible
2. the concentrations of hydroxy or other anion coordination complexes are negligible
3. the reagent HR and the chelate MR_n exist as simple undissociated molecules in the organic phase, and $[MR_n]_{aq}$ is negligible

it can be shown that

$$D = \frac{K_f K_{D_C} K_a^n}{K_{D_R}^n} \frac{[HR]_o^n}{[H^+]_{aq}^n} \tag{4.25}$$

or

$$D = K^*[HR]_o^n[H^+]_{aq}^{-n} \tag{4.26}$$

(Substituting for each equilibrium constant in (4.25) gives equation (4.24).)

Thus, for a given reagent and solvent, the extraction of the metal chelate is dependent only upon pH and the concentration of reagent in the organic phase and is *independent* of the initial metal concentration. In practice, a constant and large excess of reagent is used to ensure that all the complexed metal exists as MR_n and D is then dependent only on pH, i.e.

$$D = K^{*\prime}[H^+]_{aq}^{-n} \qquad (4.27)$$

or

$$\log D = \log K^{*\prime} + n\text{pH} \qquad (4.28)$$

where

$$K^{*\prime} = (K_f K_{D_C} K_a^n [HR]_o^n)/K_{D_R}^n \qquad (4.29)$$

The relation between D and E, the percentage extracted, is

$$D = E/(100 - E) \qquad (4.30)$$

for equal volumes of the two phases

$$\therefore \log D = \log E - \log(100 - E) = \log K^{*\prime} + n\text{pH} \qquad (4.31)$$

Equation (4.31), which defines the extraction characteristics for any chelate system, is represented graphically in figure 4.2 for a mono-, di- and tri-valent metal, i.e. $n = 1, 2$ and 3 respectively, and shows the pH range over which a metal will be extracted. No significance should be attached to the positioning of the curves relative to the pH scales as these are determined by the value of $K^{*\prime}$. Thus the more acidic the reagent or the stronger the metal

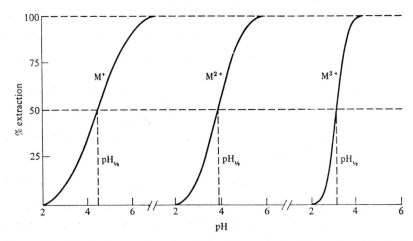

Figure 4.2.   Extraction as a function of pH for metals of different formal valencies.
(Note: the position of each curve along the pH abscissa is not significant)

complex, the lower the pH range over which the metal will be extracted. Increased reagent concentration has a similar effect.

The pH at which 50% of the metal is extracted, $pH_{\frac{1}{2}}$, can be used to assess the degree of separability of two or more metals. At $E = 50$, equation (4.31) reduces to

$$\log K^{*\prime} = -npH_{\frac{1}{2}} \tag{4.32}$$

Substituting this value for $\log K^{*\prime}$ in equation (4.28)

$$\log D = n(pH - pH_{\frac{1}{2}}) \tag{4.33}$$

For two metals, the separation factor β is defined as D'/D'', where $D' > D''$, or

$$\log \beta = \log D' - \log D'' \tag{4.34}$$

Therefore, for extraction at a specified pH,

$$\log \beta = n'(pH - pH_{\frac{1}{2}}') - n''(pH - pH_{\frac{1}{2}}'') \tag{4.35}$$

If $n' = n''$, i.e. the metals have the same formal valency,

$$\log \beta = n\Delta pH_{\frac{1}{2}} \tag{4.36}$$

Assuming that $\log \beta$ should be at least 5 for an essentially quantitative separation by a single extraction (see p. 59), $\Delta pH_{\frac{1}{2}}$ should be 5, 2.5 and 1.7 respectively for pairs of mono-, di- and tri-valent metals. Selectivity by pH control is greatest, therefore, for trivalent metals and least for monovalent. This is reflected in the slopes of the curves which are determined by n and decrease in the order $M^{3+} > M^{2+} > M^+$.

It can be seen from equation (4.25) that the value of β is determined by the formation constants and distribution coefficients of the two chelates, i.e.

$$\beta = D'/D'' = (K_f' K_{D_C}')/(K_f'' K_{D_C}'') \tag{4.37}$$

If β is insufficiently large to enable a quantitative separation to be made by pH control alone, the addition of a masking agent which forms a water-soluble complex more strongly with one metal than the other will shift the extraction curve for the former to a higher pH range with a consequent increase in β, which is now given by

$$\beta = (K_f' K_{D_C}' K_{f_m}'')/(K_f'' K_{D_C}'' K_{f_m}') \tag{4.38}$$

where K_{f_m}' and K_{f_m}'' are the formation constants of the metal complexes with the masking agent and $K_{f_m}'' > K_{f_m}'$. Masking reactions (chapter 5) play an important role in extraction procedures involving metal chelates. Some examples of neutral chelate extraction systems are given in table 4.3.

Table 4.3   Typical chelate extraction systems

CHELATING AGENT	METALS EXTRACTED
β-diketones e.g. acetylacetone, thenoyltrifluoro-acetylacetone	react with over 50 metals; especially useful for alkali metals, Be, Sn, Cr, Mn, Mo
8-hydroxyquinoline (oxine) and its derivatives	react with over 50 metals, especially useful for Al, Mg, Sr, V, W
α-dioximes e.g. dimethylglyoxime, α-furildioxime	Ni, Pd
di-alkyldithiocarbamates	react with many metals including Bi, Tl(III), Sb(III), Te(IV), As(III), Se(IV), Sn(IV), V(V)
dithizone (diphenyldithiocarbazone)	Pb, Hg, Cu, Pd, Pt, Ag, Bi, Zn and others
cupferron (ammonium salt of N-nitrosophenylhydroxylamine)	reacts with many metals, including Fe(III), Ga, Sb(III), Ti(IV), Sn(IV), Hf, Zr, V(V), U(IV), Mo(VI)
PAN (1-(2-pyridylazo)-2-naphthol	reacts with over 50 metals, including U(VI), In, V(V), Pd, Zn, Cd, Mn, Y
1-nitroso-2-naphthol	Co(III) (Cu, Ni)

Extraction of Ion-association Complexes

The extraction of charged species from an aqueous solution is not possible unless the charge can be neutralized by chelation, as described in the previous section, or by association with other ionic species of opposite charge to form a complex that is electrically neutral. A further requirement to aid extraction is that at least one of the ions involved should contain bulky hydrophobic groups. Metals and mineral acids can both be extracted as cationic or anionic complexes, chelated or otherwise, and often solvated by the organic solvent. The Nernst partition law is obeyed by ion-association systems but the number of equilibria involved is greater than for neutral chelates and the mathematical treatment, which is correspondingly more involved, will not be covered in this book. Salting-out agents are often used to increase the distribution ratio. These are electrolytes, such as di- and tri-valent metal nitrates, with a pronounced tendency to hydration. They bind large numbers of water molecules thereby lowering the dielectric of the solution and favouring ion-association.

Ion-association complexes may be classified into three types: non-chelated complexes; chelated complexes; oxonium systems.

Non-chelated Complexes

These include the simplest ion-association systems in which bulky cations and anions are extracted as pairs or aggregates without further coordination by solvent molecules. An example of this type of system is the extraction of manganese or rhenium as permanganate or perrhenate into chloroform by association with the tetraphenylarsonium cation derived from a halide salt

$$Ph_4As^+Cl^- + MnO_4^- = \underbrace{Ph_4As^+, MnO_4^-}_{\text{ion-pair}} + Cl^-$$

Anionic metal complexes such as $ZnCl_4^{2-}$, $GaCl_4^-$ and $Co(CN)_6^{3-}$ can be extracted with tetraalkylammonium salts, e.g.

$$(Bu)_4N^+Cl^- + GaCl_4^- = \underbrace{(Bu)_4N^+, GaCl_4^-}_{\text{ion-pair}} + Cl^-$$

Certain long-chain alkylammonium salts, notably tricaprylmethyl-ammonium chloride (Aliquat 336-S) and tri-*iso*-octylamine hydrochloride (TIOA) are liquids, sometimes referred to as liquid anion exchangers, which can form extractable ion-pairs or aggregates with anionic metal complexes in the same way, e.g. in sulphuric acid solution uranium is extracted as $2(TIOA—H^+)$, $UO_2(SO_4^{2-})_2$.

Alkyl esters of phosphoric acid and phosphine oxides will extract metals and mineral acids by direct solvation. Tri-*n*-butyl phosphate (TBP) and tri-*n*-octylphosphine oxide (TOPO)

$$(C_4H_9O)_3P{=}O \qquad\qquad (C_8H_{17})_3P{=}O$$
$$\text{TBP} \qquad\qquad\qquad \text{TOPO}$$

are both used for this purpose and will extract uranium, actinides and lanthanides as well as many other metals. The extractable species, such as $UO_2^{2+}(TBP)_2$, $2(NO_3^-)$ and $H^+(TBP)_2$, $UO_2(NO_3^-)_3$ in the case of uranium, vary in composition depending on acidity and total electrolyte concentration, but direct solvation of the metal ion or protons always plays an important role. TOPO is a better extractant than TBP, particularly for mineral acids, forming more definite solvates. Table 4.4 includes some of the more important non-chelated systems.

Chelated Complexes

Many cationic and anionic chelates which are not extractable by the usual organic solvents due to residual charge can be extracted in the presence of a suitable counter-ion. Two examples of charged chelates extractable by

Table 4.4 Typical ion-association extraction systems

SYSTEM	METALS EXTRACTED
NON-CHELATED ION-ASSOCIATION SYSTEMS	
tetraphenylarsonium and tetraalkyl-ammonium salts	ReO_4^-, MnO_4^-, chloro, cyano and thiocyanato complexes of Bi, Ga, Zn, Cd, Ir(IV), Zn, Co(II)
Rhodamine–B—H^+ liquid anion exchangers e.g. Aliquat 336-S, TIOA	$SbCl_6^-$, $GaCl_4^-$, $AuCl_4^-$, $TlCl_4^-$, $FeCl_4^-$ extract many metals including U(VI), Co, Fe(III), Mo(VI), Ta, Ti(III), Zn as halide, sulphate or nitrate complexes
alkyl esters of phosphoric acid and phosphine oxides e.g. TBP and TOPO	U(VI), Pu(IV), Th(IV), Sc, Y, Zr, Nb, Mo, Sb, actinides, lanthanides, mineral acids
CHELATED ION-ASSOCIATION SYSTEMS	
o-phenanthroline, ClO_4^-	Fe(II)
biquinolyl, Cl^-	Cu(I)
EDTA, liquid anion exchangers	many metals can be extracted
oxine, tetraalkylammonium salts	U(VI)
acidic alkylphosphoric esters (liquid cation exchangers) e.g. HDBP, HDEHP	metals in higher valency states, i.e. actinides, U(VI), Pu(VI)
OXONIUM SYSTEMS	
$(C_2H_5)_2O$, HCl	$FeCl_4^-$, $SbCl_6^-$, $GaCl_4^-$, $TlCl_4^-$, $AuCl_4^-$ and others
$C_2H_5COCH_3$, HF	NbF_6^-, TaF_6^-
$(C_2H_5)_2O$, HI	Sb(III), Hg(II), Cd, Au(III), Sn(II)
$(C_2H_5)_2O$, NH$_4$SCN	Sn(IV), Zn, Ga, Co, Fe(III)
$(C_2H_5)_2O$, HNO$_3$	Au(III), Ce(IV), U(VI), Th(IV)

chloroform are the Fe(II)-o-phenanthroline cation using perchlorate as a counter-ion,

$$Fe(o\text{-phen})_3^{2+},\ 2ClO_4^-$$

and the UO_2(II)-8-hydroxyquinoline anion using a tetraalkylammonium cation as the counter ion,

$$(Bu)_4N^+,\ UO_2(Ox)_3^-$$

EDTA complexes of trivalent metals can be extracted successively with liquid anion exchangers such as Aliquat 336-S by careful pH control. Mixtures of lanthanides can be separated by exploiting differences in their EDTA complex formation constants.

Acidic alkyl esters of phosphoric acid, of which dibutyl-phosphoric acid (HDBP) and di(2-ethylhexyl) phosphoric acid (HDEHP) are typical,

form extractable complexes by chelation and solvation, the acidic hydrogen being replaced by a metal, e.g. La(DBP, HDBP)$_3$. Metals in high valency states, such as tetravalent actinides are the most readily extracted. The di-alkyl phosphoric esters are liquids and are sometimes known as liquid cation exchangers.

Table 4.4 includes some of the more important chelated systems.

Oxonium Systems

Oxygen-containing solvents with a strong coordinating ability, such as diethyl ether, methyl *iso*-butyl ketone and *iso*-amyl acetate, form oxonium cations with protons under strongly acidic conditions, e.g. $(R_2O)_nH^+$. Metals which form anionic complexes in strong acid can be extracted as ion-pairs into such solvents. For example, Fe(III) is extracted from 7 M hydrochloric acid into diethyl ether as the ion-pair

$$[(C_2H_5)_2O]_3H^+(H_2O)_x, FeCl_4^-$$

The efficiency of the extraction depends on the coordinating ability of the solvent, and on the acidity of the aqueous solution which determines the concentration of the metal complex. Coordinating ability follows the sequence ketones > esters > alcohols > ethers. Many metals can be extracted as fluoride, chloride, bromide, iodide or thiocyanate complexes. Table 4.5 shows how the extraction of some metals as their chloro complexes into diethyl ether varies with acid concentration. By controlling acidity and oxidation-state and choosing the appropriate solvent, useful separations can be achieved. As, for example, the number of readily formed fluoride complexes is small compared with those involving chloride, it is evident that a measure of selectivity is introduced by proper choice of the complexing ion. The order of selectivity is $F^- > Br^- > I^- > Cl^- > SCN^-$. Examples of oxonium systems are included in table 4.4.

The use of oxonium and other non-chelated systems can be advantageous where relatively high concentrations of metals are to be extracted as solubility in the organic phase is not likely to be a limiting factor. Metal chelates, on the other hand, have a more limited solubility and are more suited to trace-level work.

Table 4.5   Extraction of metal chloro complexes into diethyl ether

| METAL | HCl (M) | | | | |
	0.3 M	1.4 M	2.9 M	4.4 M	6 M
		% extracted			
Au(III)	84	98	98		95
Fe(III)	trace	0.1	8	92	99
Tl(III)		~98		~99	~98
Sb(III)	0.3	8	22	13	6
Ge					~50
As(III)	0.2	0.7	7	37	68
Te(IV)	trace	0.2	3	12	34
Ga					~97
Sn(IV)	0.8	10	23	28	17
Hg(II)	13		0.4		0.2
Cu(II)	trace		0.05		0.05
Zn	trace		0.03		0.2
Ir(IV)	trace		0.02		5

The following metals are not extracted: Al, Be, Bi, Cd, Cr, Co, Fe(II), Pb, Mn, Ni, Pd, Os, Pt, rare earths, Ag, Ti, Th, W, U, Zr.

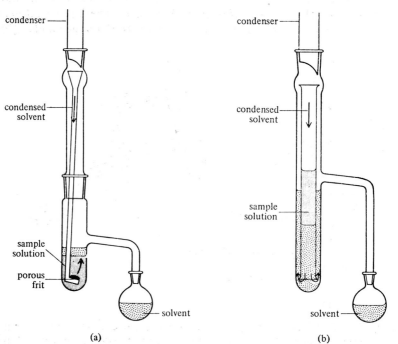

Figure 4.3.   Continuous extraction apparatus. (a) Extraction with a solvent lighter than water. (b) Extraction with a solvent heavier than water

METHODS OF EXTRACTION

Batch extraction is the simplest and most useful method, the two phases being shaken together in a separatory funnel until equilibrium is reached and then allowed to separate into two layers. If the distribution ratio is large, a solute may be transferred essentially quantitatively in one extraction, otherwise several may be necessary. The optimum conditions for quantitative extraction have been discussed on p. 58. If several extractions are required, it is advantageous to use a solvent more dense than water, e.g. carbon tetrachloride or chloroform, so that the aqueous phase can be left in the separatory funnel until the procedure is complete.

Continuous extraction consists of distilling the organic solvent from a reservoir flask, condensing it and allowing it to pass through the aqueous phase before returning to the reservoir flask to be recycled. Figure 4.3 illustrates two types of apparatus used for this purpose. The method is particularly useful when the distribution ratio is small, i.e. $D < 1$, and where the number of batch extractions required for quantitative transfer would be inconveniently large.

Discontinuous counter-current distribution is a method devised by Craig,[1] which enables substances with similar distribution ratios to be separated. The method involves a series of individual extractions performed automatically in a specially designed apparatus. This consists of a large number (50 or

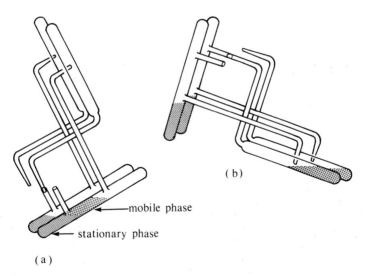

—mobile phase

— stationary phase

(a)

(b)

Figure 4.4. Two interlocking glass units for Craig counter-current distribution. (a) Position during extraction. (b) Position during transfer. (Note: by returning the apparatus from (b) to (a) the transfer is completed. The mobile phase moves on to the next unit and is replaced by a fresh portion)

more) of identical interlocking glass extraction units (figure 4.4) mounted in a frame which is rocked and tilted mechanically to mix and separate the phases during each extraction step. Initially equal volumes of the extracting solvent, which should be the more dense phase, are placed in each of the extraction units. This can be termed the *stationary phase* as each portion remains in the same unit throughout the procedure.

A solution of the mixture to be separated, dissolved in the less dense phase, is placed in the first unit and the phases mixed and allowed to separate. The upper layer is transferred automatically to the second unit whilst a fresh portion, not containing any sample, is introduced into the first unit from a reservoir. By repeating the extraction and transfer sequence as many times as there are units, the portions of lighter phase, which may be termed the *mobile phase*, move through the apparatus until the initial portion is in the last unit, and all units contain portions of both phases. A schematic representation of the first four extractions for a single solute is shown in figure 4.5

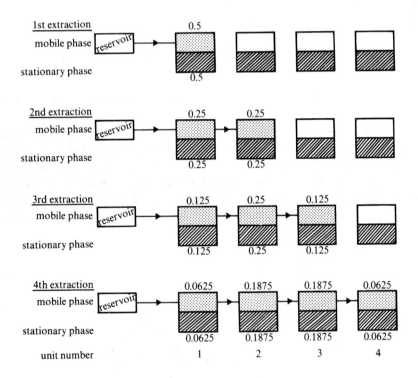

Figure 4.5. Extraction scheme for a single solute by Craig counter-current distribution. (Figures represent the proportions in each phase for $D = 1$ and equal volumes. Only the first four extractions are shown)

where it is assumed that $D = 1$ and equal volumes of the two phases are used throughout. It can be seen that the solute is distributed between the units in a manner which follows the coefficients of the binomial expansion of $(x + y)^n$ (table 4.6) where x and y represent the fractions of solute present in the mobile and stationary phases and n is the number of extractions. The values of x and y are determined by D and the proportions of mobile and

Table 4.6   Proportional distribution of a solute between extraction units for Craig counter-current distribution

EXTRACTION	UNIT NUMBER						
	1	2	3	4	5	6	7
1st	1						
2nd	1	1					
3rd	1	2	1				
4th	1	3	3	1			
5th	1	4	6	4	1		
6th	1	5	10	10	5	1	
7th	1	6	15	20	15	6	1

stationary phases used. For large values of n the distribution approximates to the normal error or Gaussian curve (chapter 2), and the effects of n and D are shown in figures 4.6 and 4.7 respectively. Thus, as the number of extractions n is increased, the solute moves through the system at a rate which is proportional to the value of D. With increasing n the solute is spread over a greater number of units, but separation of two or more components in a

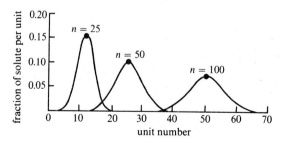

Figure 4.6.   Effect of the number n of equilibrations on solute distributions for a distribution ratio $D = 1$

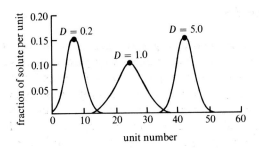

Figure 4.7. Effect of the distribution ratio D on solute distribution after 50 equilibrations

mixture will be improved. It should be noted that a 100% separation can never be achieved as the extremities of a Gaussian curve approach a baseline asymptotically. However, many separations are essentially quantitative within the context of a particular problem, e.g. 95, 99 or 99.9%.

Theoretically, any number of solutes can be separated in this manner and the method has been applied, for example, to the separation of fatty acids, amino acids, polypeptides and other biological materials with distribution ratios in some cases differing by less than 0.1. However, the procedure can be lengthy and consumes large volumes of solvents. It is frequently more convenient to use one of the chromatographic techniques described later in this chapter. These can be considered as a development of the principle of countercurrent distribution.

APPLICATIONS OF SOLVENT EXTRACTION

The technique is used predominantly for the isolation of a single chemical species prior to a determination and to a lesser extent as a method of concentrating trace quantities. The most widespread application is in the determination of metals as minor and trace constituents in a variety of inorganic and organic materials, e.g. the selective extraction and spectrometric determination of metals as coloured complexes in the analysis of metallurgical and geological samples as well as for petroleum products, foodstuffs, plant and animal tissue and body fluids.

Separation procedures for purely organic species do not possess the same degree of selectivity as systems involving metals because of a general lack of suitable complexing and masking reactions. Nevertheless, classes of compounds such as hydrocarbons, acids, fats, waxes, etc. can often be isolated prior to analysis by other techniques.

4.2 Chromatography

Historically, the word chromatography was used by Tswett[2] in 1903 to describe the separation of plant pigments by percolating a petroleum–ether extract through a glass column packed with powdered calcium carbonate. Coloured zones were produced by the various pigments migrating through the column at different rates, the components being isolated by extrusion and sectioning of the calcium carbonate packing. Modern chromatographic techniques are more complex and are used for a wide variety of separations frequently involving colourless substances, but the original term is retained. All the techniques depend upon the same basic principle, i.e. variation in the rate at which different components of a mixture migrate through a stationary phase under the influence of a mobile phase. Rates of migration vary because of differences in distribution ratios. Chromatography therefore resembles Craig counter-current distribution which has been described in the previous section. In the Craig process, individual equilibrations are performed in a series of separate vessels. If the walls of these vessels are imagined to be non-existent so that the stationary phase is continuous, and the mobile phase is allowed to move continuously rather than stepwise, the situation would be closely analogous to that found in chromatographic separations. In practice the liquid stationary phase is coated on to an inert, granular or powdered solid support which is either packed into a column or spread on a supporting sheet in the form of a thin layer. The solid stationary phases used in some chromatographic techniques have no need of a support if packed into a column but still require a supporting sheet for thin-layer operation. As the distributing components of a mixture are moved down a column or across a surface by the mobile phase, they assume a Gaussian concentration profile as they do in the Craig process. In addition, because both phases are continuous, diffusion and other kinetic effects play a significant role in determining the width of the profile. This will be discussed in more detail later.

CHROMATOGRAPHIC MECHANISMS

During a chromatographic separation solute molecules are continually moving back and forth between the stationary and mobile phases. While they are in the mobile phase, they are carried forward with it but remain virtually stationary during the time they spend in the stationary phase. The rate of migration of each solute is therefore determined by the proportion of time it spends in the mobile phase, or in other words by its distribution ratio.

The process whereby a solute is transferred from a mobile to a stationary phase is called *sorption*. Chromatographic techniques are based on four different sorption mechanisms, namely *surface adsorption, partition, ion-exchange* and *exclusion*. The original method employed by Tswett involved *surface adsorption* where the relative polarities of solute and solid stationary phase determine the rate of movement of that solute through a column or across a surface. If a liquid is coated on to the surface of an inert solid support, the sorption process is one of *partition*, and movement of the solute is determined solely by its relative solubility in the two phases or by its volatility if the mobile phase is a gas. Both adsorption and partition may occur simultaneously, and the contribution of each is determined by the system parameters, i.e. the nature of the mobile and stationary phases, solid support and solute. For example, a stationary phase of aluminium oxide is highly polar and normally exhibits strong adsorptive properties. However, these may be modified by the presence of adsorbed water which introduces a degree of partition into the overall sorption process by acting as a liquid stationary phase. Conversely, paper (cellulose) is relatively non-polar and retains a large amount of water which functions as a partition medium. Nevertheless, residual polar groups in the structure of the paper can lead to adsorptive effects.

The third sorption phenomenon is that of *ion-exchange*. Here, the stationary phase is a permeable polymeric solid containing fixed charged groups and mobile counter-ions which can exchange with the ions of a solute as the mobile phase carries it through the structure.

The fourth type of mechanism is *exclusion* although perhaps 'inclusion' would be a better description. Strictly, it is not a true sorption process as the separating solutes remain in the mobile phase throughout. Separations occur because of variations in the extent to which the solute molecules can diffuse through an inert but porous stationary phase. This is normally a gel structure which has a small pore size and into which small molecules up to a certain critical size can diffuse. Molecules larger than the critical size are excluded from the gel and move unhindered through the column or layer whilst smaller ones are retarded to an extent dependent on molecular size.

In each chromatographic technique, one of the four mechanisms predominates, but it should be emphasized that two or more may be involved simultaneously. Partition and adsorption frequently occur together and in paper chromatography, for example, ion-exchange and exclusion certainly play minor roles also.

SORPTION ISOTHERMS

Ideally the concentration profile of a solute in the direction of movement of the mobile phase should remain Gaussian at all concentrations as it moves through the system. However, sorption characteristics often change at high concentrations resulting in changes in the distribution ratio. If no such changes occurred, a plot of the concentration of solute in the mobile phase as a function of that in the stationary phase at constant temperature would be linear and the concentration profile symmetrical, figure 4.8(a). Plots of this type, known as *sorption isotherms*, can show curvature towards either axis under which circumstances the concentration profiles will show *tailing*, figure 4.8(b), or *fronting*, figure 4.8(c). Both these effects are undesirable as

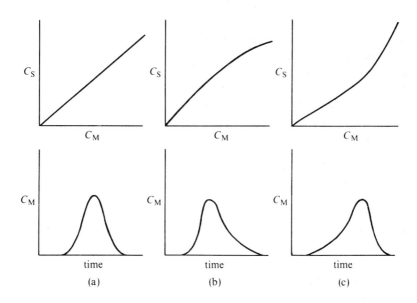

Figure 4.8. Sorption isotherms and concentration profiles. (a) Linear isotherm; Gaussian profile. (b) Curved isotherm; 'tailing'. (c) Curved isotherm; 'fronting'

they lead to poor separations and unreliable quantitative data. Fronting, which produces peaks with sloping front and sharp rear boundaries is more likely to occur in systems where partition forms the basis of the separation process, and where the solute has a small distribution ratio. Tailing produces peaks with sharp leading edges and long sloping rear boundaries. It is particularly likely to occur where adsorption is involved in the separation process. The diagrams show that both effects become more pronounced at high concentrations and are therefore symptomatic of *overloading* the column or surface with sample. Restricting the sample size so as to operate over the linear region of the sorption isotherm is a recognized means of preventing fronting and tailing.

ADSORPTION SYSTEMS

Separations in which surface adsorption is the predominant sorption process depend upon *polarity* differences between solute molecules. Those which are highly symmetrical or consist of atoms with similar electronegativities are relatively non-polar, e.g. C_5H_{12}, C_6H_6, CCl_4. The presence of functional groups leads to an increase in polarity, thus, $C_5H_{11}OH$ is more polar than C_5H_{12}, $C_6H_5NO_2$ is more polar than C_6H_6, and $CHCl_3$ is more polar than CCl_4. The more polar the molecule, the more tenaciously it will be adsorbed by a polar surface. The approximate order of increasing strength of adsorption is: paraffins < olefins < ethers < esters < ketones and aldehydes < amines < alcohols and phenols < acids. During the separation process there is competition for adsorption sites between solute molecules and those of the mobile phase. Solute and solvent molecules are continually being adsorbed and desorbed as the mobile phase travels through the system. Solutes of low polarity spend proportionately more time in the mobile phase than those that are highly polar. Consequently, the components of a mixture are eluted in order of increasing polarity (increasing distribution ratio). Adsorption isotherms often show curvature (p. 77), unless the sample loading is very low. This leads to pronounced tailing of solute peaks and a loss of resolution.

Stationary Phase

Almost any polar solid can be used, the most common choices being silica gel or alumina. A selection of stationary phases is listed in table 4.7 in order of decreasing adsorptive power. Silica gel and alumina are highly polar materials that adsorb molecules strongly. They are said to be *active* adsorbents. Activity is determined by the overall polarity and by the number

Table 4.7   Adsorbents used in
column chromatography

DECREASING ACTIVITY	
	alumina
	charcoal
	silica gel
	magnesium carbonate
	calcium carbonate
	talc
	starch
	sucrose
	cellulose

Note: The positions of alumina
and silica gel are dependent on
water content.

of adsorption sites. In silica gel, the adsorption sites are the oxygen atoms and silanol groups (—Si—OH) which readily form hydrogen bonds with polar molecules. Adsorption sites of different types are present on the surface of alumina, but, like silica gel, a proportion of them are hydroxyl groups. The amount of water present on the surface has a profound effect on activity by blocking adsorption sites. If the water is progressively removed by oven-drying, the material becomes correspondingly more active.

The choice of stationary phase and its degree of activity is determined by the nature of the sample. If sample components are adsorbed too strongly, they may be difficult to elute or chemical changes may occur. Weakly polar solutes should be separated on highly active adsorbents otherwise they may elute rapidly with little or no resolution. Strongly polar solutes are better separated on adsorbents of low activity. Silica gel can be prepared with a wider range of activities than alumina and is less likely to induce chemical changes. The latter is available in a neutral form for general use, and in a basic form (containing sodium carbonate) for use with acid-sensitive compounds. An acidic form behaves as an anion exchanger and not as a true adsorbent.

Mobile Phase

The eluting power of a solvent is determined by its overall polarity, the polarity of the stationary phase and the nature of the sample components.

Table 4.8   An example of an eluotropic series

SOLVENT	UV CUT OFF nm	RI 25°C	VISCOSITY Cp 25°C	SOLVENT POLARITY (p', parti-tion-based)	SOLVENT POLARITY (ε^0, adsorp-tion-based)
n-hexane	190	1.372	0.30	0.1	0.01
cyclohexane	200	1.423	0.90	−0.2	0.04
carbon tetrachloride	265	1.457	0.90	1.6	0.18
toluene	285	1.494	0.55	2.4	0.29
benzene	280	1.498	0.60	2.7	0.32
methylene chloride	233	1.421	0.41	3.1	0.42
n-propanol	240	1.385	1.9	4.0	0.82
tetrahydrofuran	212	1.405	0.46	4.0	0.57
ethyl acetate	256	1.370	0.43	4.4	0.58
iso-propanol	205	1.384	1.9	3.9	0.82
chloroform	245	1.443	0.53	4.1	0.40
acetone	330	1.356	0.3	5.1	0.56
ethanol	210	1.359	1.08	4.3	0.88
acetonitrile	190	1.341	0.34	5.8	0.65
methanol	205	1.326	0.54	5.1	0.95
water		1.333	0.89	10.2	

Table 4.8 lists some widely used solvents in order of their eluting power, this being known as an *eluotropic series*. In practice, better separations are achieved with the least polar solvent possible and mixtures of solvents are often used to achieve optimum separation conditions.

It is important that a given solvent should not contain impurities of a more polar nature, e.g water or acids; alcohol in chloroform, aromatics in saturated hydrocarbons, as resolution may be impaired. Certain solvent–adsorbent combinations can be chemically unstable. For example, acetone is polymerized by basic alumina.

PARTITION SYSTEMS

In a partition system the stationary phase is a liquid coated on to a solid support (p. 92). Silica gel, kieselguhr (diatomaceous earth) or cellulose powder are the most frequently used. Conditions closely resemble those of counter-current distribution so that in the absence of adsorption by the solid support, solutes move through the system at rates determined by their relative solubilities in the stationary and mobile phases. Partition isotherms usually have a longer linear range than adsorption isotherms, so tailing or fronting of elution peaks is not a particular problem, except at high concentrations.

Stationary and Mobile Phases
There is a very wide choice of pairs of liquids to act as stationary and mobile phases. It is not necessary for them to be totally immiscible, but a low mutual solubility is desirable. A hydrophilic liquid may be used as the stationary phase with a hydrophobic mobile phase or vice versa. The latter situation is sometimes referred to as a 'reversed phase' system as it was developed later. Water, aqueous buffers and alcohols are suitable mobile phases for the separation of very polar mixtures, whilst hydrocarbons in combination with ethers, esters and chlorinated solvents would be chosen for less polar materials.

In principle the wide range of stationary phases used in glc can be used in liquid–liquid partition systems, but problems can arise from mutual solubility effects which may result in the stationary phase being stripped from the column.

CHARACTERIZATION OF SOLUTES

As already described, the rate of movement of a solute is determined by its distribution ratio defined as

$$D = \frac{C_{\text{stationary phase}}}{C_{\text{mobile phase}}}$$

The larger the value of D, the slower will be the progress of the solute through the system, and the components of a mixture will therefore reach the end of a column or the edge of a surface in order of increasing value of D. In column methods, a solute is characterized by the volume of mobile phase required to move it from one end of the column to the other. Known as the *retention volume* V_R it is defined as the volume passing through the column between putting the sample on the top of the column and the emergence of the solute peak at the bottom. It is given by the equation

$$V_R = V_M + k' V_M \qquad (4.39)$$

where V_M is the volume of mobile phase in the column (the *dead* or *void volume*) and k' is the *capacity factor* which is directly proportional to D but takes account of the volume of each phase. Sometimes k' is used to characterize a solute rather than V_R.

If $k' = 0$, then $V_R = V_M$ and the solute is eluted without being retarded or retained by the stationary phase. Large values of k', which reflect large values of D, result in very large retention volumes and hence long retention times. At a constant rate of flow of mobile phase F, V_R is related to the retention time t_R, by the equation

$$V_R = F t_R \qquad (4.40)$$

If the flow of mobile phase is monitored by a detector and chart-recorder system, such as is used in gas chromatography, then V_R, t_R and the distance moved by the chart are all directly proportional, and either of the last two can be used as a measure of V_R.

In paper and thin-layer chromatography, the separation process is halted at a stage which leaves the separated components *in situ* on the surface in the form of spots. The rate at which a solute has moved is then determined by its *retardation factor* R_f which is defined as

$$R_f = \frac{\text{distance travelled by the centre of the solute spot}}{\text{distance travelled by the front of the mobile phase}}$$

It is inversely related to D and clearly cannot be greater than 1. Distances are measured from the point of application of the sample. As both V_R and R_f are related to D they will depend on the conditions under which a chromatogram is run. Valid comparisons between samples and between samples and standards can be made only if experimental conditions are identical. In many cases this is difficult to achieve and it is common practice to run samples and standards sequentially or simultaneously to minimize the effects of variations.

EFFICIENCY AND RESOLUTION

The ideal chromatographic process is one in which the components of a mixture form narrow bands which are completely resolved from one another. The narrowness of a band or peak is a measure of the *efficiency* of the process whilst *resolution* is assessed by the ability to resolve the peaks of components with similar t_R or R_f values.

Efficiency N for column separations is related to retention volume and peak width measured in terms of the standard deviation, assuming an ideally Gaussian-shaped peak (p. 17), i.e.

$$N = \left(\frac{t_R}{\sigma}\right)^2 \tag{4.41}$$

In practice it is easier to measure baseline width or the width at one half of the peak height, so N is generally calculated using one of the alternative forms:

$$N = 16\left(\frac{t_R}{W_B}\right)^2 \tag{4.42}$$

or

$$N = 5.54\left(\frac{t_R}{W_{h/2}}\right)^2 \tag{4.43}$$

where W_B is the baseline peak width and $W_{h/2}$ is the peak width measured at half of the peak height. Valid comparisons of efficiencies can be made only if the same formula is used throughout, as the computed values of N using each of the above formulae may differ considerably.

The parameter N is universally referred to as the *plate number*, but an alternative means of quoting efficiency is in terms of a *plate height*, H or HETP*. Plate number and plate height are inversely related by the equation

$$N = L/H \tag{4.44}$$

where L is the length of the column.

Values of N may be many thousands for columns having high efficiencies, the corresponding values of H being less than 1 mm.

The ultimate width of a peak is determined by the total amount of *diffusion* occurring during movement of the solute through the system, and on the rate

---

* HETP = height equivalent to a theoretical plate. It is derived from the plate theory of distillation which is a confusing concept having no basis in fact in the context of modern chromatographic separations. Nevertheless the terms *plate number* and *plate height* are still very widely used.

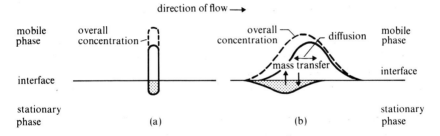

Figure 4.9.   Effects of diffusion and mass transfer on peak width. (a) Concentration
profiles of a solute at the beginning of a separation. (b) Concentration profiles of a
solute after passing some distance through the system

of *mass transfer* between the two phases. These effects are shown diagrammatically in figure 4.9. Both diffusion and mass transfer effects are interdependent and complex, being made up of a number of contributions from
different sources. Because they are kinetic effects, their influence on efficiency
is determined by the rate at which the mobile phase travels through the
system. Attempts to define efficiency in terms of diffusion and mass transfer
effects are numerous, the most useful being those of van Deemter[3] and of
Giddings[4]. Based on their approach, the following simplified equation can be
derived:

$$H = A + B/\bar{u} + C\bar{u} \qquad (4.45)$$

where $\bar{u}$ is the mean linear flow rate of the mobile phase, and A, B and C are
terms involving diffusion and mass transfer.

A is the 'multiple path' term which accounts for different portions of the

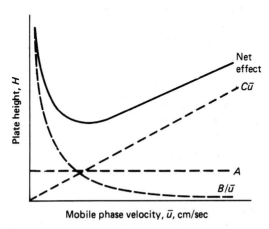

Figure 4.10.   Efficiency as a function of mobile phase velocity and the effect of each
term in equation 4.45

mobile phase, and consequently the solute, travelling different total distances because of the various routes taken around the particles of stationary phase. The effect is minimized by reducing particle size but increases with length of column or surface.

$B/\bar{u}$ is the 'molecular diffusion' term and relates to diffusion of solute molecules within the mobile phase caused by local concentration gradients. Diffusion within the stationary phase also contributes to this term, which is significant only at low flow rates and increases with column length. As B is proportional to the diffusion coefficient in the mobile phase, the order of efficiency at low flow rates is liquids > heavy gases > light gases.

$C\bar{u}$ is the 'mass transfer' term and arises because of the finite time taken for solute molecules to move between the two phases. Consequently, a true equilibrium situation is never established as the solute moves through the system, and spreading of the concentration profiles results. The effect is minimal for small particle size and thin coatings of stationary phase but increases with flow rate and length of column or surface.

Experimental values of H, obtained from equation (4.45) and plotted against the rate of flow of mobile phase for a given solute and set of conditions, produce a hyperbolic curve showing an optimum flow rate for maximum efficiency, figure 4.10. The position of the maximum varies with the solute, and a family of curves can be derived for the components of a mixture. The most efficient flow rate for a particular sample is, therefore, a matter of compromise. The equation also indicates that the highest efficiencies are achieved with stationary phases of small particle size (small values of A and $C\bar{u}$) and thin coatings of liquid (small value of $C\bar{u}$). In practice, the choice of operating conditions is often a semi-empirical one based on previous experience. Equation (4.45) applies strictly to packed column gas chromatography (p. 92) but similar equations have been derived for capillary column gas chromatography (p. 94) and for high performance liquid chromatography (p. 112).

Resolution, R_s, which is a measure of the degree of separation of two solutes, is given by the expression

$$R_s = \frac{\sqrt{N_2}}{4} \cdot \frac{(\alpha - 1)}{\alpha} \cdot \frac{k_2'}{1 + k_2'} \qquad (4.46)$$

where N_2 is the efficiency (plate number) measured for the second solute, k_2' is its capacity factor and α is the *separation factor* defined as k_2'/k_1'. An R_s value of 0 indicates that the two solutes are completely unresolved and would occur if α were one (i.e. when $k_1' = k_2'$), or if k_2' were zero (i.e. if the second solute eluted on the solvent front, p. 81). Improvement in the resolution of two solutes can be achieved by increasing the magnitude of one or more of the three terms in equation (4.46) which are essentially independent variables.

The first term is affected by particle size, column length and flowrate (p. 82 *et seq.*) but it should be noted that a doubling of resolution requires a fourfold increase in efficiency. The second and third terms are affected by the nature of the mobile and stationary phases, temperature and pressure. Very significant improvements in resolution can often be achieved by changing one or both of the phases as this can have a large effect on the value of α, especially if it is initially very close to one. Although resolution increases significantly with increasing k'_2 for early-eluting peaks, once k'_2 gets much larger than ten, further improvement is negligible because the third term rapidly approaches one.

In practice, R_s is measured from a chromatogram by relating the peak-to-peak separation to the average peak width. This is expressed by the equation

$$R_s = 2\Delta V_R/(W_1 + W_2) \tag{4.47}$$

where ΔV_R is the separation of the peak maxima and W_1 and W_2 are the respective peak widths (figure 4.11). Because of the Gaussian profile of the peaks, a 100% separation is never attainable (see counter-current distribution, section 4.1) but an R_s value of 1.5 or more indicates cross-contamination of 0.1% or less.

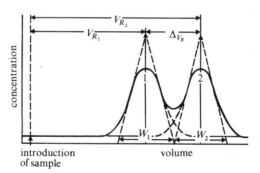

Figure 4.11. Resolution of adjacent peaks

TEMPERATURE EFFECTS

The choice of operating temperature can have a profound effect on a chromatographic separation due to the temperature dependence of the distribution ratio D of each solute or to be strict, of the distribution coefficient K_D (cf. solvent extraction, p. 57). The relation is an exponential one,

$$K_D \propto \exp\left(- \Delta H^\oplus/RT\right)$$

and a change in temperature of 20 K can result in a twofold change in K_D. As $\Delta H^\oplus$, the enthalpy of sorption, is usually negative increasing the temperature decreases K_D, or D, which leads to a corresponding decrease in V_R, the retention volume, or an increase in R_f, the retardation factor. Therefore, in the interests of speed, a higher temperature may be considered desirable but this can be at the expense of resolution, which will be affected by the increased rates of diffusion and mass transfer. At lower temperatures resolution may be better, but the time required may be unacceptably long and a compromise is usually sought. Whatever the ultimate choice of operating temperature it is important that it is reproducible if valid comparisons are to be made.

ISOLATION OF SEPARATED COMPONENTS

One aspect in which column chromatography differs from counter-current distribution is in the recovery of the separated components. In Tswett's original method, the individual pigments were recovered from the column itself rather as they would be collected from the glass separation units of the Craig apparatus. Nowadays, the usual procedure is to remove the components of a mixture from the column sequentially by sweeping them through with the mobile phase, a process known as *elution*. A sample is introduced on to the top of the column and the pure mobile phase passed through it continuously until all the components have been eluted. The order of elution depends on the individual distribution ratios for each solute, and incomplete separation may occur where these are not sufficiently different. A plot of volume of eluting agent or time against concentrations of the eluted species provides an *elution profile* from which both qualitative and quantitative information can be obtained, figure 4.12.

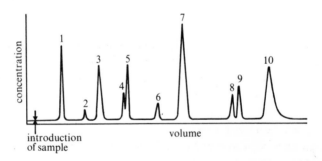

Figure 4.12.   A typical elution profile; the separation of aliphatic esters by gas–liquid chromatography

1. methyl formate	6. *iso*-propyl acetate
2. methyl acetate	7. *n*-butyl formate
3. ethyl formate	8. *sec*-butyl acetate
4. ethyl acetate	9. *iso*-butyl acetate
5. *n*-propyl formate	10. *n*-butyl acetate

Gradient elution is a procedure whereby the conditions under which the sample is eluted are progressively varied throughout the separation so as to speed up the process. This can be achieved by altering the composition of the mobile phase or increasing the temperature or flow rate. The effect is to elute components more rapidly in the latter stages and sharpen their elution profiles. *Stepwise elution* is a similar procedure in which elution conditions are changed at predetermined times rather than continuously.

Two other means of separating and removing components from a column are *frontal analysis* and *displacement development*, but these are of secondary importance. In frontal analysis sample is continuously applied to the top of the column. Eventually, as the stationary phase becomes saturated with sample, the component with the smallest distribution ratio begins to emerge from the column followed by others in order of increasing distribution ratio but not separated as well as in the elution method. Displacement development involves the movement of a sample down the column by introducing a displacing agent which has a larger distribution ratio than any of the sample components. Again solutes leave the column in order of their distribution ratios, but with much less overlap than in frontal analysis. Neither of these methods offers any advantages over elution, except in preparative work where large quantities are to be handled.

4.2.1. GAS CHROMATOGRAPHY

SUMMARY

Principles
> Separation of mixtures in microgram quantities by passage of the vaporized sample in a gas stream through a column containing a stationary liquid or solid phase; components migrate at different rates due to differences in boiling point, solubility or adsorption.

Apparatus and Instrumentation
> Injection port, heated metal, or fused quartz glass column, detector and recorder, regulated gas supply.

Applications
> Very widespread use, almost entirely for organic materials; technique is rapid, simple and can cope with very complex mixtures (100 or more components) and very small samples (nanograms); useful for both qualitative and quantitative analysis. Relative precision 2 to 5%.

Disadvantages

   Samples must be volatile and thermally stable below about 400°C; most commonly-used detectors are non-selective; published retention data is not always reliable for qualitative analysis.

   Gas chromatography, so called because the mobile phase is a gas, comprises *gas–liquid chromatography* (*glc*) and *gas–solid chromatography* (*gsc*). For glc the stationary phase is a high-boiling liquid and the sorption process is predominantly one of partition. For gsc the stationary phase is a solid and adsorption plays the major role. Samples, which must be volatile and thermally stable at the operating temperature, are introduced into the gas flow via an injection port located at the top of the column. A continuous flow of gas elutes the components from the column in order of increasing distribution ratio from where they pass through a detector connected to a recording system. A schematic diagram of a gas chromatograph is shown in figure 4.13 and details of the components are discussed below.

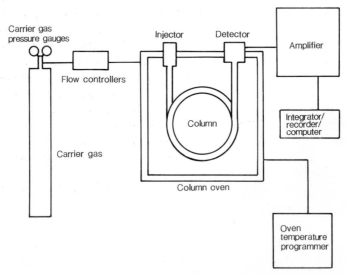

Figure 4.13.   Schematic diagram of a gas chromatograph (from R. M. Smith, *Gas and Liquid Chromatography in Analytical Chemistry*, Wiley, 1988)

1. *Mobile Phase and Flow Control*

The mobile phase or *carrier-gas* is supplied from a cylinder via a pressure-reducing head at a pressure of 10 to 40 psi giving a flow-rate of 2 to 50 cm^3 min^{-1}. Fine control of carrier-gas pressure is achieved either by a needle-valve or by a mass flow controller. The latter enables a constant flow-rate to be maintained when the temperature is increased during a separation.

With a simple needle-valve, the flow-rate will decrease with a rise in temperature due to an increase in the viscosity of the carrier-gas. Water vapour, hydrocarbons and other impurities in the gas affect column performance and detector response, but they can be removed by passing it through a trap containing a molecular sieve. Carrier-gases commonly used are nitrogen, helium and hydrogen, the choice depending on type of column (packed or capillary), cost and the detector to be used. Helium and hydrogen are the preferred gases for capillary columns because chromatographic efficiency diminishes more slowly with increasing flow rate above the optimum (figure 4.10, equation 4.45) with these gases than with nitrogen thus facilitating faster separations.

2. Sample Injection System

To ensure the best possible efficiency and resolution, the sample should be introduced into the carrier-gas stream in as narrow a band as possible. Liquids, diluted if necessary with a volatile solvent, and solids in solution, are injected through a self-sealing silicone-rubber septum using a 1 to 10 μl capacity microsyringe. Gas samples require a larger volume gas-tight syringe or gas-sampling valve as they are much less dense than liquids.

For packed columns, 0.1 to 10 μl of a liquid sample or solution may be injected into a heated zone or *flash vaporizer* positioned just ahead of the column and constantly swept through with carrier-gas (figure 4.14a). The zone is heated some 20 to 50°C above the column temperature to ensure rapid volatilization of the sample. Alternatively, to minimize the risk of decomposing thermally sensitive compounds and to improve precision, samples can be deposited directly on to the top of the packed bed of the column (*on-column injection*).

Several techniques are available for introducing samples into capillary columns which generally have a much lower sample capacity than packed columns.

Split Injection involves an inlet stream splitter incorporating a needle valve that enables most of the injected sample to be vented to the atmosphere whilst allowing only a small fraction (2% or less) to pass into the column (figure 4.14b). Split ratios between 50:1 and 500:1 are common. A disadvantage of split injection is that samples with components that vary widely in their boiling points tend to be split in differing proportions; relatively more of the lower boiling components entering the column than the high boiling ones. However, this *discrimination* effect can be assessed by chromatographing standard mixtures. Split injection is not suitable when the highest sensitivity is required as most of the sample is vented to the atmosphere.

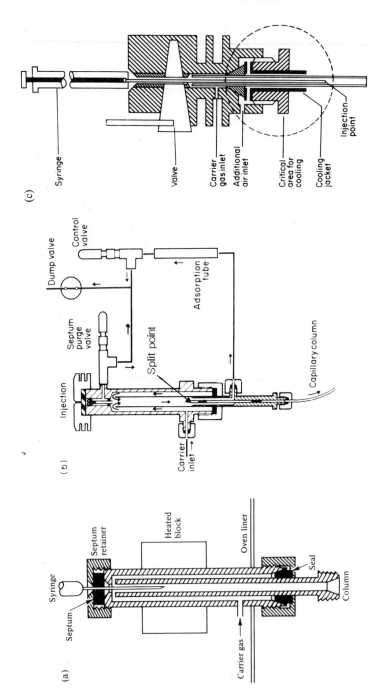

Figure 4.14   Sample injection ports. (a) Flash-vaporizer. (b) Split injector with septum purge for capillary columns. (c) Direct cold on-column injection onto a capillary column showing rotating valve and insertion of needle into the base of the column. (Reproduced by permission of Dr Alfred Hüthig Verlag from *J. High Res. Chromatogr., Chromatogr. Commun.*, **2**, 358 (1979).)

Splitless Injection avoids the problems and several variations of this technique are used. In one system, up to several microlitres of injected sample are collected in a cold trap at the top of the column initially held at over 100°C below the boiling point of the most volatile sample component. The trap is then heated to boil off the sample components sequentially. This method is suitable only if all the sample components have relatively high boiling points. In a variation devised by Grob, the sample is mixed with a high-boiling solvent, e.g. octane (b.pt. 126°C), which, on injection, condenses in a thick layer on the cooled wall at the top of the column. This layer retains and concentrates the sample components before the temperature is raised thereby, in effect, increasing the sample capacity of the column.

On-column Injection allows very small liquid samples to be placed directly into the cooled top of the column which is then heated to volatilize the components. A syringe fitted with a very fine quartz needle is used in conjunction with a specially designed septumless valve through which cooled or heated air can be passed (figure 4.14c). Air cooled to about 20°C below the boiling point of the sample solvent is passed through the valve while the sample is being injected, then warmer air is circulated to volatilize the sample *in situ* on the column. This technique avoids discrimination and reduces the risk of decomposition of thermally sensitive compounds.

Automatic Injectors, which eliminate variations due to the analyst, improving reproducibility, and which can be controlled by computer, are of value where large numbers of samples are to be analysed or unattended operation is required. Samples and standards loaded into racks or turntables can be run in a pre-determined sequence and under different operating conditions. Such devices can also be used for single samples to improve injection precision.

Solid samples can be introduced as a solution or in a sealed glass ampoule which is crushed in the gas stream by means of a gas-tight plunger. Only solids which have appreciable vapour-pressures at the operating temperature of the column can be successfully chromatographed.

3. *The Column*

The column is the heart of the gas chromatograph in which the separation process occurs. It consists of a coil of stainless steel, glass or fused silica (quartz) tubing which may be one metre to one hundred metres long and have an internal diameter of between 0.1 mm and about 3 mm.

To ensure operation under reproducible conditions, the column is enclosed in a thermostatically controlled oven whose temperature can be held constant to within ±0.1°C. Operating temperatures range from

ambient to over 400°C and may remain constant during a separation—
isothermal operation—or automatically increased at a predetermined rate to
speed the elution process—*temperature programming*. The latter is a form of
gradient elution. Rapid temperature equilibration and changes in operating
temperature are achieved by circulating the oven air with a fan.

(a) Packed Columns

These rarely exceed 2 to 3 metres in length with an internal diameter of 2 to 3
mm and are made of stainless steel or glass, the latter being less reactive to
thermally sensitive compounds and facilitating visual inspection of the
packed bed. For glc they are completely filled with an inert porous granular
solid support which is coated with a thin film of a *liquid* or *semi-liquid*
stationary phase (*vide infra*). Gsc columns are filled with a solid stationary
phase which has adsorbent properties. Packed columns do not have the very
high resolving power of capillary columns but, compared to the latter, they
are cheap, robust and have a high sample capacity which allows the use of a
simpler injection system. Their popularity has diminished steadily in
recent years as capillary columns have improved, but for some applications
they may still be preferred.

Solid Support

The function of a solid support is to hold the liquid phase used for packed
column glc immobile during the separation process. It should be inert, easily
packed and have a large surface area. Calcined diatomaceous earth and
firebrick, both mainly silica, are commonly used, being marketed under
various trade names such as Celite, Chromosorb and Stermachol. The
materials must be rendered chemically inert before coating with stationary
phase because trace metal impurities and surface silanol (Si—OH) groups
produce surface active sites which promote undesirable adsorption effects.
Adsorption causes tailing (p. 77) and may result in catalysed decomposition
or rearrangements of the solutes passing through the column. Pretreatment
consists of acid- or alkali-washing to remove the trace metals, and silanizing
to convert the Si—OH groups to silyl ethers, e.g. Si—O—Si(CH$_3$)$_3$.
Dimethyl dichlorosilane or hexamethyldisilazane are frequently used for
this purpose. If highly polar compounds are to be separated, silanized glass
beads or granular PTFE are supports less likely to cause tailing. Because
they are non-porous, they can support a maximum of only about 3% of
stationary phase, and the size of sample that can be chromatographed is
smaller than with the silaceous solid supports.

The particle size of a solid support is critical in striking a compromise
between column efficiency and speed of separation. Both the multiple path
term A and the mass transfer term ($C\bar{u}$ of equation (4.45) (p. 83)) are reduced

by reducing particle size thus leading to increased efficiency. However, as particle size is reduced, the pressure drop across the column must be increased if a reasonable flow-rate is to be maintained. The optimum particle sizes for $\frac{1}{8}$ in columns are 80/100 or 100/120 mesh and for $\frac{1}{4}$ in columns 40/60 or 60/80 mesh.

Stationary Phase

The number of stationary phases suitable for gas chromatography is quite extensive, and choice is dictated largely by the nature of the sample. A liquid stationary phase should be non-volatile and thermally stable at the operating temperature of the column, otherwise it will 'bleed' during operation and cause a drifting baseline on the recorder. In addition it should be chemically stable and inert towards samples to ensure reliable results. Stationary phases are described as non-polar or polar according to their structure and separating abilities. Non-polar types include hydrocarbon and silicone oils and greases. Polar types cover a wide range of polarity and include high molecular weight polyesters, ethers, carbowaxes, amines, etc. Solid adsorbents used in gsc, e.g. silica and alumina, are among the most polar of stationary phases.

In general, the most suitable stationary phase for a given sample is that which is chemically similar to it. Thus, a mixture of saturated hydrocarbons is best separated on a non-polar hydrocarbon-type liquid such as squalane, Apiezon-L grease or silicone oil DC200. If there is an appreciable difference in polarity between the sample components and the stationary phase, elution profiles may show tailing or fronting because of non-linearity of the sorption isotherms. Where a sample contains unknowns or compounds of varying polarity, a compromise stationary phase must be chosen, usually by trial and error. The order of elution can be altered by changing the liquid phase. For example, on a non-polar paraffin-type column, t-butyl alcohol (b.pt. 82.6°C) will elute before cyclohexane (b.pt. 80.8°C). If a more polar liquid phase containing hydroxyl groups is used, the cyclohexane elutes first because of hydrogen-bonding between the alcohol and the stationary phase.

Stationary phases can be made highly selective by adding compounds to them which have affinities for certain chemical species. For example, silver nitrate, incorporated into a polar liquid preferentially retards the elution of olefins by formation of weak π-complexes. A selection of stationary phases with their maximum operating temperatures and useful applications is given in table 4.9.

The amount of stationary phase or 'loading' carried by a solid support affects the efficiency of the column and the size of sample that can be injected. A high loading, 15 to 30% by weight, produces a thick liquid film which impairs efficiency by increasing the mass transfer term $C\bar{u}$ in equation (4.45). Loadings of 1 to 10% are to be preferred, although at the lower end of this

Table 4.9   Some stationary phases used in gas chromatography

STATIONARY PHASE	OPERATING TEMPERATURE (°C)	TYPICAL APPLICATIONS
squalane ⎱ high M.W.	0–130	saturated hydrocarbons
Apiezon-L ⎰ hydrocarbons	50–280	high-boiling hydrocarbons
Porapak-Q solid aromatic polymer	200	water, light hydrocarbons, permanent gases
silicone gum rubber SE 30	50–350	general, steroids, pesticides
silicone oil DC 550	20–250	general, aromatics
di-nonyl phthalate	20–150	esters, alcohols
di-ethyleneglycol succinate	20–200	fatty acid esters
carbowax 20 M (polyethylene glycol)	60–300	alcohols, amines, halogen and sulphur compounds, essential oils
Bonded phases		
dimethylpolysiloxane	-60–325	amines, hydrocarbons, pesticides, PCBs, phenols, sulphur compounds
phenyl/methylpolysiloxane	-60–280	glycols, drugs, pesticides, steroids
polyethylene glycol	60–220	alcohols, free acids, aromatics, essential oils
PLOT (Al_2O_3)	200 max.	C1–C10 hydrocarbons
PLOT (molecular sieve)	350 max.	permanent gases
PLOT (carbon)	115 max.	He, N_2, O_2, CO, CO_2, CH_4, C_2H_6
Dexsil 300 (carborane/methylsilicone)	450 max.	general use; low bleed gc–ms

(Left margin, vertical: INCREASING POLARITY, with downward arrow)

range the sample size may have to be restricted to prevent overloading the column. In general lower loadings allow the use of a lower operating temperature which is an important factor when handling thermally sensitive compounds. However if the liquid film is too thin, adsorption of the solutes by the solid support may cause tailing, decomposition or rearrangements.

(b) Capillary (open tubular) Columns
This type of column has become the most widely used because of its superior resolving power for complex mixtures compared to that of a packed column. Capillary columns are typically 5 to 50 metres long and between 0.1 mm and 0.60 mm internal diameter. A thin film (0.1 to 5 μm thick) of the liquid stationary phase is coated or bonded onto the inner wall of the tube (hence the alternative description of *wall-coated open tubular* or WCOT columns)

which is made of high-purity fused silica (quartz). Keeping the total metallic impurities in the silica to around 1 ppm or less renders it extremely inert thereby minimizing peak tailing and the possibility of thermal decomposition of chromatographed substances. The exterior of the tube is coated with a layer of a polyimide or aluminium as a protection against cracking or scratching.

The unrestricted flow of carrier-gas through the centre of capillary columns results in a much smaller pressure drop per metre than for packed columns. They can therefore be made very much longer and will generate many more theoretical plates, i.e. up to about 150,000 plates per 25 metres compared with a few thousand for a 2-metre packed column. A narrow bore and thin layer of stationary phase are essential to promote rapid mass transfer between the phases as this increases efficiency by reducing the $C\bar{u}$ term in equation (4.45). The most efficient capillary columns are those with the narrowest bore (0.1 mm) and the thinnest liquid coating (0.1 μm) but they have a very low sample capacity ($<< 0.1 \mu$l) which necessitates the use of a sample inlet stream splitter, cold trap or special on-column injection system (p. 90). Sample capacity can be progressively increased at the expense of reduced efficiency by increasing both column diameter and liquid phase thickness. Wide-bore capillary columns (> 0.5 mm id) with a relatively thick layer (1 to 5 μm) have sample capacities approaching those of packed columns so they can be used with a simple packed column injection system. Furthermore, although their efficiencies are much lower than those of the narrowest bore columns, they are nevertheless considerably more efficient than packed columns and have become popular replacements for them. Very rapid separations can be achieved on short wide-bore capillary columns (5 to 10 metres), with efficiencies at least as good as those of packed columns.

Capillary columns are much more expensive than packed columns but their working life can be extended and performance improved by chemically bonding the stationary phase to the wall of the tubing. This greatly reduces 'column bleed', especially at high operating temperatures, and is particularly advantageous in minimizing the contamination of detectors and gc–ms systems. Bonded-phase columns, which can also be washed with solvents to remove strongly-retained material accumulating from samples in the first few metres over a period of time, are becoming the most popular type of capillary column for routine work. Stationary phases of various polarities are available and the range of applications includes petrochemicals, essential oils and biomedical samples. For gas–solid chromatography (gsc), porous-layer open-tubular (PLOT) columns are used. These have a thin porous layer of a finely-divided solid, usually alumina or molecular sieve, deposited on the inside wall of the tube. They are used to separate mixtures of low RMM hydrocarbons and the permanent gases. A capillary column separation of organic acids in human urine is shown in figure 4.19.

4. Detectors

The purpose of a detector is to monitor the carrier-gas as it emerges from the column and respond to changes in its composition as solutes are eluted. Ideally a detector should have the following characteristics: rapid response to the presence of a solute; a wide range of linear response; high sensitivity; stability of operation.

Most detectors are of the differential type, that is their response is proportional to the concentration or mass flow rate of the eluted component. They depend on changes in some physical property of the gas stream, e.g. thermal conductivity, density, flame ionization, electrolytic conductivity, β-ray ionization, in the presence of a sample component. The signal from the detector is fed to a chart recorder, computing integrator or VDU screen via suitable electronic amplifying circuitry where the data are presented in the form of an elution profile. Although there are a dozen or more types of detector available for gas chromatography, only those based on thermal conductivity, flame ionization, electron-capture and perhaps flame emission and electrolytic conductivity are widely used. The interfacing of gas chroma-

Table 4.10   GC detector characteristics

DETECTOR	MINIMUM DETECTABLE QUANTITY $(g\,sec^{-1})$	LINEAR RANGE	TEMPERATURE LIMIT (°C)	REMARKS
Thermal conductivity (tcd)	10^{-9}	10^4	450	non-destructive, temperature and flow sensitive
Flame ionization (fid)	10^{-12}	10^7	400	destructive, excellent stability
Electron capture (ecd)	10^{-13}	10^2 to 10^3	350	non-destructive, easily contaminated, temperature-sensitive
Phosphorus	10^{-14}	10^5	400?	similar to flame ionization
Nitrogen	10^{-13}	10^5	400?	similar to flame ionization
Flame photometric (fpd)				
(P cmpds)	10^{-12}	10^4	~ 250	signal approximately proportional to square of S concn.
(S cmpds)	10^{-10}	—	~ 250	

tographs with infrared and mass spectrometers, so-called 'hyphenated' techniques, is described on p. 108 *et seq.* Some detector characteristics are
summarized in table 4.10.

Thermal Conductivity Detector (tcd)

This detector is based on the principle that a hot body loses heat at a rate
which depends on the thermal conductivity and therefore the composition of
the surrounding gas. Sometimes called a katharometer, it consists of two
heated filaments of a metal which has a high coefficient of resistance, e.g.
platinum, and which form two arms of a Wheatstone bridge circuit. The two
filaments are situated in separate channels in a heated metal block, figure 4.15.
Pure carrier gas flows through one channel and the effluent from the column
through the other. The rate of heat loss from each filament determines its
temperature and therefore its resistance. A change in thermal conductivity
of the gas flowing through the sample channel arising from elution of a
sample component alters the temperature and hence the resistance of the
filament in that channel and this produces an out-of-balance signal in the

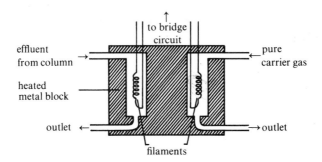

Figure 4.15.   Thermal conductivity detector (tcd)

bridge circuit. The imbalance is registered as a deflection of the recorder pen.
Sensitivity, which is determined by the difference in thermal conductivity
between the reference and sample streams, is highest when hydrogen or
helium is used as the carrier-gas. The thermal conductivity detector is robust
and reliable but has only moderate overall sensitivity which varies widely
from compound to compound. Response is non-linear and sensitive to
changes in temperature and flow-rate so that it is not particularly suitable for
quantitative work.

Ionization Detectors

Ionization detectors depend on the principle that the electrical conductivity
of a gas is directly proportional to the concentration of charged particles
within it. Effluent gas from the column passes between two electrodes across

which a dc potential is applied. An ionizing source partially ionizes the carrier gas allowing a steady current to flow between the electrodes and through a resistor where a corresponding voltage drop is amplified and fed to a recorder. When a sample component is eluted from the column, it is also ionized in the electrode gap thereby increasing the conductivity and producing a response in the recorder circuit. Ionization detectors are very sensitive, respond rapidly, usually linearly, and are mostly stable to variations in temperature and flow-rate.

Flame Ionization Detector (fid)

A schematic view of a flame ionization detector (fid) which is one of the most widely used, is shown in figure 4.16. Effluent gas from the column is mixed with hydrogen and air and burned at a small metal jet. The jet forms the negative electrode of an electrolytic cell, the positive or collector electrode being a loop of wire or short tube placed just above the flame. The potential difference applied across the electrodes is about 200 V. The fid responds to virtually all organic compounds except formic acid, air and other inorganic gases. Its response to water is very low. It has a very high sensitivity and the widest linear range (10^7) of any detector in common use (table 4.10).

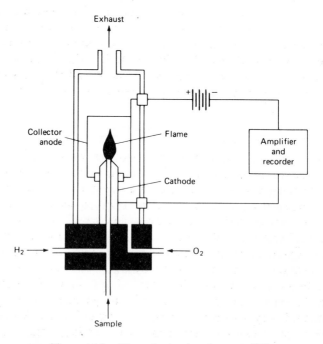

Figure 4.16.   Flame ionization detector (fid)

Nitrogen/Phosphorus Detector (NP)

This is a more recent introduction which is again basically an fid, and can in fact be used as such, if a selective response to phosphorus and/or nitrogen compounds is not required. Nitrogen and phosphorus response is achieved through the incorporation of an electrically-heated glass bead containing a rubidium silicate which is positioned a few millimetres above the burner jet. By heating the bead to 600–800°C and applying a negative potential (-180 V) to it, a reaction cycle occurs involving the vaporization, ionization and recapture of rubidium by the bead. During this recycling of rubidium, which prevents its long-term depletion in the bead, an electron flow to the positive collector electrode occurs. This flow or background current is enhanced when nitrogen or phosphorus compounds are eluted, as they form radicals in the flame which participate in the reaction cycle and accelerate the rate of rubidium recycling. Thus, nitrogen compounds are believed to form mainly cyanide radicals which then abstract electrons from vaporized neutral rubidium atoms to form CN^- ions. The CN^- ions are captured by the collector electrode, thereby increasing the detector current, whilst the Rb^+ ions are recaptured by the bead. The phosphorus response is thought to be due to similar processes involving PO_2^* and PO_2^- species.

The detector can be made to respond to phosphorus compounds only by earthing the jet, which is at a negative potential for simultaneous nitrogen and phosphorus detection, and altering the flow rates of the flame gases. If the bead is not electrically heated the response is the same as a conventional fid. The detector is thus a three-in-one device which can be very easily switched between the three modes, fid, N and P, and P only. Sensitivity for nitrogen and phosphorus compounds exceeds 10^{-13} g sec^{-1} and the linear range is about 10^5.

Electron Capture Detector (ecd)

This is the most widely used of several detectors which employ a β-ray ionizing source. A schematic diagram is shown in figure 4.17. Unlike the fid the electron capture detector depends on the recombination of ions with free electrons and therefore measures a reduction in signal. As the nitrogen carrier-gas flows through the detector a tritium or ^{63}Ni source ionizes the gas forming 'slow' electrons which migrate towards the wire anode under an applied potential difference of 20 to 50 V. The drift of 'slow' electrons constitutes a steady current while only carrier-gas is present. If a solute with a high electron affinity is eluted from the column, some of the electrons are 'captured' thereby reducing the current in proportion to its concentration. The detector is very sensitive to compounds containing halogens and sulphur, anhydrides, peroxides, conjugated carbonyls, nitrites, nitrates and organometallics, but is virtually insensitive to hydrocarbons, alcohols, ketones and

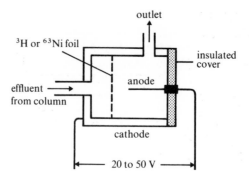

Figure 4.17.   Electron capture detector (ecd)

amines. Additional selectivity can be obtained by increasing the applied potential when the response of weakly electron-capturing compounds is eliminated. Application of a pulsed dc potential rather than a continuous one improves sensitivity and linear range and reduces contamination of the anode by deposition of negatively-charged solute species. However, this mode of operation is more costly and requires the use of an argon/methane carrier gas mixture. The electron capture detector is particularly useful in the analysis of halogen-containing pesticides which can be detected in the sub-picogram range. Although it is the most sensitive available, its linear range is restricted to only 10^2 or 10^3 and it is sensitive to temperature changes. The carrier-gas must be exceptionally pure as oxygen, air and water at levels exceeding about 10 ppm affect performance as does column bleed. Halogenated solvents should be avoided in sample preparation as residual traces can de-activate the detector.

Flame Photometric Detector (fpd)

This is a selective detector for phosphorus and sulphur-containing compounds which makes use of the characteristic emission of radiation by $S_2^{\cdot}$ and HPO$^{\cdot}$ species in a suitable flame. Earlier versions incorporated special burners to produce a flame with hydrogen enveloping the air/carrier gas mixture (inverted or inside-out flame). More recently, modified fid burners have proved satisfactory (figure 4.18). The $S_2^{\cdot}$ emission at 394 nm and the HPO$^{\cdot}$ emission at 526 nm are selected by means of appropriate narrow bandpass filters and the lower half of the flame is shielded to reduce background emission which is due largely to $C_2^{\cdot}$ and CH$^{\cdot}$. The emission intensity can be monitored with a fibre optics light pipe connected to a remote photomultiplier tube, as the response of the tube is temperature-sensitive. An alternative design utilizes heat filters, a long tube and a metal

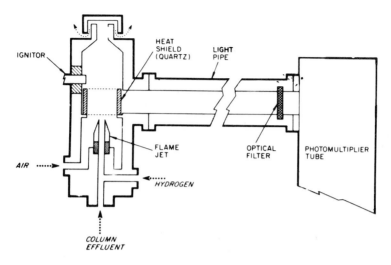

Figure 4.18. Schematic cross-section of a flame photometric detector (fpd)

block to act as a heat sink. Sensitivity is high (table 4.10) but the response to phosphorus is linear over a range of only 10^4, while for sulphur it is proportional to the square of the sulphur concentration, varies with the type of sulphur compound and is linear over an even smaller range.

SPECIALIZED TECHNIQUES USED IN GAS CHROMATOGRAPHY

Non-volatile and Thermally Sensitive Compounds
Paints, plastics, polymers, ionic and many biologically important compounds fall into this category. They can either be *pyrolysed* under controlled conditions to produce characteristic lower molecular mass and therefore volatile products or, in some cases, converted into related and more volatile *derivatives*.

Pyrolysis gc involves the rapid and controlled thermal decomposition of a few milligrams of the sample in the injection part of the chromatograph, the volatile products then being swept onto the column in a narrow band of carrier gas. This technique is discussed more fully in section 11.5.

Derivatization of non-volatile polar or thermally sensitive compounds to enhance their volatility and stability prior to chromatography is a well-established technique. Compounds containing hydroxyl, carboxyl and amino functional groups can be readily reacted with appropriate reagents to convert these polar groups into much less polar methyl, trimethylsilyl or trifluoroacetyl derivatives of greater volatility. Fatty acids, carbohydrates,

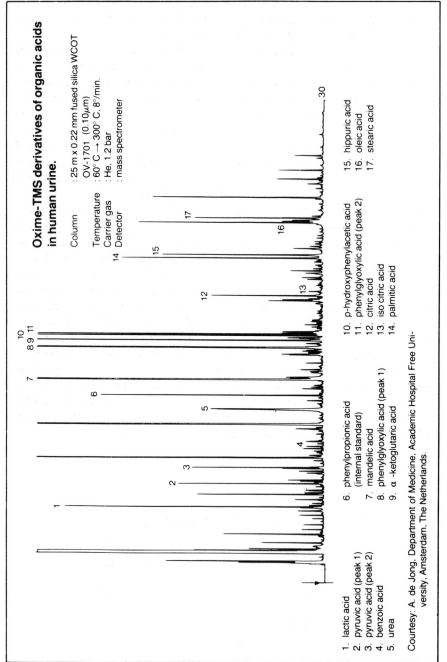

Oxime-TMS derivatives of organic acids in human urine.

Column   : 25 m x 0.22 mm fused silica WCOT
                  OV-1701 (0.10μm)
Temperature   : 60° C → 300° C, 8°/min.
Carrier gas   : He, 1.2 bar
Detector   : mass spectrometer

1. lactic acid
2. pyruvic acid (peak 1)
3. pyruvic acid (peak 2)
4. benzoic acid
5. urea

6. phenylpropionic acid (internal standard)
7. mandelic acid
8. phenylglyoxylic acid (peak 1)
9. α-ketoglutaric acid

10. p-hydroxyphenylacetic acid
11. phenylglyoxylic acid (peak 2)
12. citric acid
13. iso citric acid
14. palmitic acid

15. hippuric acid
16. oleic acid
17. stearic acid

Courtesy: A. de Jong, Department of Medicine, Academic Hospital Free University, Amsterdam, The Netherlands.

Figure 4.19.   Capillary column separation of some constituents of human urine

phenols, amino acids and other compounds of biological interest are the most frequently derivatized although liquid chromatography (hplc) is very often the preferred technique for these types of compound (p. 112).

Thermal desorption is a technique that involves the pre-concentration of substances prior to chromatography. For example, it enables the build-up of volatile materials such as toxic solvent vapours in an industrial or laboratory environment to be monitored. The vapours are allowed to pass through a small tube containing an adsorbent such as *Tenax* or *Poropak* by atmospheric diffusion for a prescribed period of time. The tube is subsequently connected to the injection port of the chromatograph and purged with carrier-gas whilst being rapidly heated. This causes any previously adsorbed substance to be *thermally desorbed* and swept onto the column in a narrow band to be separated in the normal way.

Headspace analysis involves examination of the vapours derived from a sample by warming in a pressurized partially-filled and sealed container. After equilibration under controlled conditions, the proportions of volatile sample components in the vapours of the *headspace* are representative of those in the bulk sample. The system, which is usually automated to ensure satisfactory reproducibility, consists of a thermostatically heated compartment in which batches of samples can be equilibrated, and a means of introducing small volumes of the headspace vapours under positive pressure into the carrier-gas stream for injection into the chromatograph (figure 4.20). The technique is particularly useful for samples that are mixtures of volatile and non-volatile components such as residual monomers in polymers, flavours and perfumes, and solvents or alcohol in blood samples. Sensitivity can be improved by combining headspace analysis with thermal desorption whereby the sample vapours are first passed through an adsorption tube to pre-concentrate them prior to analysis.

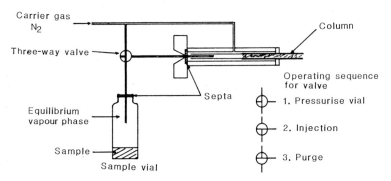

Figure 4.20.   Headspace sample injection system (positioned in a heated oven)

Identification of the component peaks of a chromatogram, which may be numerous, can be achieved in two ways: comparison of retention times (discussed below); trapping the eluted components for further analysis by other analytical techniques such as infrared and mass spectrometry or by direct interfacing of these techniques with a gas chromatograph. This latter approach is discussed on p. 108.

Comparison of Retention Times

Provided that operating conditions remain constant and are reproducible, the retention times of the components of a sample can be compared directly with those of known materials and synthetic mixtures. An unfamiliar peak can sometimes be identified by 'spiking' a sample with a pure substance whose presence is suspected. An increase in the size of the unknown peak is good evidence for it being the substance added. As two materials may have the same retention time for a given stationary phase, this method is not infallible. It is advisable, therefore, to run unknowns on two different stationary phases.

In cases where a mixture has a large number of components, or pure standards are not available, published retention data must be consulted. The uncorrected retention time, t_R (p. 81) is not suitable for this purpose because it cannot be compared with data from different columns and instruments. Valid comparisons can be made using relative retention data which are dependent only on column temperature and type of stationary phase. An *adjusted retention time*, t'_R is first obtained by subtracting from t_R the time required to elute a non-retained substance such as air, figure 4.21.

Thus,

$$t'_R = t_R - t_{air} \qquad (4.48)$$

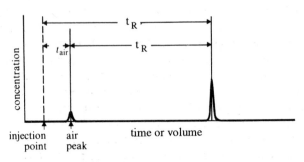

Figure 4.21. Measurement of adjusted retention time, t'_R

where t_{air} denotes the time taken to elute the air peak. The term t_{air} is related to the volume of the injection port, column void space and detector, known as the *dead volume* V_{DV}, by the expression $t_{air} = V_{DV}/F$ where F is the volumetric flow rate. Values of t'_R are obtained for the sample components to be identified and also for a standard substance chromatographed under identical conditions, preferably by adding it to the sample. The *relative adjusted retention time* α_x for component x is given by

$$\alpha_x = \frac{t'_R(x)}{t'_R \text{ (standard)}} \tag{4.49}$$

The value of α_x is then compared with published values of known compounds for the appropriate temperature and stationary phase used.

If a sample contains one or more members of a homologous series, identifications can be made using a plot of log t_R against the number of carbon atoms, previously prepared from standards. The plot, which is valid for one temperature only, is linear and can be used for alkanes, alkenes, alcohols, aldehydes, ketones, esters and ethers.

A universal system for qualitative identification was devised by Kováts some years ago. Based on the linear relation between log t'_R and the number of carbon atoms for a homologous series, Kováts selected n-alkanes as standards for the following reasons:

(i) they cover a very wide range of boiling points
(ii) they are readily separated on virtually any column
(iii) they are chemically very stable and non-toxic
(iv) they are easily obtained and relatively cheap

The Kováts retention index for each *n*-alkane is defined as 100 times the number of carbon atoms in the chain at all temperatures and for any column, e.g. pentane is 500 and octane is 800. A plot of log t'_R against retention index for a series of *n*-alkanes is linear as shown in figure 4.22. A substance with a retention time between an adjacent pair of *n*-alkanes will have a retention index that can be determined graphically or calculated by linear interpolation. For example, using the plot in figure 4.22, the graphical method for an unknown substance that elutes with a retention time a little more than halfway between those of *n*-octane (C8) and *n*-nonane (C9), will have a retention index in the region of 860. The possible identity of unknowns can be ascertained from tables (published or previously compiled in house) or by using a computer data base. Kováts indices are also used as a basis for comparing gc stationary phases by computing a set of values from the retention times of a selected group of test compounds on each phase.

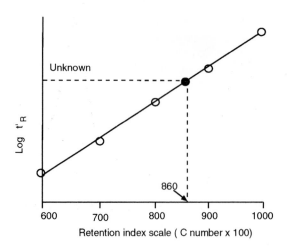

Figure 4.22.   Graph of log adjusted retention times on *n*-alkanes compared to carbon number. The open circles show log retention times of *n*-alkane standards. The retention of an unknown (full circle) can be determined from the graph as shown

QUANTITATIVE ANALYSIS

Gas chromatography has become one of the most useful separation techniques because quantitative information can so readily be obtained from it. Standardization of operating conditions is of prime importance and detector response factors must be known for each compound to be determined. The integrated area of a peak is directly proportional to the amount of solute eluted. Peak height can be used as a measure of peak area, but compared to an area measurement the linear range is less and it is more sensitive to changes in operating conditions. A more reliable method is to multiply the peak height by the retention time as these parameters are inversely related. Heights are measured by drawing perpendiculars from the peak maxima to the baseline or projected baseline if there is drift or the peaks to be measured overlap. Measurement of area is accomplished by one of the following methods, which vary considerably in precision (table 4.11).

Table 4.11   Precision of methods of peak area measurement

METHOD OF MEASUREMENT	RELATIVE PRECISION, %
computing integrator	0.4
cutting out and weighing peaks	1.7
height × weight at $\frac{1}{2}$-height	2.6
$\frac{1}{2}$-base × height	4.0

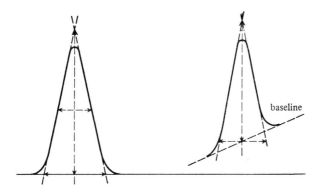

Figure 4.23. Measurement of peak height and area

Geometric Methods

As normal peaks have a Gaussian profile, which approximates to an isosceles triangle, their area can be estimated by multiplying the height by the width at half height or by calculating the area of a triangle formed by the baseline and the sides of the peak produced to intersect above the maximum (figure 4.23) i.e. $\frac{1}{2}$ base × height. The methods are simple and rapid but are unreliable if peaks are narrow or asymmetrical. Precision is only moderate.

Cutting-out and Weighing

The peak is cut from the original chart paper or from a photocopy and weighed on an analytical balance. This method is fairly precise and particularly useful for unsymmetrical peaks but is subject to errors arising from variation in thickness and moisture content of the paper.

Automatic Integration

Electronic integrators are the most rapid and precise means of determining peak areas. They have a digital output derived by feeding the detector signal into a voltage-to-frequency converter which produces a pulse-rate proportional to the input signal. The total number of pulses is a measure of the peak area and this can be printed out directly or stored until required. Electronic integrators have a wide linear range, a high count-rate and may automatically correct for baseline drift. In addition, the more expensive versions will print retention data alongside peak areas. Computing integrators, based on a microcomputer, are now widely available and are discussed in chapter 13.

CALCULATION OF QUANTITATIVE RESULTS

Internal Normalization

If the components of a mixture are similar in chemical composition and if all are detected, the percentage weight of each is given by

$$\text{percentage } x = \frac{\text{area for component } x}{\text{total area for all components}} \times 100 \qquad (4.50)$$

The formula assumes that the detector sensitivity is the same for each component. If this is not the case, the response of each must first be determined using a set of standards. Areas are then multiplied by correction factors obtained by setting the response of one component equal to unity.

Internal Standardization
An accurately known amount of a standard is added to the sample before it is chromatographed. The ratio of peak area of standard to that of the component of the sample to be determined is calculated. This ratio is converted to weight of component using a previously prepared calibration curve (p. 10). The internal standard should have a retention time close to those of the components being determined but well resolved from them. Preferably it should be present at a similar concentration level.

Standard Addition
If a pure sample of the component to be determined is available, the sample can be chromatographed before and after the addition of an accurately known amount of the pure component. Its weight in the sample is then derived from the ratio of its peak areas in the two chromatograms.

The advantages of internal standardization are that the quantities of sample injected need not be measured accurately and the detector response need not be known, as neither affect the area ratios. Standard addition is particularly useful in the analysis of complex mixtures where it may be difficult to find a suitable internal standard which can be adequately resolved from the sample components.

COMBINATION OF GAS CHROMATOGRAPHY WITH OTHER ANALYTICAL TECHNIQUES

The identification of gc peaks other than through retention data, which are sometimes ambiguous or inconclusive, can be facilitated by the direct inter-facing of gc with infrared spectrometry (p. 377 *et seq.*) or mass spectrometry (p. 425 *et seq.*), so called 'coupled' or 'hyphenated' techniques. The general instrumental arrangement is shown in figure 4.24 (a).

GC—Mass Spectrometry
Identification of separated components can be achieved by feeding the effluent gases from a gc column directly to a mass spectrometer. Where a

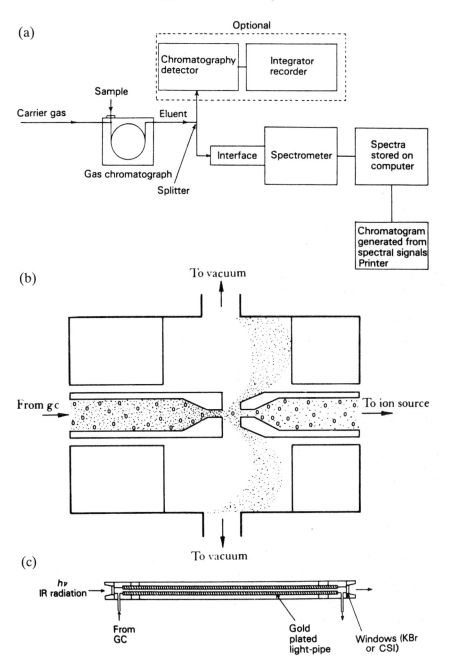

Figure 4.24.   (a) GC-spectrometer system. (b) Jet-orifice separator for GC-MS.
(c) Flow-through light pipe cell for GC-IR

packed column is used, the interfacing of the two instruments presents difficulties in that the mass spectrometer operates at very low pressures (10^{-7} to 10^{-9} Nm^{-2}) whereas gas emerging from the column is at atmospheric pressure. A jet-orifice separator (figure 4.24b) enables the carrier gas (usually helium) to be pumped away whilst allowing the sample vapours to pass on into the mass spectrometer. It is constructed of glass and consists of two fine-bore jets separated by a 0.5 to 1 mm gap and surrounded by an evacuated tube. As the gc effluent is passed through the separator, the momentum of the relatively large and heavy sample molecules carries them across the gap and on into the mass spectrometer whereas the lighter helium atoms are deflected outwards by the vacuum and are pumped away. Increased enrichment of the sample vapour can be achieved by incorporating a second pair of jets into the device. The diameter and hence the volumetric gas flow through a capillary column is much lower than that through a packed column which obviates the need for a separator as the carrier gas can be readily pumped away by the vacuum system of the spectrometer. Furthermore, the column can be inserted directly into the spectrometer reducing dead volume. .

In many gc separations, a large solvent peak may precede those of sample components, and minor components sometimes elute on the tails of major component peaks. It is advantageous to be able to divert solvent peaks and sometimes other large peaks from the mass spectrometer to avoid swamping it. This is easily accomplished by means of a *solvent dumping valve* positioned between the end of the column and the separator; carefully timed operation of this valve enables selected peaks to be vented to the atmosphere. As in other instances where gc is interfaced with another technique (e.g. gc-ir) any transfer tubing used between the end of the column and the spectrometer must be heated to prevent condensation of samples during the transfer stage. In the case of gc-ms, the dumping valve and separator must be heated also, either with a separate heater or by enclosing them within the gc oven. The total volume of the separator, dumping valve and transfer line (dead volume) must be kept to a minimum so as not to degrade chromatographic resolution, and glass or glass-lined metal components are preferable as hot metal surfaces can catalyse the decomposition of thermally sensitive compounds.

A stream-splitter may be used at the end of the column to allow the simultaneous detection of eluted components by destructive gc detectors such as an fid. An alternative approach is to monitor the *total ion current* (*tic*) in the mass spectrometer which will vary in the same manner as the response of an fid detector. The *total ion current* is the sum of the currents generated by all the fragment ions of a particular compound and is proportional to the instantaneous concentration of that compound in the ionizing chamber of the

mass spectrometer.\By monitoring the ion current for a selected mass fragment (m/e) value characteristic of a particular compound or group of compounds, detection can be made very selective and often specific. Selected ion
monitoring (sim) is more sensitive than tic and is therefore particularly useful
in trace analysis.

As early gc peaks elute in a few seconds or less, rapid scanning of the mass
range of interest is necessary. Fast scanning also allows partially resolved gc
peaks to be sampled several times, *peak slicing*, to facilitate identification of
the individual components (figure 12.5) provided that the dead volume of the
interface is small compared to peak volumes. For the speedy interpretation
of spectral data from complex chromatograms a computerized data
processing system is essential. Quantitative analysis can be accomplished by
monitoring standards and samples at a selected mass fragment (m/e) value.
The technique of gc-ms is now well-established as one of the most powerful if
somewhat costly analytical tools available for the study of complex samples.

GC-Infrared Spectrometry
Effluent gas emerging from a gas chromatograph at atmospheric pressure can
be led directly into a heated infrared gas cell via a heated transfer line.
Vapour-phase infrared spectra of eluting components can be recorded as they
pass through a cell by a Fourier transform (FT) infrared spectrometer
enabling a full-range spectrum to be collected and stored in a second or less.

For maximum sensitivity, the volume of the gas cell should be similar to
the volume of the eluting gc peaks, e.g. 50 to 300 μl for capillary columns.
This is achieved with a *light pipe*, a glass tube 50 cm by 2 mm i.d. and coated
on the inside with gold to maximize transmission of the radiation. The ends
are sealed with an ir transparent material such as potassium bromide or
caesium iodide (figure 4.24c). If the cell volume is appreciably less than the
peak volumes, good spectra from partially resolved peaks can be obtained
by careful sampling, i.e. *peak slicing*. Conversely, if the cell volume is
appreciably more than the peak volumes, cross-contamination of one peak
with another may occur. Wide-bore capillary columns or packed columns
having higher sample capacities but poorer resolution are used where
increased sensitivity is required. As in the case of gc-ms, a stream-splitter
facilitates the simultaneous use of an fid detector. Alternatively, if a non-
destructive thermal conductivity detector is used no stream splitting is
necessary.

Compared to gc-ms, gc-ir is much less sensitive. Whereas a mass spectrum
can be recorded from as little as 10^{-10} g of sample, at least 10^{-6} g is required
for an ir spectrum. Care is required in the interpretation of vapour phase ir
spectra as they differ in certain respects from the corresponding liquid or solid
phase spectra. Rotational fine structure may appear, band positions may be

shifted slightly and hydrogen bonding effects are non-existent. The availability of libraries containing thousands of digitized vapour-phase infrared spectra that can be searched in seconds by computer and compared with those collected from a chromatographic run has increased the power of gc-ir for qualitative analysis. The full range of computerized spectral enhancement facilities enables the quality of work or noisy spectra to be improved and the spectra of contaminants to be subtracted.

Gc-ir is becoming as widely used as gc-ms because FT spectrometers (p. 278) have virtually replaced the older dispersive types and even with computerized enhancements are much cheaper than mass spectrometers.

APPLICATIONS OF GAS CHROMATOGRAPHY

Along with high performance liquid chromatography, gas chromatography is the most widely used of the chromatographic techniques. Some applications are included in table 4.9 but the list is far from exhaustive. It is particularly suited to the rapid analysis of volatile mixtures containing dozens or even hundreds of components and as a result is much used by the food and petroleum industries. Specialized techniques such as pyrolysis derivatization, head-space analysis and thermal desorption have extended the range of applications so that it is also used widely by the plastics, paints and rubber industries and for the monitoring of toxic atmospheric pollutants. The overall relative precision of quantitative analysis is 2 to 5%.

4.2.2. HIGH PERFORMANCE LIQUID CHROMATOGRAPHY

SUMMARY

Principles

Separation of mixtures in microgram to gram quantities by passage of the sample through a column containing a stationary solid by means of a pressurized flow of a liquid mobile phase; components migrate through the column at different rates due to different relative affinities for the stationary and mobile phases based on adsorption, size or charge.

Apparatus and Instrumentation

Solvent delivery system; stainless steel columns; injection port; flow-through detector, recorder.

Applications

Used largely for the separation of non-volatile substances including ionic and polymeric samples; complementary to gas chromatography.

Disadvantages
> Column performance very sensitive to settling of the packed bed or the accumulation of strongly adsorbed materials or particulate matter at the top; universal detection system not available.

High performance liquid chromatography (hplc) has its origins in classical column chromatography although in both theory and practice it is similar to gas chromatography. In column chromatography the sample is introduced into a liquid mobile phase which flows through a column of relatively coarse particles of the stationary phase, usually silica or alumina, under the influence of gravity. Flow rates are of the order of $0.1\,cm^3\,min^{-1}$ which results in extremely lengthy separation times and quite inadequate efficiencies and separations of multicomponent mixtures. The poor performance is largely due to very slow mass transfer between stationary and mobile phases and poor packing characteristics leading to a large multiple path effect (p. 83). It was recognized that much higher efficiencies and hence better resolution could be achieved through the use of smaller particles of stationary phase, and that rapid separations would require higher flow rates necessitating the pumping of the mobile phase through the column under pressure. The means of meeting these two basic requirements were developed during the 1960s together with suitable pumps, injection systems and low dead-volume detectors and the new technique, which is now at least as extensively used as gc, became known as *'high-performance' liquid chromatography* (hplc) or simply *'liquid chromatography'* (lc), The mobile phase is typically pumped at pressures up to about 3 000 psi (200 bar), and flow rates of 1 to 5 $cm^3\,min^{-1}$ can be achieved through 10–25 cm columns packed with particles as small as 3 μm in diameter. At its best, hplc is comparable to gc for speed, efficiency and resolution and it is inherently more versatile. It is not limited to volatile and thermally stable samples and the choice of stationary phase includes solid adsorbents, chemically modified adsorbents, ion-exchange and exclusion materials thus allowing all four sorption mechanisms (p. 76) to be exploited. A much wider choice of mobile phases than in gc facilitates a very considerable variation in the selectivity of the separation process.

A schematic diagram of a high performance liquid chromatograph is shown in figure 4.25 and details of the components are discussed below. All materials which come into contact with the mobile phase are manufactured from stainless steel, PTFE, sapphire, ruby or glass for inertness.

1. *Solvent Delivery Systems*
These include solvent reservoirs and inlet filters, solvent degassing facilities and one or more pumps with associated pressure and flow controls. Most systems are microprocessor or computer controlled enabling parameters to

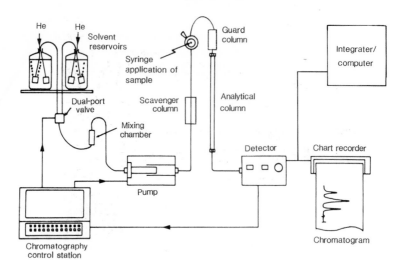

Figure 4.25.   Schematic diagram of a binary (2-solvent) hplc system

be selected and monitored during operation using simple keypad dialogues. The ability to store sets of parameters as method files and to run diagnostic tests of the system are also often available. A single solvent may be used as the mobile phase for *isocratic* elution or mixtures of two to four solvents (*binary*, *ternary* and *quaternary*) blended together under microprocessor or computer control for *gradient* elution, i.e. where the composition of the mobile phase is altered during the chromatographic run so as to optimize the separation. Pumps for hplc should be capable of delivering a constant, reproducible and pulse-free supply of mobile phase to the column at flow-rates between 0.1 and at least 5 cm^3 min^{-1} and at operating pressures up to about 3 000 psi (200 bar). They should be chemically inert to the various solvents that may be used and preferably have a very small hold-up volume to facilitate rapid changes of mobile phase and for gradient elution. A number of types of pump have been developed and these can be classified according to whether they function at *constant flow*, which is desirable for reproducible retention data, or *constant pressure*. The latter will deliver a constant flow only if column backpressure, solvent viscosity and temperature also remain constant.

   Constant flow reciprocating pumps are now the most widely used type (figure 4.26a), but because their mechanical action inherently produces a pulsating delivery of the mobile phase the flow must be smoothed so as to eliminate the pulsations. This can be achieved in several ways, the simplest being the incorporation of a *pulse damper* in the flow to the column. One

(a)

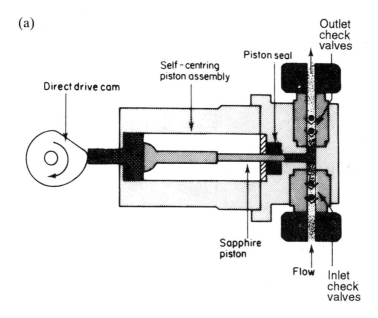

(b)

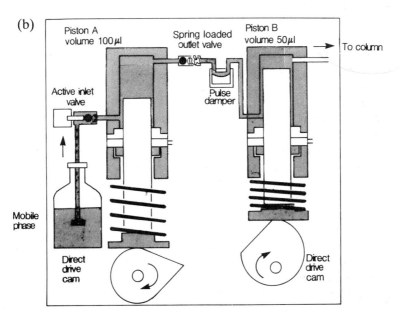

Figure 4.26. (a) Single-head reciprocating pump (Reproduced by permission of Gilson Medical Electronics, Inc.). (b) Double-head 'in-series' reciprocating pump

(c)

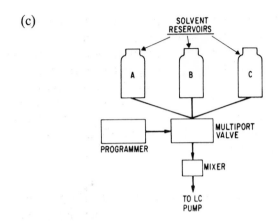

Figure 4.26(c).   A low-pressure gradient former with three solvents

such device is a flexible bellows or diaphragm enclosed in a small oil-filled chamber which absorbs the pulsations. Alternative designs of pump include a double-headed arrangement where two pistons operate in parallel but with delivery strokes 180° out-of-phase and sharing common solvent inlets and outlets. Another is an in-line double-headed pump with one piston (A), having twice the capacity of the other and delivering solvent both to the column and to the chamber of the second piston simultaneously (figure 4.26b). The smaller piston (B) then takes over the delivery whilst the larger one is refilling.

With both designs, residual pulsations can be virtually eliminated with a pulse damper or by suitable cam design that varies the speed of the pistons during the fill and delivery parts of the cycle so as to maintain a constant flow. The flow rates of reciprocating pumps are varied by altering the length of stroke of the piston(s) or through the use of a variable speed stepper-motor. Constant flow can be ensured by incorporating flow or pressure sensors into the design which automatically adjust pumping stroke or motor speed by means of a feed-back system. Automatic compensation for solvent compressability can therefore be achieved.

Two alternative but much less common types of constant flow pump are the screw-driven syringe and the hydraulic amplifier. The former consists of a variable-speed stepping motor which drives a plunger into a stainless-steel cylinder of large capacity (up to 500 cm^3) while in the latter a variable speed gear pump supplies oil under pressure to a pressure intensifier which in turn acts upon the mobile phase contained in a cylinder and delivers it to the column at a greatly increased pressure. Disadvantages of these types of pump are cost and the need for frequent refilling especially when solvents need to be changed, although both achieve constant and pulseless flows without the need for pulse-dampers or feed-back controls.

Where binary, ternary or quaternary gradient elution (p. 87) is required, a microprocessor controlled low-pressure gradient former is the most suitable, figure 4.26c. The solvents from separate reservoirs are fed to a mixing chamber via a multiport valve, the operation of which is pre-programmed via the microprocessor, and the mixed solvent is then pumped to the column. For the best reproducibility of solvent gradients small volume pumps ($< 100\ \mu l$) are essential.

2. Sample Injection System

Sample injection in hplc is a more critical operation than in gc. Samples may be injected either by *syringe* or with a *valve injector* although the former is now rarely used. Valves, which can be used at pressures up to about 7 000 psi (500 bar), give very reproducible results for replicate injections (<0.2% relative precision) and are therefore ideal for quantitative work (p. 129). They consist of a stainless steel body and rotating central block into which are cut grooves to channel the mobile phase from the pump to the column (figure 4.27). The sample is loaded into a stainless steel loop incorporated

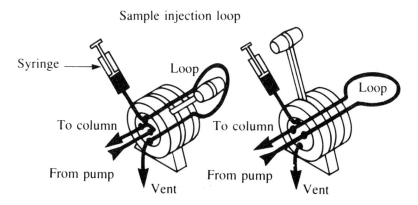

Sample injection loop

Figure 4.27. Valve injector with external loop

into the valve body or attached externally whilst the mobile phase is passed directly to the column. By rotating the central block, the flow can be diverted through the loop thereby flushing the sample onto the column. Returning the block to its original position enables the next sample to be loaded ready for injection. Although the sample injected is generally a fixed volume as determined by the size of the loop, these are interchangeable and range from 2 μl to over 100 μl. *Multiport valves* which can accommodate several loops of different sizes are available, and some loops can be used partially filled. Automated injection systems that allow a series of samples and standards to be injected over a period of time whilst the instrument is unattended and under variable chromatographic conditions are frequently used in industrial laboratories.

3. *The Column*

Columns are made from straight lengths of precision-bore stainless-steel tubing with a smooth internal finish. Typically they are 10 to 25 cm long and 4 to 5 mm i.d. Microbore columns, 20 to 50 cm long and with an i.d. of 1 to 2 mm, are sometimes used where sample size is limited and to minimize solvent consumption because the volumetric flow rate through them is less than a quarter of that through conventional columns. The stationary phase or packing is retained at each end by thin stainless-steel frits or mesh discs of 2 μm porosity or less. Columns are packed by a *slurry method* which involves suspending the particles of packing in a suitable solvent and 'slamming' it into the column rapidly and at pressures in excess of 3 000 psi (200 bar). The choice of slurry solvent depends upon the nature of the packing and many solvents have been investigated. *Balanced-density* slurries, in which solvent density is matched with particle density, are sometimes used to minimize settling out of larger particles during the packing procedure or for packings which are larger than 10 μm.

Hplc columns need more careful handling and storage than gc columns to avoid disturbance of the packed bed. They should be kept sealed at both ends when not in use and flushed with methanol prior to sealing. Column life is generally six months or more and can be prolonged by use of a *guard column* and a *scavenger column*. The former consists of a very short length of column placed between the injection port and the analytical column to trap strongly retained species or particulate matter originating in the mobile phase, the samples or from wearing of the injection valve. It is packed with relatively large particles (~30 μm) of the same or a similar stationary phase to that used in the analytical column and require periodic renewal. Scavenger columns are short lengths of tube packed with large particle silica and positioned between the pump and the injection valve with the principal object of saturating an aqueous mobile phase with silica to reduce attack on the packing in the analytical column, especially by high or low pH buffers.

4. *Stationary Phase* (*column packing*)

Unmodified or chemically modified microparticulate silicas (3, 5 or 10 μm) are preferred for nearly all hplc applications. The particles, which are totally porous, may be spherical or irregular in shape but it is essential that the size range is as narrow as possible to ensure high column efficiency and permeability. For separations based on adsorption, an unmodified silica, which has a polar surface due to the presence of silanol (Si-OH) groups, is used. Appropriate chemical modification of the surface by treatment with chloro- or alkoxy-silanes, e.g. $R(CH_3)_2SiCl$, produces *bonded-phase* packings which are resistant to hydrolysis by virtue of forming siloxane (Si—O—Si—C) bonds. The reactions are similar to those used to silanize gc supports (p. 92). Materials with different polarities and chromatographic characteristics can be prepared. The most extensively used are those with a non-polar hydrocarbon-like surface, the modifying groups, R, being octadecyl (C_{18} or ODS), octyl or aryl. More polar bonded-phases, e.g. aminopropyl, cyanopropyl (nitrile) and diol, and cation and anion exchange materials are also available. Mixed ODS/aminopropyl and ODS/nitrile phases having enhanced selectivity for certain classes of compound have also been produced. Ion-exchange chromatography is discussed more fully in 4.2.5. Hplc separations based on exclusion (gpc) are best achieved with microparticulate silicas which are sometimes chemically modified with such groups as trimethylsilyl to eliminate or minimize adsorption effects. Exclusion separations using polymeric gels are discussed in 4.2.6.

Chiral stationary phases for the separation of *enantiomers* (optically active isomers) are becoming increasingly important. Among the first types to be synthesized were chiral amino acids ionically or covalently bound to amino-propyl silica and named *Pirkle* phases after their originator. The ionic form is susceptable to hydrolysis and can be used only in normal phase hplc whereas the more stable covalent type can be used in reverse phase separations but it less stereoselective. Polymeric phases based on chiral peptides such as bovine serum albumin or α_1-acid glycoproteins bonded to silica, esterified cellulose and chiral cyclodextrins can all be used in reverse phase separations. The latter, α-, β- or γ-cyclodextrins form barrel-shaped cavities into which enan-tiomers can fit. Chiral phases form complexes with analytes by binding them at specific sites through H-bonding, π–π and dipolar interactions. In addition, steric repulsion, solvent, pH, ionic strength and temperature all affect chroma-tographic retention. If the total adsorption energies of two enantiomers differs then a separation is possible.

The recently introduced graphitized carbon and a new generation of rigid porous polymeric micro-beads based on styrene/divinyl benzene as alterna-tives to silica, can both be used over an extended pH range between 1 and 13. These materials have increased the choice of stationary phases and the scope of hplc, particularly for highly polar and basic substances where peak tailing on silica-based columns is a frequent occurrence. Some examples of column packings used in hplc and their applications are given in table 4.12.

Table 4.12   Some column packings used in hplc

PACKING	MODE OF HPLC*	APPLICATIONS
microparticulate silicas; spherical or irregular particles; mean particle size $3\,\mu m, 5\,\mu m, 10\,\mu m$	LSC (adsorption)	non-polar to moderately polar mixtures, e.g. poly-aromatics, fats, oils, mixtures of isomers
chemically modified versions of the above (bonded-phase packings): octadecyl (ODS or C_{18})	BPC, IPC	wide range of moderately polar mixtures, e.g. pharmaceuticals and drugs, amino acids
octyl (C_8)	BPC, IPC	more polar mixtures, e.g. pesticides, herbicides,
short chain (C_3 or less)	BPC, IPC	peptides, metabolites in body fluids IPC applications of above three packings include bases, dyestuffs and other multiply charged species; can be used instead of IEC
diol	BPC	very polar and water-soluble compounds, e.g. food and drink additives
nitrile	normal phase and BPC	alternative to silica and can give better results
aminoalkyl	BPC	carbohydrates including sugars
anion and cation exchangers (tertiary amine or sulphonic acid)	IEC	ionic and ionizable com-pounds, e.g. vitamins, water-soluble drugs, amino acids, food and drink additives
controlled porosity silicas (may be chemically modified to reduce adsorption effects)	SEC	polymer mixtures, screen-ing of unknown samples. Increasing use for separ-ating mixtures of smaller molecules before other modes of hplc
chiral amino acids bound to aminopropyl	CC	mixtures of enantiomers especially of drugs
chiral peptides	CC	
cyclodextrins	CC	

* LSC = liquid-solid chromatography   IEC = ion-exchange chromatography
  BPC = bonded-phase chromatography   SEC = size exclusion chromatography
  IPC = ion-pair chromatography   CC = chiral chromatography

5. *Mobile Phase*

Unlike gc, in hplc appropriate selection of the mobile phase composition is crucial in optimizing chromatographic performance. The eluting power of the mobile phase is determined by its overall polarity, the polarity of the stationary phase and the nature of the sample components. For 'normal phase' separations (polar stationary phase/non-polar mobile phase) eluting power *increases* with increasing solvent polarity whereas for 'reverse phase' separations (non-polar stationary phase/polar mobile phase) eluting power *decreases* with increasing solvent polarity. Some examples of solvents suitable for hplc are given in table 4.8 together with their polarities as measured by a solubility parameter, P' (applicable to partition-based separations) and an adsorption parameter, ε^0 (applicable to adsorption-based separations). Such a list is often called an *eluotropic series*. Other properties of solvents which must be taken into account include boiling point and viscosity, detector compatibility, flammability and toxicity. Generally, lower boiling and hence less viscous solvents give higher chromatographic efficiencies and lower back pressures. The most commonly used detectors are based on absorbance of UV radiation and on refractive index (see below). The UV cut-off and refractive indices of solvents therefore need to be known. These are included in table 4.8.

Often, optimum retention and resolution are achieved by using a mixture of two solvents. The solubility-based parameter, P', varies linearly with the proportion of the two solvents, being given by the weighted arithmetic mean of the two individual values. The adsorption-based parameter, ε^0, however, increases rapidly with small additions of a more polar solvent to a less polar one and levels off as the proportion increases. Elution with a single solvent or mixed solvent of fixed composition is called *isocratic* as opposed to *gradient* elution (p. 87). Gradient elution is sometimes employed where sample components vary widely in polarity. Gradient formers (p. 116) enable binary, ternary and quaternary mixtures of solvents to be blended reproducibly. In reverse phase separations the most widely-used mobile phases are mixtures of aqueous buffers with methanol, or water with acetonitrile. In normal phase work pentane or hexane with dichloromethane, chloroform or an alcohol are frequently used.

Many solvents for hplc require purification before use as the impurities may either be strongly UV absorbing, e.g. aromatic or olefinic impurities in *n*-alkanes, or they may be of much higher polarity than the solvent itself, e.g. traces of water or acids, or ethanol in chloroform etc. All mobile phases should be filtered and degassed before pumping through the column, the former to prevent contamination and clogging of the top of the column and the latter to prevent noise in the detector from the formation of air bubbles due to the pressure dropping to atmospheric at the column exit.

Table 4.13 Characteristics of liquid chromatography detectors*

DETECTOR BASIS	TYPE†	MAXIMUM SENSITIVITY‡	FLOW RATE SENSITIVE?	TEMPERATURE SENSITIVITY	USEFUL WITH GRADIENT?
UV absorption	S	5×10^{-10}	No	Low	Yes
IR absorption	S	10^{-6}	No	Low	Yes
Fluorimetry	S	10^{-10}	No	Low	Yes
Refractive index	G	5×10^{-7}	No	$\pm 10^{-4}\,°C$	No
Conductometric	S	10^{-8}	Yes	$\pm 1°C$	No
Mass spectrometry	S	10^{-10}	No	None	—
Amperometric	S	10^{-10}	Yes	$\pm 1°C$	—
Radioactivity	S	—	No	None	Yes

*Most of these data were taken from: L. R. Snyder and J. J. Kirkland, *Introduction to Modern Liquid Chromatography*. New York: Wiley-Interscience, 1964, p. 165. With permission.
† G = general; S = selective.
‡ Sensitivity for a favourable sample in grams per millilitre.

6. *Detectors*

The ideal hplc detector should have the same characteristics as those required for gc detectors, i.e. rapid and reproducible response to solutes, a wide range of linear response, high sensitivity and stability of operation. No truly universal hplc detector has yet been developed but the two most widely applicable types are those based on the absorption of UV or visible radiation by the solute species and those which monitor refractive index differences between solutes dissolved in the mobile phase and the pure mobile phase. Other detectors which are more selective in their response rely on such solute properties as fluorescence, infrared absorption, electrical conductivity, diffusion currents (amperometric) and radioactivity. The characteristics of the various types of detector are summarized in table 4.13.

UV/Visible Photometers and Dispersive Spectrophotometers

These detectors respond to UV/visible absorbing species in the range 190 to 700 nm and their response is linear with concentration, obeying the Beer–Lambert law (p. 355). They are not appreciably flow or temperature sensitive, have a wide linear range and good but variable sensitivity.

Photometers are designed to operate at one or more fixed wavelengths only, e.g. 220, 254, 436 and 546 nm, whereas *spectrophotometers* facilitate monitoring at any wavelength within the operating range of the instrument. Both types of detector employ low-volume (10 μl or less) flow-through cells fitted with quartz windows. Careful design of the cell, which should be of

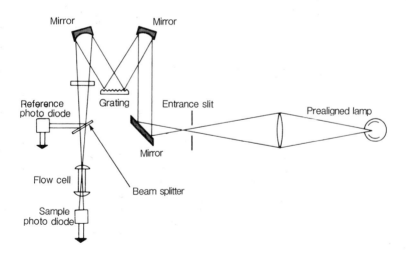

Figure 4.28. Variable-wavelength detector, showing deuterium lamp, optical path, reference photodiode and monochromator

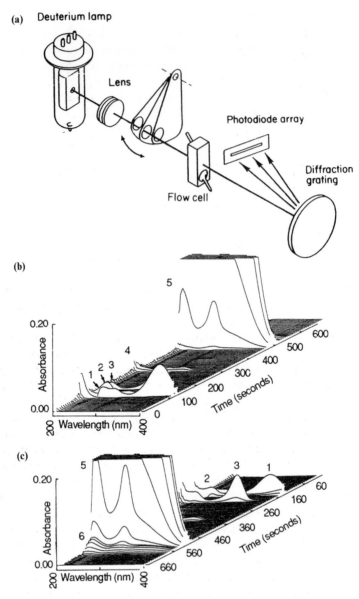

Figure 4.29.   (a) Optical path in a UV photodiode array detector. (b and c). Three-dimensional display from UV diode array detector; azathioprine and impurities separated by reversed-phase (RP) HPLC with HP 1040A detector coupled to an HP 85 microcomputer: (b) isometric projection of data showing components 1–5; (c) reversed projection of data showing components 1–3, 5 and 6. *Key*: (1) 1-methyl-4-nitro-5-hydroxyimidazole; (2) 1-methyl-4-nitro-5-thioimidazole; (3) 6-mercaptopurine; (4) 1-methyl-4-nitro-5-chloroimidazole; (5) azathioprine; (6) process impurity

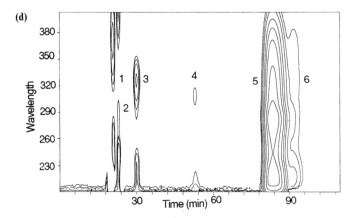

Figure 4.29(d).   Contour plot of data for azathioprine and impurities in (b) and (c), showing the isoabsorbance contours (from 1 000 to 5 m AU) plotted in the wavelength–time plane. Key to the peaks, as in (b) and (c)

minimal volume to reduce band-spreading and maximal path length for high sensitivity, is necessary to reduce undesirable refraction effects at the cell walls as solutes pass through.

Although many substances absorb appreciably at 254 nm or one of the other fixed wavelengths available with a photometer, a much more versatile detection system is based on a spectrophotometer fitted with a grating monochromater and continuum source, e.g. a deuterium lamp for the UV region and a tungsten-halogen lamp for the visible region (figure 4.28). They have double-beam optics (p. 275), stable low-noise electronics and are often microprocessor controlled. Some can be programmed to select a sequence of optimum monitoring wavelengths during or between chromatographic runs, and the recording of a complete UV spectrum after stopping the flow with a selected peak in the detector cell is a feature of other designs. The most recent development is a rapid scanning capability that allows a complete spectrum to be recorded in a fraction of a second without the need to stop the flow, therefore rivalling diode array detectors. Like the latter, full computer control and high resolution colour graphics enable chromatograms to be displayed in 3D and other formats and peak purity assessed. Sensitivity and resolution are better than some diode array detectors.

Photometers are more sensitive than spectrophotometers, are cheaper and more robust and are well suited to routine work where monitoring at 254 nm or some other fixed wavelength is acceptable. Spectrophotometers, however, allow 'tuning' to the most favourable wavelength either to maximize sensitivity for a particular solute or to 'detune' the response to other solutes. By allowing monitoring down to 190 nm, weakly absorbing or saturated compounds can be detected.

Diode Array Spectrophotometers

These can provide more spectral information than photometers or conventional dispersive spectrophotometers but are much more expensive and generally less sensitive (figure 4.29(a) and p. 353).

However, they enable sets of complete UV or UV and visible spectra of all the sample components to be recorded as they elute from the column. The stored spectral information can be processed in several ways by the microcomputer and displayed using sophisticated colour graphics software packages. The most usual is a 3D chromatogram of time/absorbance/ wavelength as shown in figure 4.29(b). This can be rotated on the screen to allow examination of otherwise hidden regions behind major peaks, figure 4.29(c), or viewed from directly above and shown as an absorbance contour map to provide useful information on peak purity, figure 4.29(d). The comparison of spectra selected from any points on the time axis by overlaying them in various colours is an alternative assessment of peak purity and spectra can be matched with a library of standards for identification purposes. Some software packages also include the facility for calculating peak purity factors from absorbance ratios at two or more wavelengths for different time slices of an eluting peak, values close to unity indicating a high degree of purity.

Fluorescence Detectors

These are highly selective and among the most sensitive of detectors. They are based on filter fluorimeters or spectrofluorimeters (p. 376) but are usually purpose-designed for hplc or capillary electrophoresis (p. 168). The optical arrangement of a typical detector using filters is shown in figure 4.30. Excitation and emission wavelengths are selected by narrow bandpass filters,

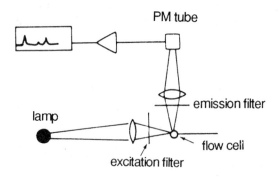

Figure 4.30.   Schematic diagram of a simple dual-filter fluorimeter. Excitation using a xenon lamp. Filters used to select the wavelength in both the excitation and emission beams

and the flow cell has a capacity of 10 to 25 μl. The optical paths of the excitation and fluorescent emission beams are at 90° to each other. This provides an extremely low background signal whilst only mobile phase is flowing through the cell. Sensitivity can be improved further by focusing the fluorescence onto the photomultiplier tube with a curved parabolic mirror. Analytes that do not fluoresce naturally can be derivatized with commercially available fluorescent reagents prior to chromatography or between the column and the detector (pre- or post-column derivatization). Fluorescence detectors are relatively insensitive both to pulsations in the mobile phase flow and to temperature fluctuations.

Refractive Index (RI) Monitors

There are several types of RI detector, all of which monitor the difference between a reference stream of mobile phase and the column effluent. Any solute whose presence alters the refractive index of the pure solvent will be detected, but sensitivity is directly proportional to the difference between the refractive index of the solute and that of the solvent. At best they are two orders of magnitude less sensitive than UV/visible detectors. All RI detectors are highly temperature-sensitive, and some designs incorporate heat exchangers between column and detector to optimize performance. They cannot be used for gradient elution because of the difficulty in matching the refractive indices of reference and sample streams.

The most common type of RI monitor is the *deflection refractometer* (figure 4.31). A visible light source is directed through a two-compartment cell divided by a diagonal piece of glass. The light is refracted on the way through the cell, reflected back from a mirror behind the cell then refracted again before being focused on a phototube detector. Whilst only solvent passes through each half of the cell, the phototube signal is constant, but when a solute passes through the sample compartment, the light beam is deflected and the change in intensity of radiation falling on the phototube is registered

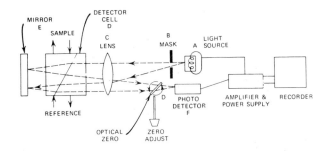

Figure 4.31. Refractive index (RI) detector

by the recorder. This type of detector has a wide linear range of response but (like other RI monitors) only moderate sensitivity.

Two other types of RI monitor are based on Fresnel's laws of reflection and the principle of interferometry respectively. The former utilizes a very small volume (3 μl) sample cell and is therefore useful for highly efficient columns, but linearity is limited and the cell windows need to be kept scrupulously clean for optimum performance. The main advantages of the interferometric design are improved sensitivity and a wide linear range.

Electrochemical Detectors

Conductance monitors can be used where the sample components are ionic and providing that the conductivity of the mobile phase is very low. They are used extensively in *ion chromatography* (p. 139) for the detection of inorganic unions, some inorganic cations and ionized organic acids.

Amperometric detectors which are based on the principle of polarography, rely on measuring the current generated in an electrochemical cell at a fixed applied potential by the facile oxidation or reduction of an eluted compound at the surface of a glassy carbon, gold or platinum micro-electrode. The cell is completed with a calomel reference electrode (p. 226) and an auxiliary electrode, the purpose of the latter being to enable the applied potential to be stabilized when the cell resistance alters by virtue of the currents generated. The mobile phase acts as a supporting electrolyte for the redox

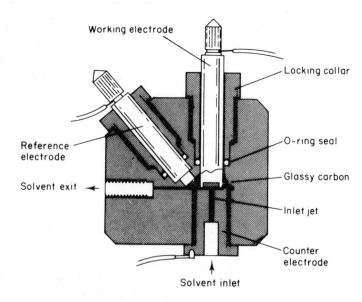

Figure 4.32. 'Wall-jet' amperometric detector

reactions and its composition is therefore restricted to predominantly aqueous solvent mixtures. Several designs have been produced, some with internal cell volumes as little as 1 μl; one type is shown in figure 4.32. Amperometric detectors are amongst the most sensitive available but they are susceptible to noise, caused by any residual pulsations from the pump affecting the flow of mobile phase, and to surface contaminations of the micro-electrode due to the build-up of electrode reaction products which impairs reproducibility. However, their high sensitivity and selectivity (through variation of the applied potential) enhances their value for the trace analysis of certain types of compound, e.g. phenols and arylamines such as catecholamines (by oxidation) and nitro or azo compounds (by reduction) although in the latter cases air must be purged from the mobile phase with nitrogen to eliminate interference by the reduction of dissolved oxygen.

Of the remaining types of detector, those based on fluorimetry are both selective and particularly useful for trace analysis as they can be orders of magnitude more sensitive than UV/visible photometers.

QUALITATIVE AND QUANTITATIVE ANALYSIS

Methods similar to those used in gc are applicable to hplc. Thus, comparison of retention data is the most useful means of qualitative identification, the capacity factor k' generally being used in preference to retention time, t_R (p. 81)—compare gas chromatography (p. 104). As in gc, members of a homologous series show a linear relation between log t_R and the number of carbon atoms. The trapping of hplc peaks for examination by other techniques such as infrared, nuclear magnetic resonance and mass spectrometry is straightforward, providing that the time taken for the peak to reach a sample vial via a transfer line from the detector can be accurately measured. Complete UV/visible spectra are recorded by diode array detectors or the flow can be stopped with the peak in the cell of a conventional dispersive spectrophotometric detector while the complete spectrum is scanned.

HPLC–MASS SPECTROMETRY

Direct interfacing of hplc with mass spectrometry is a 'coupled' or 'hyphenated' technique similar to GC–Mass Spectrometry (p. 108) which provides structural information on separated sample components. Its development has been slow because of difficulties inherent in removing the liquid mobile phase whilst allowing only the analytes to pass into the mass spectrometer, particularly as reverse phase hplc often employs mobile phases containing aqueous

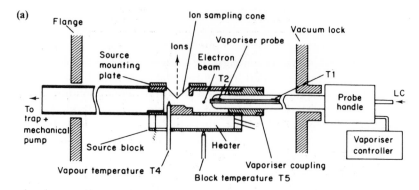

Figure 4.33(a). Thermospray interface for hplc-ms. T1, inlet temperature probe; T2 outlet temperature probe

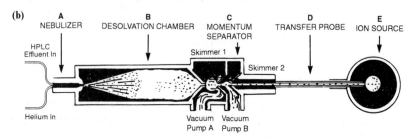

Figure 4.33(b). Particle-beam interface in hplc-ms

buffers and inorganic salts. Several designs of interface have been developed, the main differences between them being the means of separating analytes from the mobile phase and the method of ionization employed.

With the *thermospray* interface (figure 4.33(a)), the mobile phase, usually containing an ammonium ethanoate buffer, is passed through a heated probe (350–400°C) into an evacuated source chamber where it forms a supersonically expanding mist of electrically charged droplets. The liquid evaporates to leave charged solid particles which then release molecular ions such as MH^+ and MNH_4^+ by an ammonia chemical ionization (CI) process. The analyte ions are 'skimmed off' into the mass spectrometer whilst the vaporized solvent is pumped away. An electron beam is also employed to enhance the production of ions by CI.

A *particle-beam* interface (figure 4.33(b)) employs helium to nebulize the mobile phase, producing an aerosol from which the sample is evaporated at near ambient temperature and pressure. The mixture of helium, solvent vapour and analyte molecules is accelerated into a low-pressure two-stage 'momentum separator' where it expands supersonically. The helium and solvent are pumped away whilst the relatively heavy analyte molecules,

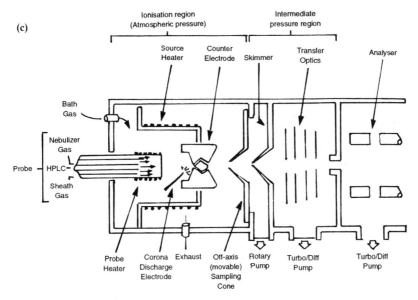

Figure 4.33(c).   APCI interface for hplc-ms

having much greater momentum, pass straight through two skimmer plates and along a narrow probe into a heated ionization chamber where electron impact (EI) ionization occurs.

An atmospheric pressure chemical ionization (APCI) interface uses nitrogen as a nebulizing gas, and a heated nebulizer probe where the mobile phase is vaporized and reactant ions are formed by a corona discharge (figure 4.33(c)). The ions and analyte molecules are accelerated through skimmers into a low-pressure region where the solvent is pumped away. Collisions between analyte molecules and reactant ions cause chemical ionization (CI) as the analytes pass into the mass spectrometer.

Only the particle-beam interface produces EI spectra for direct comparison with computerized library spectra, the other two systems enabling the RMM of analytes up to several thousand to be established but providing little or no structural information through fragmentation. New or modified lc-ms interfaces are likely to be developed in the future.

Quantitative analysis using valve injection, has a relative precision of 0.2% or better. Precision is highest where 50–200 μl sample loops are used. Isocratic elution is preferable to gradient elution because conditions are more reproducible. Internal standards are not always required, the concentration of samples being determined from previously prepared calibration curves or by using a factor if the curve is linear. For routine analyses, automatic sampling systems are better because they eliminate operator error. Either peak height or peak area can be measured, the former being suitable for sharp, early-eluting peaks, where peaks are not fully resolved, and for trace analysis. Peak area measurements give better precision and are more reliable if peaks are asymmetrical; they are less susceptible to changes in chromatographic conditions.

OPTIMIZING A SEPARATION

Adequate resolution of the components of a mixture in the shortest possible time is nearly always a principal goal. Establishing the optimum conditions by trial and error is inefficient and relies heavily on the expertise of the analyst. The development of computer-controlled hplc systems has enabled systematic automated optimization techniques, based on statistical experimental design and mathematical resolution functions, to be exploited. The basic choices of column (stationary phase) and detector are made first followed by an investigation of the mobile phase composition and possibly other parameters. This can be done manually but computer-controlled optimization has the advantage of releasing the analyst for othes duties or running overnight. Gradient elution (p. 87) is sometimes used as a preliminary step for unknown samples to indicate possible *isocratic* (constant mobile phase composition) conditions.

A simple means of optimizing a reverse phase separation is to obtain a series of six chromatograms with binary mixtures of an organic solvent, such as acetonitrile or methanol, and water ranging from 0 to 100 per cent water in steps of 20 per cent. Assessment of each chromatogram, followed if necessary by additional ones with intermediate proportions of the organic solvent and water, then enables the best composition to be selected. An extension of this procedure is to use ternary mixtures of two organic solvents and water. An example of the separation of a six component mixture using five different proportions of methanol, tetrahydrofuran and water is shown in figure 4.34.

More sophisticated, computerized and usually automated methods of mobile phase optimization are commercially available. Most software packages are designed to use one of two alternative approaches. In the first, a predetermined series of chromatograms is recorded followed by evaluation to find the optimum composition. In the second a *directed search* routine is used whereby the computer evaluates each chromatogram in turn according to specified criteria before selecting a new composition for the next run. With the former or 'simultaneous' technique, evaluation of each chromatogram is based on the computation of a single numerical value called a *chromatographic response or optimization function* (CRF or COF). This represents the quality of the chromatogram as a function of the peak overlap (resolution) for each pair of adjacent peaks in the sample and the total elution time. Contour maps are then plotted using the individually computed CRF or COF values,

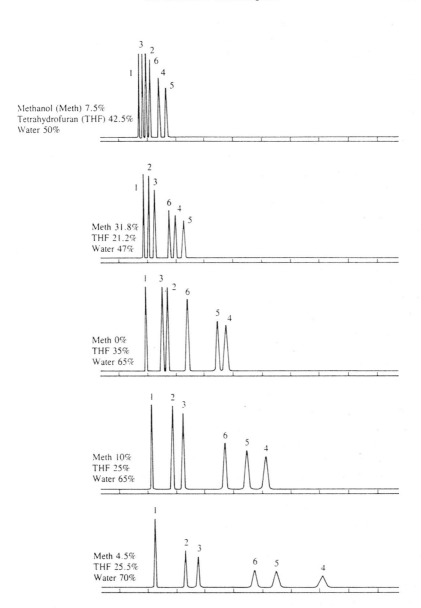

Figure 4.34.   Optimizing an hplc separation using five ternary mobile phases. Peaks:
(1) benzyl alcohol, (2) phenol, (3) 3-phenylpropanol, (4) 2,4-dimethylphenol, (5) benzene, and (6) diethyl o-phthalate. (After R. D. Conlon, 'The Perkin-Elmer Solvent
Optimization System,' *Instrumentation Research*, p. 95 (March 1985). Courtesy of
Instrumentation Research.)

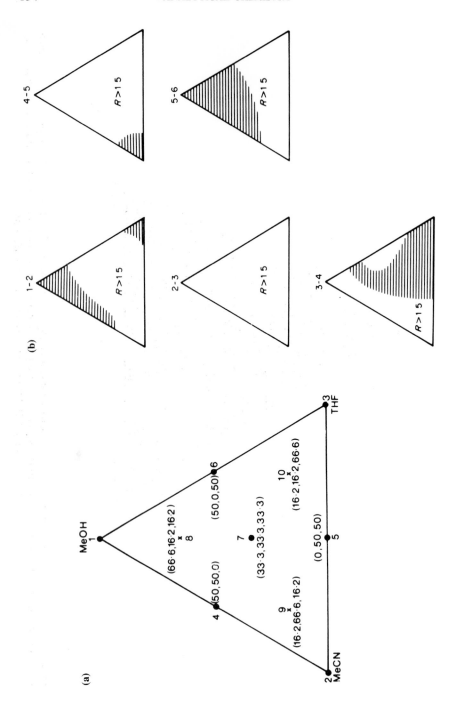

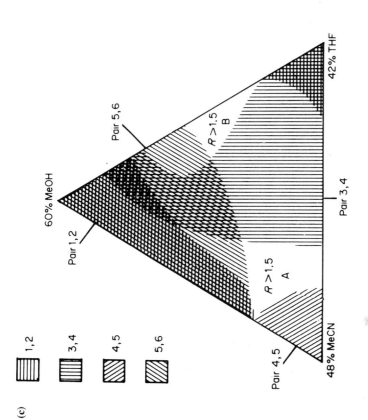

Figure 4.35. (a) Simplex lattice design for reversed-phase chromatographic optimization showing relative proportions of each solvent to be used. (b) Individual resolution maps for the five pairs of solutes in a 6-component test mixture. (c) Overlapping resolution map (ORM) for the 6-component test mixture.

and additional values are derived by interpolation. Directed searches offer greater flexibility and may require a smaller number of chromatograms, but they may select a localized optimum composition rather than a global one. However, they are *adaptive* and have the ability to cope with unexpected outcomes during the optimization process. A problem with both approaches is caused by 'peak crossovers', i.e. changes in elution order. Keeping track of each peak in a chromatogram represents a major challenge for optimization software.

The *simplex lattice design* is a straightforward example of the simultaneous technique and is based on a *solvent selectivity triangle* originally proposed by Snyder and involving the plotting of an *overlapping resolution map* (ORM). Three solvents of differing chromatographic selectivity are mixed with a fourth (solvent strength adjuster) to provide three mobile phases of equal solvent strength (*isoeluotropic*), e.g. methanol, acetonitrile, tetrahydrofuran and water would be suitable for a reverse phase separation. The three mobile phases form the apices of a solvent selectivity triangle. At least four other compositions are made by blending the initial three in various proportions; these lie along the sides or within the triangle. The mixture is then chromato-graphed using each of the seven (or more) compositions in turn, and a resolution contour map generated from the COF values for each pair of adjacent components in the mixture for all compositions. The superimposi-tion of the individual contour maps produces an ORM which is used to identify the optimum mobile phase composition for separating the mixture with a specified resolution. The provision of a computer system with high-resolution colour graphics greatly enhances the value of this approach. The solvent composition triangle and set of resolution maps for a 6-component mixture is shown in figure 4.35.

Other more mathematical techniques, which rely on appropriate computer software and are examples of chemometrics (p. 33), include the generation of one, two or three dimensional *window diagrams, computer directed searches* and the use of *expert systems* (p. 533). A discussion of these is beyond the scope of this text.

APPLICATIONS OF HIGH PERFORMANCE LIQUID CHROMATOGRAPHY

The use of hplc in all its forms is growing steadily and may eventually exceed that of gc. This is because all four sorption mechanisms can be exploited and the technique is well suited to a very wide range of compound types including ionic, polymeric and labile materials. The most appropriate choice of mode of hplc for a given separation problem is based on the molecular weight, solubility characteristics and polarity of the compounds to be separated and a guide to this is given in figure 4.36.

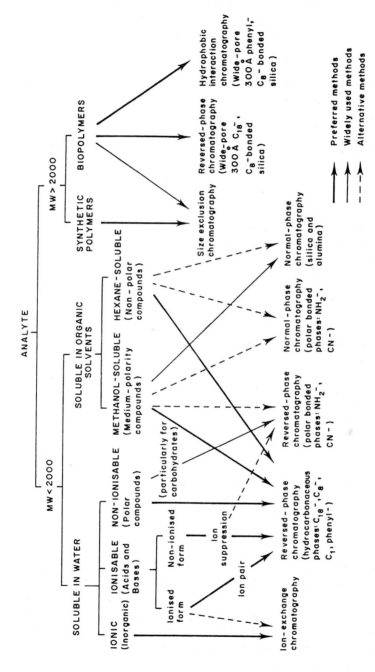

Figure 4.36. Selection of suitable liquid chromatographic methods for analysis depending on solute structure and properties

Adsorption chromatography on silica is well suited to the separation of less polar compounds such as polyaromatic hydrocarbons, fats and oils, and for the separation of isomers or compounds with differing functional groups.

Bonded-phase chromatography (BPC) in one form or another is suitable for most hplc separations ranging from mixtures of weakly polar to highly polar and ionizable compounds. Reverse phase chromatography using octadecyl (ODS or C18) columns and methanol/aqueous buffers or acetonitrile/ water mobile phases is by far the most widely used. The mechanism of separation is not clear and may involve adsorption or partition depending on the solutes and particular phases used. For weakly acidic or basic solutes, pH control is very important as retention times vary considerably with degree of dissociation or protonation, the non-ionic form of a solute having a greater affinity for a non-polar stationary phase. Figure 4.37 shows the relation

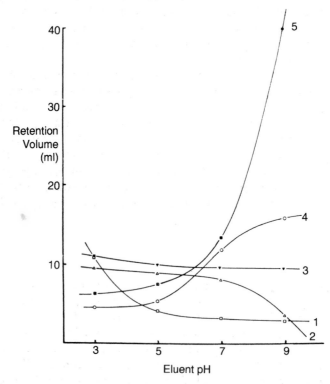

Figure 4.37.   Effect of eluent pH on selectivity in reverse-phase *BPC*. Column, 30 × 0.4 cm, μ-Bondapak-C$_{18}$, 10 μm; mobile phase, 0.025 M NaH$_2$PO$_4$–Na$_2$HPO$_4$ with 40% v methanol; flowrate, 2 ml/min; temp., ambient; detector, UV, 220 nm; samples, 1–10 μg. (1) Salicylic acid; (2) phenobarbitone; (3) phenacetin; (4) nicotine; (5) methylamphetamine

between pH and retention time for five weakly acidic or basic drugs chromatographed on an ODS stationary phase. Ion-pair or paired-ion chromatography (IPC or PIC) on non-polar bonded-phase columns is often used for the separation of ionic or ionizable solutes in place of conventional ion-exchange resins or bonded-phase ion-exchangers. Mobile phases used in IPC are buffered and contain counter-ions of opposite charge to the solute ions, e.g. tetrabutylammonium ions for the separation of ionized carboxylic acids at pH 7 or 8, or alkyl sulphonates for the separation of protonated weak bases at pH 3 or 4. Under the appropriate conditions, solute ions form ion-pairs with the counter-ions, which partition into or are adsorbed selectively onto the stationary phase. IPC offers the advantages of greater efficiency and column stability and more selectivity in the separation of ionic or ionizable compounds compared to bonded-phase or conventional ion-exchangers.

Size exclusion (gel filtration or permeation) chromatography (gpc) is suitable for solutes with molecular weights of 2 000 or more and is also useful for the preliminary investigation of unknown samples. Separated fractions can then be subjected to one of the other modes of hplc. Exclusion chromatography is discussed in 4.2.6.

Hplc has had considerable success in separating compounds as diverse as steroids, carbohydrates, vitamins, dyestuffs, pesticides and polymers. It is used routinely for the assay of pharmaceutical products, the monitoring of drugs and metabolites in body fluids and for other biomedical, biochemical and forensic applications, such as the detection of drugs of abuse. The determination of additives in foodstuffs and beverages including sugars, vitamins, flavourings and colourings and the quality control of polymers, plastics and resins are further examples of the very wide and growing scope of this technique. Figure 4.40 illustrates a number of separations employing different modes of hplc.

ION CHROMATOGRAPHY

For the separation of inorganic and some organic anions and cations a form of hplc known as *ion chromatography* (ic) has been developed. The technique involves an ion-exchange column and a means of *suppressing* (removing) ionized species other than the sample ions in the eluting mobile phase to facilitate detection of the sample ions by a conductivity monitor (figure 4.38(a)). Separation of the analytes is achieved using a *pellicular* (*vide infra*) cation- or anion-exchange resin (p. 156) and eluting with sodium hydroxide or sodium hydrogen carbonate/sodium carbonate (for anionic analytes) or methane sulphonic acid (for cationic analytes). The eluent from the column passes through a *suppressor* where the eluting or 'background' electrolyte is effectively removed by converting it into water or water and

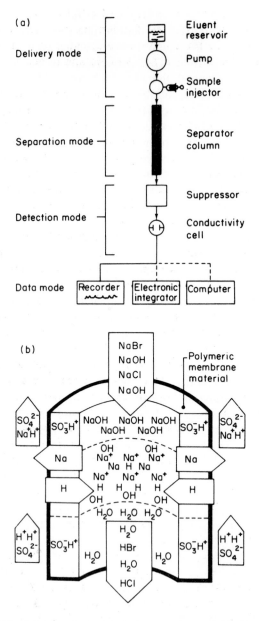

Figure 4.38.   (a) Configuration of ion chromatograph with ion-exchange separation column and conductometric detector. (b) Ion chromatography suppressor fibre exchanging hydrogen ions for sodium ions during an anion separation. The halide ions remain as the mineral acids

carbon dioxide, i.e. sodium ions are replaced with hydronium ions or methyl-sulphonate ions with hydroxyl ions. This is a necessary step which leaves the analytes (all now in the same ionic form) as the only ionized species and thereby easily detected down to very low (sub ppm) levels by a conductivity monitor.

To separate anionic analytes the analytical column is packed with a pellicular strong-base anion-exchange resin in the hydrogen carbonate (HCO_3^-) form. The particles have an inert non-porous central core surrounded by a surface layer of anionic resin micro-beads. Mass transfer in pellicular materials is very fast and, although the columns are of limited capacity, they have very high efficiencies. The suppressor is a cartridge containing a porous polymeric cation-exchange membrane in the H^+ form, which enables hydronium ions to replace sodium ions in the eluent and to form water by combination with hydroxyl ions (figure 4.38(b)). As the membrane becomes depleted of H^+ ions it is continually replenished from an external acidic solution whilst the Na^+ ions passing into the external solution are removed. For cationic analytes, a pellicular cationic resin in the H^+ form is used, the suppressor utilizing an anionic membrane.

A miniaturized 'self-regenerating' suppressor cartridge incorporating an electrolysis cell has recently become available. Hydronium ions and oxygen are continually formed by the electrolysis of a stream of deionized water passing through an anode compartment whilst hydroxyl ions and hydrogen are similarly formed in a cathode compartment. Both compartments are separated from the eluent by either cation- or anion-exchange membranes depending on whether anionic or cationic analytes are to be separated. The process is shown diagrammatically in figure 4.39. The oxygen and hydrogen generated by the electrolysis are vented to the atmosphere while the hydronium or hydroxyl ions pass through the membrane into the eluent in the central compartment to form water by replacing Na^+ ions or methyl sulphonate anions (MSA), respectively. These pass outwards into the cathode or anode compartment where they are removed, along with the oxygen and hydrogen, in the regenerant stream of deionized water. The device is essentially maintenance free and provides superior sensitivity and operational stability compared to earlier designs.

Ion chromatography has been used mainly for the separation of inorganic anions such as chloride, bromide, fluoride, sulphate, nitrate, nitrite and phosphate at ppm levels in surface waters, industrial effluents, food products, pharmaceuticals and clinical samples. However, separations of organic acids and bases, alkali, alkaline earth and transition metal cations is becoming more widespread. Organic solvents such as methanol or acetonitrile can be added to the mobile phase when organic species, e.g. low molecular weight amines or acids, are to be separated. Ion-exchange resins which have a

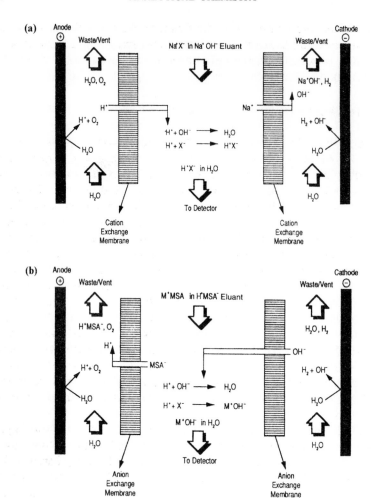

Figure 4.39.   Self-regenerating suppressor cartridge incorporating an electrolysis cell. (a) For anionic analytes. (b) For cationic analytes

proportion of the ionic sites replaced with hydrophobic reversed phase groups commonly used in hplc columns, typically octadecylsilane (ODS), enable mixtures of both non-ionic and ionic species to be separated. Some examples of ic separations are included in figure 4.40. Like hplc, ic has become micro-computer-controlled and fully automated.

An alternative means of detection involves UV spectrophotometry, the mobile phase containing the strongly-absorbing phthalate ion which gives a constant high absorbance baseline signal that displays negative peaks as the sample components elute from the separator column. No suppressor column is therefore needed and the sensitivity is comparable to conducto-metric detection.

4.2.3. SUPERCRITICAL FLUID CHROMATOGRAPHY

Supercritical fluid chromatography (sfc) is a recently developed technique that is a hybrid between gas and liquid chromatography employing a supercritical fluid as the mobile phase. Supercritical fluids are produced when a gas or liquid is subjected to a temperature and pressure that both exceed critical values. Under these conditions the properties of the fluid, including density, viscosity and solute diffusion coefficients are intermediate between those of a gas and a liquid and vary with pressure. Increasing the pressure further makes a supercritical fluid more like a liquid, i.e. higher density and viscosity, smaller solute diffusion coefficients, whilst reducing the pressure causes it to become more like a gas, i.e. lower density and viscosity and larger solute diffusion coefficients. The only substance to be extensively used for sfc so far is carbon dioxide, for which the critical temperature is 31.1°C and the critical pressure is 72.9 bar. Xenon, with values of 16.5°C and 57.6 bar respectively, may prove useful, if expensive, for sfc-FT-ir (cf. gc-FT-ir, p. 111) as it is completely transparent to ir radiation. Other substances such as ammonia, hexane and nitrous oxide have critical values that could easily be exceeded in practice but are hazardous, toxic and/or corrosive.

The instrumentation developed for sfc is based on both gas and liquid chromatographs. The supercritical fluid is delivered to the column by a modified hplc pump. Columns are either very narrow bore capillaries of the type used in gc with a chemically bonded stationary phase (p. 94) or packed reverse phase hplc columns (p. 118). They are enclosed in a constant temperature oven and, although *temperature programming* is not used, *pressure programming* is often employed as a form of gradient elution. Increasing the pressure throughout a separation speeds up the elution process by increasing the density and hence the solubilizing power of the supercritical fluid. Carbon dioxide is a non-polar mobile phase so small amounts of more polar compounds such as methanol or glycol ethers can be added if polar mixtures are to be separated as this improves peak shapes.

The nature of a supercritical fluid enables both gas and liquid chromatographic detectors to be used in sfc. Flame ionization (fid), nitrogen phosphorus (npd), flame photometric (fpd) gc detectors (p. 96 et seq) and UV and fluorescence hplc monitors are all compatible with a supercritical fluid mobile phase and can be adapted to operate at the required pressures

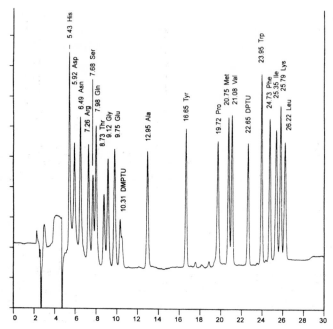

Narrow-bore column separation of PTH-amino acids. Column 25 × 0.21 cm, Zorbax SB-C18; gradients: 12%–38% solvent B in 18 min, hold at 38% B solvent; solvent A: 0.04 M acetic acid with 5% tetrahydrofuran/triethylamine to pH 4.10; solvent B: acetonitrile; flow rate: 0.21 mL/min; 55 °C; 50-µL sample (50 pmol).

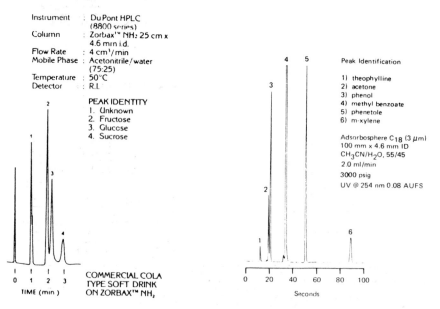

Instrument   : DuPont HPLC
               (8800 series)
Column   : Zorbax™ NH₂ 25 cm x
               4.6 mm i.d.
Flow Rate   : 4 cm³/min
Mobile Phase : Acetonitrile/water
               (75:25)
Temperature : 50°C
Detector   : R.I.

PEAK IDENTITY
1. Unknown
2. Fructose
3. Glucose
4. Sucrose

COMMERCIAL COLA
TYPE SOFT DRINK
ON ZORBAX™ NH₂

TIME (min)

Peak Identification

1) theophylline
2) acetone
3) phenol
4) methyl benzoate
5) phenetole
6) m-xylene

Adsorbosphere C₁₈ (3 µm)
100 mm x 4.6 mm ID
CH₃CN/H₂O, 55/45
2.0 ml/min
3000 psig
UV @ 254 nm 0.08 AUFS

Seconds

Figure 4.40.   Examples of separation by hplc and ic

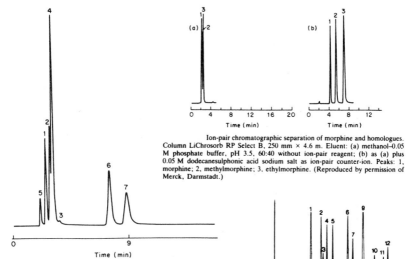

Ion-pair chromatographic separation of morphine and homologues. Column LiChrosorb RP Select B, 250 mm × 4.6 m. Eluent: (a) methanol–0.05 M phosphate buffer, pH 3.5, 60:40 without ion-pair reagent; (b) as (a) plus 0.05 M dodecanesulphonic acid sodium salt as ion-pair counter-ion. Peaks: 1, morphine; 2, methylmorphine; 3, ethylmorphine. (Reproduced by permission of Merck, Darmstadt.)

Separation of polar purine and pyrimidine bases on a polystyrene–divinylbenzene column at pH 8.5. Column PLRP-S, 150 mm × 4.6 mm. Eluent acetonitrile–ammonium formate, pH 8.5, 1:100 at 1 ml min⁻¹. Peaks: 1, cytosine; 2, uracil; 3, guanine; 4, hypoxanthine; 5, xanthine; 6, adenine; 7, thymine. (Reproduced by permission of Polymer Laboratories Ltd from *PL applications note*, No. 314.)

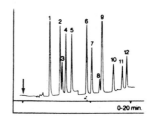

Antiepileptic drugs

Column	: 250 × 4.6 mm
Packing	: Spherisorb 5 ODS
Flow rate	: 1 ml/min.
Eluent	: Solvent A: methanol/water 20 : 80 Solvent B: acetonitril
Gradient	: 17.5% B → 45% B in 15 minutes
Detector	: UV 210 nm
Injection	: Valco 7000 psi valve 5 µl

Organic and Inorganic Anions

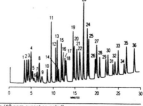

Separation of tricyclic antidepressants on a cyano-bonded silica column. Column Supelcosil LC-PCN, 5 µm, 150 mm × 4.6 mm. Eluent acetonitrile–methanol–0.01 M K₂HPO₄ buffer, pH 7.0, 60:15:25 at 2 ml min⁻¹. Ultraviolet spectrophotometric detection at 215 nm. Peaks: 1, trimipramine (internal standard); 2, doxepin; 3, amitriptyline; 4, imipramine; 5, desmethyldoxepin; 6, nortriptyline; 7, desipramine; 8, proptriptyline (internal standard). (Reproduced with permission of Supelco, Inc., Bellefonte, PA 16823.)

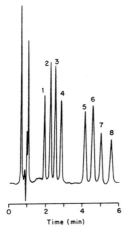

Peaks (10 ppm except as noted)

1. F⁻ (1.5 ppm)	13. Cl⁻ (3 ppm)	25. SeO₄²⁻
2. α-Hydroxybutyrate	14. Galacturonate	26. α-Ketoglutarate
3. Acetate	15. NO₂⁻ (5 ppm)	27. Fumarate
4. Glycolate	16. Glucuronate	28. Phthalate
5. Butyrate	17. Dichloroacetate	29. Oxalacetate
6. Gluconate	18. Trifluoroacetate	30. PO₄³⁻
7. α-Hydroxyvalerate	19. HPO₃²⁻	31. AsO₄³⁻
8. Formate (5 ppm)	20. SeO₃²⁻	32. CrO₄²⁻
9. Valerate	21. Br⁻	33. Citrate
10. Pyruvate	22. NO₃⁻	34. Isocitrate
11. Monochloroacetate	23. SO₄²⁻	35. cis-Aconitate
12. BrO₃⁻	24. Oxalate	36. trans-Aconitate

(up to several hundred bar). A very wide range of solute types can therefore be detected in sfc. In addition the *coupled* or *hyphenated* techniques of sfc-ms and sfc-FT-ir are likely to become increasingly important in the future (cf. gc-ms and gc-ir, p. 108 et seq).

The applications of sfc include the analysis of mixtures of hydrocarbons, triglycerides, high relative molecular mass and thermally labile compounds. Its advantage over gc lies in its ability to separate mixtures at much lower temperatures and over hplc in improved efficiency due to more rapid mass transfer (larger solute diffusion coefficients) and ease of coupling to a mass spectrometer. The wide choice of detector and ease of controlling retention times by pressure programming are also features that make sfc attractive but the instrumentation is currently much more expensive than either gc or hplc. Its future growth is therefore uncertain.

4.2.4. THIN-LAYER CHROMATOGRAPHY

SUMMARY

Principles
Separation of mixtures in microgram quantities by movement of a solvent across a flat surface; components migrate at different rates due to differences in solubility, adsorption, size or charge; elution is halted when or before the solvent front reaches the opposite side of the surface and the components examined *in situ* or removed for further analysis.

Apparatus and Instrumentation
Thin layers of powdered cellulose, silica gel, alumina, ion-exchange or gel-permeation material supported on glass plates, plastic sheets or aluminium foil; development tanks; components sometimes examined by reflectance or transmittance densitometry or removed for spectrometric analysis.

Applications
Very widespread use, largely for qualitative purposes and for both organic and inorganic materials; especially useful for checks on purity, to monitor reactions and production processes and to characterize complex materials.

Disadvantages
Migration characteristics very sensitive to conditions; thin layers easily damaged; quantitative precision only moderate: 5 to 10%.

The difference between this technique and gc or hplc is that the separation process occurs on a flat essentially two-dimensional surface. The separated components are not usually eluted from the surface but are examined *in situ*. Alternatively, they can be removed mechanically for further analysis. In thin-layer chromatography (tlc), the stationary phase is usually a polar solid such as silica gel or alumina which is coated on to a sheet of glass, plastic, or aluminium. Although some moisture is retained by the stationary phase, the separation process is predominantly one of surface adsorption. Thin layers are sometimes made from ion-exchange or gel-permeation materials. In these cases the sorption process would be ion-exchange or exclusion.

Samples are applied as spots or streaks close to one edge of the plate and the mobile phase is allowed to travel from that edge over the samples and towards the opposite edge. In so doing, the components of the sample move across the surface at rates governed by their distribution ratios and therefore separate into individual spots or bands. This procedure is called *development*. After development, coloured substances are immediately visible on the surface. Those which are colourless can be visualized by treatment with a chromogenic reagent, detected by fluorescence under a uv lamp or by using radioactive tracers. A typical thin-layer chromatogram is shown in figure 4.41.

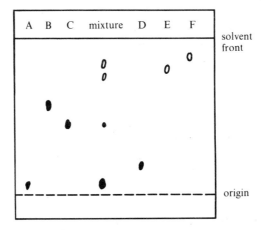

Figure 4.41.   Thin-layer chromatogram

Diffusion and mass transfer effects cause the dimensions of the separated spots to increase in all directions as elution proceeds, in much the same way as concentration profiles become Gaussian in column separations (p. 82). Multiple path, molecular diffusion and mass transfer effects all contribute to spreading along the direction of flow but only the first two cause lateral spreading. Consequently, the initially circular spots become progressively elliptical in the direction of flow. Efficiency and resolution are thus impaired. Elution must be halted before the solvent front reaches the opposite edge of the plate as the distance it has moved must be measured in order to calculate the retardation factors (R_f values) of separated components (p. 81).

FACTORS AFFECTING R_f VALUES AND RESOLUTION

Basically, the R_f value of a solute is determined by its distribution ratio which in turn is dependent on relative solubilities for partition systems or relative polarities for adsorption systems. For example, if adsorption tlc is used to separate a mixture of squalene (a hydrocarbon), methyl oleate, cholesterol and α-tocopherol (vitamin E), then squalene, being the least polar, will move furthest and the cholesterol, being the most polar, will remain close to the origin. Methyl oleate is less polar than α-tocopherol and will therefore be found between it and the squalene. The role of polarity is discussed more fully on p. 78.

The effect of temperature on distribution ratios has already been mentioned on page 85. Although the separation proceeds more quickly at elevated temperatures, resolution suffers because of increased rates of diffusion. However, in adsorption tlc only small increases in R_f values are observed even with a 20°C rise. Strict temperature control is not necessary if samples and standards are run at the same time, although large fluctuations should be avoided. The quality of the thin-layer materials, and in particular the presence of impurities in them, determine the extent to which partition, adsorption, ion-exchange and exclusion participate in the sorption process. These factors affect R_f values in an unpredictable manner. Thin layers should be of uniform thickness, between 0.2 and 0.3 mm; with thinner layers, local variations in thickness can result in appreciable variations in R_f values.

A stable atmosphere saturated with the vapour of the mobile phase is required to ensure reproducible R_f values. Unless saturation conditions prevail, solvent will evaporate from the surface of the thin layer causing an increased solvent flow but slower movement of the solvent front; R_f values consequently increase. In practice, chromatograms are best developed in a sealed glass tank in which a saturated atmosphere has been produced by lining the walls with filter-paper soaked in the appropriate solvent or solvent mixture. The tank should be prepared about 15 minutes before the chromatogram is to be developed.

The optimum amount of sample required to produce detectable spots with a minimum of spreading due to overloading is about 0.01 μg to 50 μg. With samples much larger than this, isotherms become non-linear, the R_f values alter significantly with sample size and resolution suffers because of increased tailing or fronting.

The process of placing a sample on the thin layer is called *spotting*. The sample should be dissolved in up to 10 μl of a solvent which should be volatile so as to prevent undue spreading of the spot. Accurately measured amounts of sample can be spotted using a micropipette or microsyringe but for qualitative purposes a melting-point tube drawn out to a fine tip will suffice. A faint pencil line can be drawn about 2 to 3 cm from one edge of the plate and the sample spots placed along this line 1 cm apart and not closer than 1.5 cm to the sides. A ruler or raised perspex template can be used to aid spotting the samples on to the plate accurately.

DEVELOPMENT PROCEDURES

For convenience, the ascending method is used mostly although there is no reason why descending or horizontal development cannot be used for special purposes. An alternative method consists of forming a sandwich with a glass plate the same size as the chromatographic plate. The two are clamped together with a thin spacer to protect the surface of the thin layer and the bottom side is immersed in a narrow trough containing the solvent. The small volume of air trapped between the two plates is rapidly saturated with the developing solution.

Two-dimensional Development

More effective separation of complex mixtures and those consisting of components with similar R_f values can be achieved by successive developments at 90° to each other. The sample is spotted at one corner and developed with the first solvent. After thorough drying, the paper or plate is turned through 90° and developed with the second solvent. The two-way chromatogram (figure 4.42) can be compared with a 'standard map' obtained from chromatographing known mixtures.

DETECTION OF SEPARATED COMPONENTS

Coloured substances can be observed visually *in situ* after development of the chromatogram, but those which are colourless must be *visualized* by physical or chemical means. For many samples, and especially for unknowns, the chromatogram should first be examined under a uv lamp. Many organic compounds show up as fluorescent spots when irradiated at 370 or 254 nm.

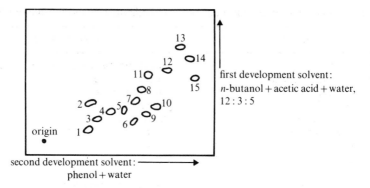

Figure 4.42.   Two-dimensional chromatogram; standard map for amino acids

1. taurine	6. lysine	11. tyrosine
2. glutamic acid	7. alanine	12. valine
3. serine	8. α-amino-*n*-butyric acid	13. leucine
4. glycine	9. arginine	14. β-phenylalanine
5. threonine	10. hydroxyproline	15. proline

An alternative method is to incorporate a fluorescent indicator into the thin layer when non-fluorescing materials show as dark spots on a fluorescent background. Radioactive counting techniques can be used for the detection of labelled compounds.

The most widely used detection methods are those in which spots are visualized chemically by spraying with a chromogenic reagent or by dipping into a reagent solution. Dipping provides a more even cover with the reagent, but may dissolve some of the separated components or cause spreading of the spots by diffusion. Solvents used in both methods should be reasonably volatile to facilitate drying after treatment. Acetone, ethanol and chloroform are commonly used. Reagents may be general in that they produce coloured spots with a wide range of compounds or they may be selective in reacting only with compounds containing certain functional groups. Metals can be visualized by using a reagent with which they form coloured insoluble salts or complexes. Some examples of chromogenic reagents and their applications are given in table 4.14. A general and very sensitive method for detecting organic substances separated by tlc is to spray the chromatogram with concentrated sulphuric acid and heat the plate to about 200°C for several minutes. The charred organic matter shows up as brown or black spots.

Chromatograms can be stored easily for future reference unless, as sometimes happens, the spots fade. More permanent less bulky records can be made by photographing the chromatogram, spraying it with a lacquer or photocopying it on to light sensitive 'diazo' paper.

Table 4.14   Chromogenic reagents for visualizing thin-layer chromatograms

REAGENT	APPLICATIONS
iodine vapour	general organic, unsaturated compounds
phosphomolybdic acid	general organic
fluorescein/bromine	general organic
sulphuric acid	general organic (tlc only)
ninhydrin or isotin	amino acids
2,4-dinitrophenylhydrazine	ketones and aldehydes
H_2S water, diphenylcarbazide or	metals
rubeanic acid	metals
aniline phthalate	sugars
antimony trichloride	steroids, essential oils
chloroplatinic acid	alkaloids
bromothymol blue	lipids

STATIONARY PHASE

Nature and Function of Thin-Layer Materials

Any of the materials used in column, ion-exchange and exclusion chromatography can be used for tlc provided that they can be obtained in the form of a homogeneous powder of fine particle size (1 to 50 μm). Coarse materials will not easily form a uniform thin layer which adheres to a glass plate or other supporting sheet. A list of some of the more important stationary phases and their uses is given in table 4.15. Pre-coated thin-layer plates are available commercially but are relatively expensive. Plates with a wide variety of stationary phases can be prepared in the laboratory using a mechanical spreading apparatus. This consists of a trough to hold a slurry of the station-

Table 4.15   Stationary phases for thin-layer chromatography

STATIONARY PHASE	PREDOMINANT SORPTION PROCESS	USE
silica gel	adsorption or partition	general
modified silica gels	adsorption or partition	similar to bonded phase hplc
alumina	adsorption or partition	general
cellulose powder	partition	inorganic, amino acids, nucleotides, food-dyes
kieselguhr	partition	sugars
modified celluloses e.g. DEAE and CM	ion-exchange	nucleotides, phospholipids
Sephadex gels	exclusion	macromolecules

ary phase and a flat bed to support a number of plates. By moving the trough across the plates, an even coating of slurry is deposited from which the solvent can be evaporated by oven-drying.

Silica gel or silicic acid, has found the most widespread use in tlc. It functions primarily as a surface adsorbent if dried above 100°C, otherwise the adsorbed water acts as the stationary phase for a partition system. Plates coated with silica gel often contain about 10% w/w calcium sulphate (plaster of Paris) as a binder to improve adherence to the plate, although this is not essential if a very finely powdered material is used. Indicators which fluoresce under a uv lamp can be incorporated into the layer when it is prepared, e.g. the sodium salt of fluorescein. Alumina and kieselguhr (diatomaceous earth) are sometimes used as alternatives to silica gel but offer no particular advantages. Chemically modified silica gels similar to the modified microparticulate silicas used in hplc (p. 118) have become available. For relatively polar compounds, octadecyl (C_{18} or ODS) modified plates are particularly useful. The production of plates with layers of very small and uniform particles, which result in more compact sample spots and improved resolution, has given rise to the term 'high performance tlc' or hptlc. Cellulose powder is used for partition tlc where it acts largely as a solid support as in paper chromatography. More compact spots are obtained than with paper chromatography and development times are faster because of the fine particle size. A number of ion-exchange cellulose powders are available for separations of ionic species (p. 151).

Reagents which selectively retard certain chemical species can be incorporated into a thin-layer plate. Thus, silver nitrate, which forms weak π-complexes with unsaturated compounds, aids their separation from saturated compounds.

The mobile phase is drawn through the thin layer by capillary action, but the rate of movement is relatively fast because of the uniform and small particle size. Tlc separations on a 20 × 20 cm plate take only 20 to 40 min compared with two hours or more for a comparable-sized paper chromatogram.

MOBILE PHASE

The choice of mobile phase is largely empirical but general rules can be formulated. A mixture of an organic solvent and water with the addition of acid, base or complexing agent to optimize the solubility of the components of a mixture can be used. For example, good separations of polar or ionic solutes can be achieved with a mixture of water and n-butanol. Addition of acetic acid to the mixture allows more water to be incorporated and increases the solubility of basic materials, whilst the addition of ammonia increases the solubility of acidic materials. If the stationary phase is hydrophobic, various mixtures of benzene, cyclohexane and chloroform provide satisfactory mobile phases. It should be emphasized that a large degree of trial and error is involved in their selection. For tlc on silica gel, a mobile phase with as low a polarity as possible should be used consistent with achieving a satisfactory separation. Polar solvents can themselves become strongly adsorbed thereby producing a partition system, a situation which may not be as desirable. An eluotropic series (table 4.8) can be used to aid the selection of the best solvent or solvent mixture of a particular sample.

QUALITATIVE ANALYSIS

Components are identified by comparison of their R_f values with those of standards run under identical conditions, or by removing the materials from the chromatogram and subjecting them to further qualitative tests, e.g. spot-tests, mass spectrometry, infrared spectrometry. Factors affecting R_f values were discussed on p. 148. Chromatographic materials and conditions are usually so variable that it is advisable to run standards with samples to ensure that comparisons are valid.

QUANTITATIVE ANALYSIS

Thin-layer chromatography does not provide quantitative information of the highest precision and accuracy. Linear relationships between the weight of a substance and the logarithm or square-root of the spot area can sometimes be established under very closely controlled conditions. The optical absorbance of a spot determined by reflectance measurements can be similarly related to weight, or the substances can be scraped from the plate and dissolved in a suitable solvent for a spectrometric determination. The main difficulties with area and density measurements lie in defining the boundaries of spots and controlling chromogenic reactions in a reproducible manner. Relative precision can be as good as 1 to 2% but is more usually 5 to 10%.

APPLICATIONS OF THIN-LAYER CHROMATOGRAPHY

Thin-layer chromatography is very widely used, mainly for qualitative purposes; almost any mixture can be at least partially resolved. Inorganic applications, such as the separation of metals in alloys, soil and geological samples, and polar organic systems, such as mixtures of amino acids or

Table 4.16   Typical applications of thin-layer chromatography

SEPARATION	TECHNIQUE	MOBILE PHASE
amino acids and peptides	cellulose tlc	2-dimensional 1. n-butanol/acetic acid/water 2. phenol/water
hydrocarbon oils, ester oils	tlc on silica gel	light petroleum/ether
simple sugars	cellulose tlc	n-propanol/ethyl acetate/water
steroids e.g. bile acids, estrogens, sterols	tlc on silica gel	chloroform/acetone
Co, Cu, Fe, Mn, Zn, Ni	cellulose tlc	acetone/conc HCl
alkaloids e.g. belladonna, morphine, opium, purine	tlc on silica gel	chloroform/ethanol
chlorinated insecticides e.g. Aldrin, DDT, Heptachlor, Endrin	tlc on silica gel	hexane
vitamins e.g. A, D, E	tlc on silica gel	hexane-acetone
B, C	tlc, on silica gel	water

sugars in urine, are particularly suited to cellulose tlc. The versatility of tlc has resulted in a rapid spread in its use in all fields especially for the separation of organic materials. It is ideally suited to following the course of complex reactions, quality control, purity checks, clinical diagnosis and forensic tests. Some typical separations by thin-layer chromatography are given in table 4.16.

4.2.5. ION-EXCHANGE CHROMATOGRAPHY

SUMMARY

Principles
    Separation of ionic materials in microgram to gram quantities by passage of a solution through a column or across a surface consisting of a porous polymeric resin incorporating exchangeable ions.

Apparatus and Instrumentation
    Glass columns for separation by gravity flow; glass, metal or nylon tubing for pressurized systems; fraction collector, detector and recorder.

Applications

Mainly inorganic, especially mixtures of metals with similar chemical characteristics, e.g. lanthanides; separation of amino acids.

Disadvantages

Gravity flow separations slow; separated components accompanied by a large excess of eluting electrolyte.

Ion-exchange separations are limited to samples containing ionized or partially ionized solutes. The stationary phase consists of an insoluble but porous resinous material which contains fixed charge-carrying groups and mobile counter ions of opposite charge. The counter ions can be reversibly exchanged for those of a solute which carry a like charge as the mobile phase travels through the system. Variations in the affinity of the stationary phase for different ionic species is responsible for differential rates of migration. Separations are often enhanced by eluting with a mobile phase containing a complexing agent.

STRUCTURE OF ION-EXCHANGE RESINS

The most widely used type of resin is a copolymer of styrene and divinyl-benzene produced in bead form by suspension polymerization in an aqueous medium.

The proportion of divinylbenzene (DVB) is 2 to 20%, which results in a three-dimensional cross-linked structure that is both rigid, porous and highly insoluble. The degree of *cross-linking*, expressed as the weight percent of DVB, affects the rigidity of the structure and the size of the pores. A low degree of cross-linking produces beads which swell appreciably when in contact with a polar solvent and have large pores enabling ions to diffuse into the structure and exchange rapidly. Resins with a high degree of cross-linking have smaller pores and are more rigid. Swelling is less, the exchange process is slower, and large ions may not be able to diffuse into the interior of the bead at all. By control of the polymerization process, the resin produced can be made gel-like (micro-reticular) having only pores of sizes comparable to the dimensions of inorganic and small organic ions. Alternatively, a more open structure can be produced (macro-reticular) in which the pores are tens of nanometres in diameter and through which bulky organic ions can diffuse readily.

Cation or anion-exchanging properties are introduced into the resin by chemical modification after polymerization. *Cation-exchangers* can be sub-divided into *strong-acid* types containing —SO₃H groups and *weak-acid* types containing —COOH groups. The former are produced by reacting the resin with chlorosulphonic acid which results in mainly *para* substitution of the benzene rings. The sulphonic acid groups are dissociated over a wide pH range and these resins will exchange their protons for other cations under both acid and alkaline conditions. The weak-acid type differs in that it is prepared by direct polymerization of DVB and methacrylic acid, $CH_3C(=CH_2)COOH$. Protons are exchanged for other cations only above pH 5 as below this value the carboxylate groups are not dissociated.

Anion-exchangers comprise *strong-base* types incorporating quaternary ammonium groups ($—N^+R_3$) and *weak-base* types incorporating primary, secondary or tertiary amines. They are prepared by chloromethylating the resin followed by treatment with the appropriate amine

The chloromethylation reaction can result in cross-linking additional to that produced by the DVB. Thus for anionic resins, the DVB content is not a reliable guide to pore size.

Resins can also be prepared with chelating functional groups which show selective affinities for certain metals. Iminodiacetic acid, vinylacetylacetone, glyoxal-thiophenol and 8-hydroxyquinoline have all been used to produce these so-called 'chelating resins'. Their selectivities are similar to those of the free reagent.

Non-Resinous Ion-Exchange Materials

Cellulose, modified by the introduction of ionic groups, is available in paper form or as a powder for use in tlc and is particularly useful for the separation of macromolecules and biological materials. Cation exchangers are produced by introducing acidic groups, e.g. $-OCH_2SO_3H$ (sulphomethyl, SM), $-OCH_2COOH$ (carboxymethyl, CM), which are bonded to the cellulose structure via ether or ester linkages. Anion exchangers are formed by reacting the cellulose with epichlorhydrin and an amine, e.g. $-OCH_2CH_2N^+(C_2H_5)_3$ (triethylaminoethyl, TEAE).

Liquid ion-exchangers have been discussed in the section on solvent extraction (p. 67). They can be used in column form by coating them on to a solid support such as cellulose powder or Kel-F (polytrifluorochloro-ethylene). Tris-*n*-octylamine (TNOA) and bis(2-ethylhexyl)phosphoric acid (HDEHP) behave as strong-base and strong-acid exchangers for anions and cations respectively.

Inorganic ion-exchange materials with similar properties to resins are sometimes used to separate mixtures at high temperatures or under conditions of high-energy radiation. These include micro-crystalline heteropolyacid salts (e.g. ammonium molybdophosphate), hydrated zirconium oxide and zirconium phosphate. Some of the materials are highly selective but tend to react with acids, bases and complexing agents. For hplc, chemically modified microparticulate silicas (p. 119) incorporating cationic or anionic ion-exchange groups are available. Being rigid and incompressible they are more suited to pressurized systems than ion-exchange resins although their *capacity* is lower. Often, however, ion-pair chromatography (p. 139) on non-polar bonded-phases proves to be superior to the use of ion-exchangers.

PROPERTIES OF ION-EXCHANGE RESINS

The *capacity* of an ion-exchange resin, i.e. the number of exchangeable ions, is determined by the degree of cross-linking and is expressed as milliequivalents per gram of dry resin. Values range between 2 and 10, depending on the resin, and are generally quoted for the hydrogen or chloride forms. In practice, the full capacity is never available due to non-ideal operating conditions. For column separation, the practical or 'break-through' capacity is reached when an exchanging ion continuously introduced at the top of the column leaks through with the column effluent.

Swelling is a phenomenon which accompanies the use of most ion-exchange resins, and is especially important in column operation. It results from the osmotic pressure set up when a resin bead, which can be considered to be a concentrated electrolyte solution, is surrounded by a more dilute polar solution. Solvent flows into the bead and distends the structure in an attempt to reduce the osmotic pressure by dilution. The change in *bed-volume*, that is the size of the resin column, can be considerable and may well alter as different sample or eluting solutions are passed through. This affects pore size and thus the ease with which exchanging ions can penetrate the beads. In non-polar solvents, microreticular resins remain almost unswollen so their exchange capacity is small and the rate of exchange slow. Macroreticular resins, however, still retain an open structure and are therefore particularly useful for separations in relatively non-polar solvents or for the separation of large organic ions. The effect of the degree of cross-linking on swelling has already been mentioned.

Selectivity

The affinity between a resin and an exchangeable ion is a function both of the resin and the ion. Ion-exchange is an equilibrium process which for a cationic

Table 4.17   Selectivity coefficients for some common cations and anions

CATIONS	4% DVB	8% DVB
Li^+	0.76	0.79
H^+	1.00	1.00
Na^+	1.20	1.56
NH_4^+	1.44	2.01
K^+	1.72	2.28
Ag^+	3.58	6.70
Mg^{2+}	0.99	1.15
Zn^{2+}	1.05	1.21
Cu^{2+}	1.10	1.35
Ni^{2+}	1.16	1.37
Ca^{2+}	1.39	1.80
Sr^{2+}	1.57	2.27
Pb^{2+}	2.20	3.46
Ba^{2+}	2.50	4.02

ANIONS	6–10% CROSS-LINKING
OH^-	0.09
F^-	0.09
CH_3COO^-	0.17
Cl^-	1.0
CN^-	1.6
Br^-	2.8
NO_3^-	3.8
I^-	8.7

resin can be represented by the equation

$$nR^-H^+ + M^{n+} = (R^-)_nM^{n+} + nH^+ \tag{4.51}$$

where R represents the resin matrix.

The equilibrium constant, also known as the *selectivity coefficient*, is given by

$$K = \frac{[M^{n+}]_R[H^+]^n}{[M^{n+}][H^+]_R^n} \tag{4.52}$$

where $[M^{n+}]_R$ and $[H^+]_R^n$ are the concentrations (strictly activities) of the exchanging ion and the hydrogen ion within the resin structure. Thus, the greater the affinity for a particular ion, relative to hydrogen, the greater the value of K. Selectivity coefficients are functions of the proportions of the exchanging ions, the total concentration of the solution and the degree of cross-linking. In dilute solutions (<0.1 M) values of K increase with increasing formal valency, i.e. $M^{4+} > M^{3+} > M^{2+} > M^+$. There is a further variation within each charge group as shown by the examples given in table 4.17.

Values are quoted relative to the hydrogen ion for cations and to the chloride ion for anions. As shown, selectivity also increases with increasing degree of cross-linking. At concentrations greater than 0.1 M selectivity for monovalent over polyvalent ions increases. Ionic properties which determine resin affinity are complex involving hydration energy, polarizability and hydrated ionic radius. The last shows an approximately inverse relationship to the selectivity coefficient.

ION-EXCHANGE SEPARATIONS

Most separations are performed using columns of resin and an elution procedure. The sample is introduced as a small band at the top of the column from where the various components are moved down the column at a rate depending on their selectivity coefficients. Sorption isotherms are approximately linear in dilute solutions so that elution peaks are symmetrical. Tailing can be expected at high concentrations as the isotherms curve towards the mobile phase concentration axis (p. 77).

The mobile phase contains an ion of low resin affinity, and the separated components collected at the bottom of the column are thus accompanied by a relatively high concentration of this ion. Procedures often adopted in ion-exchange chromatography are *gradient elution*, involving continuous variation of the composition of the eluting agent, *stepwise elution*, in which the composition is altered at specific points during the separation, and *complexing elution* where a reagent which forms complexes of varying stability with the sample components is included in the solution. Acids, bases and buffers are the most widely used eluting agents.

Separated components emerging in the column effluent can be monitored by means of a physical measurement, e.g. uv or visible absorbance, refractive index, conductivity or radioactivity. Alternatively, separate fractions can be collected automatically and subjected to further analysis.

APPLICATIONS OF ION-EXCHANGE CHROMATOGRAPHY

The applications of ion-exchange chromatography are exemplified by the selection shown in table 4.18. Among the most notable are the separation of lanthanides and actinides using a citrate, lactate or EDTA eluting agent; the separation of many metals as halide complexes on anionic resins and the separation of amino-acids with citrate buffers. The use of pressurized systems for complex mixtures is likely to become more widespread in the future.

4.2.6. GEL-PERMEATION CHROMATOGRAPHY

SUMMARY

Principles
   Separation of materials according to molecular size and shape by passage of a solution through a column or across a surface consisting of a polymeric gel.

Apparatus and Instrumentation
   Glass columns for separations by gravity flow; glass, metal or nylon tubing for pressurized systems; fraction collector, detector and recorder.

Applications
   Separation and desalting of high-molecular weight materials; determination of molecular weights.

Disadvantages
   Gravity flow separations slow; resolution of low molecular weight compounds poor or non-existent.

Molecules that differ in size can be separated by passing the sample solution through a stationary phase consisting of a porous cross-linked polymeric gel. The pores of the gel exclude molecules greater than a certain critical size whilst smaller molecules can permeate the gel structure by diffusion. The process is described as *gel-permeation, gel-filtration* or *exclusion chromato-*

Table 4.18   Some applications of ion-exchange chromatography

SEPARATIONS	RESIN USED	ELUTION METHOD
transition metals, e.g. Ni^{2+}, Co^{2+}, Mn^{2+}, Cu^{2+}, Fe^{3+}, Zn^{2+}	anion	stepwise elution 12 M to 0.005 M HCl
lanthanides	cation	stepwise or gradient elution with citrate buffers
Zr, Hf	anion	3.5% H_2SO_4
phosphate mixtures, e.g. ortho-, pyro-, tri-, tetra- etc.	anion	1 M to 0.005 M KCl
trace metals in industrial effluents	chelating reins	concentrated acids
amino acids	cation	stepwise or gradient elution with citrate buffers
aldehydes, ketones, alcohols	anion	ketones and aldehydes held as bisulphate addition compounds. Eluted with hot water and NaCl respectively
sugars	anion	as borate association complexes. Eluted by gradient pH
carboxylic acids	anion	gradient pH elution
pharmaceuticals	cation/anion	acid or alkali buffer

graphy. Excluded molecules pass through the system more rapidly than smaller ones which can diffuse into the gel. Diffusion within the gel also varies with molecular size and shape because pores of different dimensions are distributed throughout the gel structure in a random manner. These smaller molecules are eluted at rates dependent upon their degree of permeation into the gel, and components of a mixture therefore elute in order of decreasing size or molecular weight.

STRUCTURE AND PROPERTIES OF GELS

Gels used for the stationary phase can be hydrophilic, for separations in aqueous and other polar solvents, or hydrophobic, for use in non-polar or weakly-polar solvents. Agar, starch, polyacrylamide and cross-linked dextrans possess hydroxyl or amide groups and are thus hydrophilic. They swell in aqueous media and in such solvents as ethylene glycol and dimethylformamide. Bio-Gel (a co-polymer of acrylamide and *N,N'*-methylene-bisacrylamide) and Sephadex (dextran cross-linked with epichlorhydrin) are two commercially available gels made in bead form. Cross-linking produces a

Table 4.19   Fractionation ranges of some commercial gels

HYDROPHILIC GELS	FRACTIONATION RANGE
Sephadex G-10	Up to 700
G-50	500–10 000
G-100	4 000–150 000
G-200	5 000–800 000

HYDROPHOBIC GELS	MEAN PORE SIZE IN ÅNGSTROMS	FRACTIONATION RANGE
μ-Styragel	100	Up to 700
(polystyrene/DVB)	500	$0.05 \times 10^4 - 1 \times 10^4$
	10^4	$1 \times 10^4 - 20 \times 10^4$
	10^6	$5 \times 10^6 - 10 \times 10^6$
RIGID GELS		
μ-Bondagel (silica)	125	$0.2 \times 10^4 - 5 \times 10^4$
	500	$0.2 \times 10^5 - 5 \times 10^5$
	1 000	$5 \times 10^5 - 20 \times 10^5$
Zorbax PSM-60	60	$0.01 \times 10^4 - 4 \times 10^4$
PSM-500	350	$0.1 \times 10^5 - 5 \times 10^5$
PSM-1 000	750	$0.03 \times 10^6 - 2 \times 10^6$

three-dimensional network which renders them insoluble. By varying the degree of cross-linking, the range of pore sizes can be controlled so as to produce beads that will fractionate samples over different molecular weight ranges, table 4.19. The gels are chemically stable over a pH range of 2 to 11 but are attacked by strong acids, bases and oxidizing agents. Sephadex may contain a small number of carboxyl groups which act as ion-exchange or adsorption sites, but the effect on retention volumes is only slight. Gels based on agarose have exceptionally large pores which facilitate separations in molecular weight ranges up to 2×10^8. They are not cross-linked and are more chemically and thermally unstable than other hydrophilic gels.

Hydrophobic gels are made by cross-linking polystyrene with divinyl benzene and thus resemble ion-exchange resins but without the ionic groups. As a result they can absorb relatively non-polar solvents such as tetrahydrofuran, benzene, chloroform and cyclohexanone with ease but not aqueous or other highly polar liquids. Dextran-based gels can be made hydrophobic by acylation or alkylation of the hydroxyl groups. The fractionation ranges of some commercially available hydrophobic gels are also given in table 4.19.

Porous glasses and silica gels with controlled pore sizes which are rigid and non-compressible are particularly useful in separations where the mobile phase is pressurized as in hplc (p. 112, table 4.18).

THE SEPARATION PROCESS

Columns of gel beads are more often used than thin layers. The process can be considered analogous to a partition system wherein molecules which are completely excluded from the gel have a distribution ratio of zero whilst those small enough to penetrate all parts of the structure have a value of one. Adsorption or ion-exchange effects can cause the distribution ratios of polar molecules to exceed one. Components of a mixture are characterized by their retention volume V_R, which is determined by the distribution ratio.

A solute always has the same retention volume for a given gel. It is virtually independent of temperature and flow-rate because separation occurs by selective diffusion within a single liquid phase. True partition is an equilibrium process involving the transfer of a solute between two immiscible phases. For solutes whose molecular size or mass falls within the fractionation range of the gel, i.e. for which distribution ratios lie between zero and one, retention volume is approximately a linear function of the logarithm of the molecular weight, figure 4.43. Molecules that are completely excluded from the gel are eluted with a retention volume equal to the void or dead volume of the column (p. 81); molecules which freely permeate all parts of the gel network are

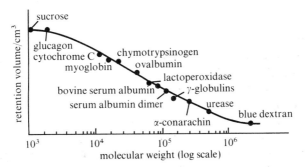

Figure 4.43. Relation between molecular weight and retention volume for proteins at pH = 7.5 (Sephadex G-200 column)

eluted by a volume equal to the dead volume + the volume of liquid within the gel particles. A column is calibrated by eluting substances of known relative molecular mass (RMM) or molecular mass range. Elution profiles are Gaussian because of diffusion effects; peak broadening can be minimized by eluting with a solvent of low viscosity, e.g. tetrahydrofuran. Efficiency is determined only by peak width, as retention volumes are constant (cf. other column techniques, p. 82). Eluted components are detected by monitoring a physical property of the column effluent such as refractive index or uv absorption or by collecting and analysing separate fractions.

APPLICATIONS OF GEL-PERMEATION CHROMATOGRAPHY

Relative molecular mass (RMM) distributions for components of bio-chemical and polymer systems can be determined with a 10% accuracy using standards. With biochemical materials, where both simple and macro-molecules may be present in an electrolyte solution, *desalting* is commonly employed to isolate the macromolecules. Inorganic salts and small molecules are eluted well after such materials as peptides, proteins, enzymes and viruses. Desalting is most efficient if gels with relatively small pores are used, the process being more rapid than dialysis. Dilute solutions of macro-molecules can be concentrated and isolated by adding dry gel beads to absorb the solvent and low RMM solutes.

4.3 Electrophoresis

SUMMARY

Principles

    Separation of charged materials by differential migration across a surface or through a column in an applied potential gradient; migration rates dependent upon size, shape and charge of species.

Apparatus and Instrumentation

    Traditional electrophoresis: paper, cellulose acetate or polymeric gels used as a supporting medium for the electrolyte solution; enclosed tank with electrodes and buffer reservoirs; dc power supply.

    Capillary electrophoresis: narrow-bore fused-silica capillary tube; injection system; detector; recorder or VDU.

Applications

    Qualitative and quantitative characterization of biologically active materials; especially useful for clinical and forensic work where small amounts of complex samples may be involved. Nanogram to 3 aptomole scale separations by capillary electrophoresis.

Disadvantages

    Mobilities very sensitive to supporting medium; precision poor for quantitative work by traditional methods (5 to 20%)

    0.5 to 3% for capillary methods.

Electrophoresis involves the differential migration of charged species in an electrolyte solution under the influence of an applied potential gradient. The rate of migration of each species is a function of its charge, shape and size.

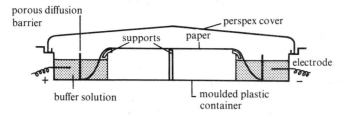

Figure 4.44. Apparatus for paper electrophoresis

In traditional *zone electrophoresis*, the electrolyte solution is retained by an inert porous supporting medium, usually paper or gel, in the form of a sheet or column. Application of a dc potential across the solution for a period of time results in the components of a mixture, originally placed at the centre, separating into individual bands or spots. Detecting techniques such as spraying with a chromogenic reagent or staining with a dyestuff are then used to visualize the developed *electropherogram*. A schematic diagram of a paper electrophoresis apparatus is shown in figure 4.44 and a typical electropherogram in figure 4.45. The technique differs from chromatography in that only a single phase is involved, i.e. the electrolyte solution which essentially remains stationary on the supporting medium. *High performance capillary electrophoresis* (hpce) is a comparatively recent development that is proving to be a very powerful separation technique of growing importance (*vide infra*).

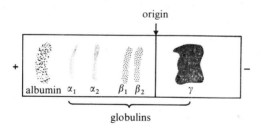

Figure 4.45. Electropherogram of blood serum on cellulose acetate. (Buffer: 0.50 M barbitone, pH = 8.6)

FACTORS AFFECTING IONIC MIGRATION

If charged species, dissolved or suspended in an electrolyte solution, are subjected to a uniform potential gradient, they rapidly assume a constant rate of migration. As in conventional electrolysis cations migrate towards the cathode (negative electrode) and anions to the anode (positive electrode). The rate of migration reaches a constant value when the attractive force exerted by the electrode is balanced by the frictional force due to movement of the species through the electrolyte solution. Each species is characterized by its *electrophoretic mobility* μ which is determined by its total charge, its overall size and shape and the viscosity of the electrolyte solution. The distance d travelled by a given species is related to its mobility μ and to the time of application and size of the potential gradient by the equation

$$d = \mu t(E/S) \tag{4.53}$$

where t denotes time, E is the applied potential and S is the distance between the electrodes. Typically, 50 to 150 V is applied across a distance of 10 to 20 cm. For two species of differing mobility, μ_1 and μ_2, the separation after time t is given by:

$$d_1 - d_2 = (\mu_1 - \mu_2)t(E/S) \tag{4.54}$$

Separation is hindered by diffusion which increases in proportion to the square root of the time taken. Resolution is therefore maximized by applying a large potential gradient E/S for a short time.

EFFECT OF TEMPERATURE, pH AND IONIC STRENGTH

As mobility and rate of diffusion both increase with increasing temperature, close control is necessary to ensure that valid comparisons can be made and that component bands are sharply resolved. The net charge and hence the mobility of many species, especially organic compounds, is pH dependent. To ensure that the migration characteristics of the components of a mixture remain constant during a separation, the electrolyte solution employed is usually a buffer. In an unbuffered medium, pH changes could lead to gross changes in the ionic concentrations of partially dissociated compounds such as weak acids and bases. Ampholytes, e.g. amino-acids, and proteins, might even reverse their direction of migration as a result of a change in pH

$$\underset{\substack{|\\R}}{H_3N^+-CH-COOH} \underset{H^+}{\overset{OH^-}{\rightleftharpoons}} \underset{\substack{|\\R}}{H_2N-CH-COOH} \underset{H^+}{\overset{OH^-}{\rightleftharpoons}} \underset{\substack{|\\R}}{H_2N-CH-COO^-}$$

Large reservoirs of buffer solution are maintained in the electrode compartments to minimize the effect of electrolysis which is to produce hydrogen ions at the anode and hydroxyl ions at the cathode. The choice of buffer is critical in its effect on the degree of separation of a mixture, and is dictated largely by the values of the various dissociation constants.

The total ionic strength should be kept fairly low (0.01 to 0.1 M) as mobilities decrease with increasing salt concentration. Furthermore, at high electrolyte concentrations, more current flows through the solution causing increasing difficulty in the dissipation of the evolved heat. Macromolecules are frequently insoluble in solutions containing a high level of salts.

ELECTROOSMOSIS

During the migration of cations and anions towards their respective electrodes, each ion tends to carry solvated water along with it. As cations are usually more solvated than anions, a net flow of water towards the cathode occurs during the separation process. This effect, known as *electroosmosis*, results in a movement of neutral species which would normally be expected to remain at the point of application of the sample. If required, a correction can be applied to the distances migrated by ionic species by measuring them relative to the position reached by a neutral molecule. Propanone, urea or glucose is sometimes added to a sample for this purpose. *Electroosmotic flow* (EOF) is particularly important in high performance capillary electrophoresis, *vide infra*.

SUPPORTING MEDIUM

The function of the supporting medium is to provide an inert porous structure for the electrolyte solution. Filter paper, cellulose acetate and various gels are used for this purpose. Although paper is cheap and convenient to use, its fibrous structure and the presence of ionic groups result in poor resolution of migrating species due to tailing. Polymerized cellulose acetate is available in sheets and has a small uniform pore size. This fact coupled with the absence of ion-exchange or adsorption sites results in separations showing better resolution than those on paper. However, electroosmosis is pronounced and the maximum sample loading is only about a quarter that of paper. Agar and polyacrylamide gels have found the most widespread use. Both are used in the form of thin flat beds or in columns. Agar is particularly useful for column work because its high mechanical strength enables it to be readily extruded for visualizing or further analysis after the separation process is complete. Polyacrylamide gels are used in gel filtration procedures where varying the degree of cross-linking produces materials fractionating over different molecular size ranges. In electrophoresis separations, fractionation according to size is therefore superimposed on the migration process. This results in enhanced degrees of separation for complex mixtures of macromolecules.

DETECTION OF SEPARATED COMPONENTS

For separations on paper, after a preliminary drying, spraying or dipping with any of the chromogenic reagents used in paper and thin-layer chromatography serves to visualize the individual components. Cellulose acetate has the added advantage that it can be made water-clear by treatment with an oil or with a mixture of acetic acid and ethanol. This facilitates subsequent measurements of spot densities. Proteins and other macromolecules are visualized by treating the electrophoretogram with a stain or dye that becomes adsorbed on to the solute molecules but which is easily washed off the solute-free areas. Gel electrophoretograms cannot be dried before spraying or dipping, and a precipitant is incorporated into the visualizing reagent to prevent materials dissolving during treatment.

APPLICATIONS OF TRADITIONAL ZONE ELECTROPHORESIS

This is considered to be largely a qualitative technique. Difficulties that arise in obtaining reproducible quantitative data are similar to those encountered in thin-layer chromatography. In addition, adsorption characteristics of dyes on macromolecules are so variable that only semi-quantitative comparisons can be made. These are, however, still very useful in detecting an abnormal distribution of components in a series of routine samples. It is used widely for the analysis of biologically active materials. Proteins, nucleic acids, enzymes, viruses and drugs are among the many classes of compound amenable to separation by this technique. A significant proportion of the applications of electrophoresis lies in the fields of clinical diagnosis and forensic tests. It is particularly useful for the characterization of body fluids such as serum, urine, gastric juices, etc.

Two variations of the basic technique are isoelectric focusing and immunoelectrophoresis. The former offers improved resolution and sharper bands in the separation of weak acids, weak bases and ampholytes through the use of pH and density gradients superimposed along the potential gradient. The latter employs specific antigen–antibody interactions (chapter 10) to visualize the separated components of serum samples

HIGH PERFORMANCE CAPILLARY ELECTROPHORESIS

The technique of hpce, or ce, involves high-voltage electrophoresis in narrow-bore fused-silica capillary tubes or columns and on-line detectors similar to those used in hplc (p. 123). Components of a mixture injected into one end of the tube migrate along it under the influence of the electric field (potential gradient) at rates determined by their electrophoretic mobilities. On passing through the detector, they produce response profiles that are similar to, but generally sharper than, chromatographic peaks. Recorded as a function of time, the peaks of a capillary *electropherogram* resemble a very high-efficiency hplc chromatogram. Efficiencies measured in plate numbers (equation 4.42 or 4.43) approach or exceed 10^6. This is firstly because the peaks are not broadened by mass transfer or multiple-path effects, molecular or longitudinal diffusion being the only significant band-spreading mechanism (p. 83) and secondly because flat flow-profiles are generated by a very pronounced electroosmotic effect (*vide infra*). The factors affecting efficiency are given by the equation

$$N = \frac{\mu E d}{2D}$$

$$(4.55)$$

where D is the diffusion coefficient of the migrating species, d the distance travelled and μ its electrophoretic mobility, E being the applied potential gradient (field). Hence, a high field, short migration time and distance, and small diffusion coefficients result in the highest efficiencies. In general, diffusion coefficients are inversely related to the size of the migrating species; some examples are given in table 4.20. A particular advantage of capillary tubes over traditional electrophoresis systems is their high electrical resistance. This facilitates the use of very high fields which generate only a modest amount of heat that is readily dissipated through the wall of the capillary. High fields result in fast rates of migration and hence short analysis times.

Electroosmosis (p. 167) plays a very significant role in hpce because the interior surface of a quartz capillary develops a negative charge when in contact with aqueous solutions due to the ionization of surface silanol groups (Si–OH) above pH 4 and the adsorption of anions. As a result, a layer of cations from the bulk solution builds up close to the wall to maintain a charge balance by forming an electrical 'double-layer'. The high fields employed

Table 4.20   Diffusion coefficients of species of different sizes

SPECIES	RMM (MW)	D $(cm^2 \, sec^{-1}) \cdot 10^5$
HCl	36	3.05
NaCl	58	1.48
β-alanine	89	0.93
1,2-aminobenzoic acid	137	0.84
glucose	180	0.67
citrate	192	0.66
cytochrome C	13 370	0.11
β-lactoglobulin	37 100	0.075
catalase	247 500	0.041
myocin	480 000	0.011
tobacco mosaic virus	40 590 000	0.0046

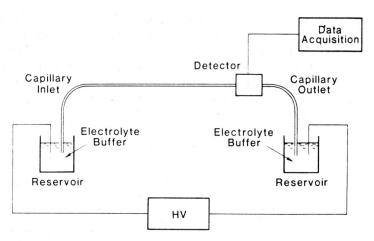

Figure 4.46.   Schematic diagram of a capillary electrophoresis system

cause a pronounced *electroosmotic flow* (EOF) as the highly solvated cations are drawn towards the cathode. All analyte species, whether cationic, anionic or neutral, are carried towards the cathode by the EOF and hence through a detector cell positioned close to the cathodic end of the tube. As the EOF originates at the wall of the capillary, an essentially flat flow-profile is produced across the tube thereby minimizing band-spreading and resulting in very high efficiencies.

A schematic diagram of an hpce system is shown in figure 4.46. The fused-quartz capillary is generally 50 to 75 cm long with an i.d. of 25 to 100 μm and an o.d. of about 400 μm. Like capillary gc columns, it is protected with an outer layer of a polyimide. Potentials of between 10 and 30 kV are applied across the capillary during electrophoresis creating fields of 100 to 500 V cm^{-1}. The on-line detector positioned close to the cathodic end of the capillary is most commonly a UV absorbance or fluorescence monitor or a diode array spectrometer providing absorbance data at multiple wavelengths. The detector cell is simply a small section of the capillary tube from which the polyimide protective coating around the outside has been removed to allow radiation from the detector lamp to pass through. The walls of the tube therefore serve as the optical windows of the cell. The optical path through the tube is only 25 to 75 μm, but it can be extended by creating an enlarged region known as a 'bubble cell' thereby increasing sensitivity by a factor of about three without any loss of efficiency. It should be noted that, although the absolute detection limits in hpce are extremely low, generally 10^{-13} to 10^{-20} g, the concentration limits are comparable to and sometimes poorer than those in hplc due to the very short optical path length through the capillary.

Fluorescence detection can be up to four orders of magnitude more sensitive than UV absorbance, especially where laser induced excitation is used, mass detection limits being as low as 10^{-20} to 10^{-21} mole. Pre- and post-column derivatization methods are being developed to extend the applicability of fluorescence detection to non-fluorescent substances. Several types of electrochemical and mass spectrometric detector have also been designed. Detector characteristics are summarized in table 4.21.

Table 4.21 Detectors for capillary electrophoresis

METHOD	MASS DETECTION LIMIT (moles)	CONCENTRATION DETECTION LIMIT (molar)[a]	ADVANTAGES/ DISADVANTAGES
UV-vis absorption	10^{-13}–10^{-16}	10^{-5}–10^{-8}	• universal • diode array offers spectral information
fluorescence	10^{-15}–10^{-17}	10^{-7}–10^{-9}	• sensitive • usually requires sample derivatization
laser-induced fluorescence	10^{-18}–10^{-20}	10^{-14}–10^{-16}	• extremely sensitive • usually requires sample derivatization • expensive
amperometry	10^{-18}–10^{-19}	10^{-10}–10^{-11}	• sensitive • selective but useful only for electroactive analytes • requires special electronics and capillary modification
conductivity	10^{-15}–10^{-16}	10^{-7}–10^{-8}	• universal • requires special electronics and capillary modification
mass spectrometry	10^{-16}–10^{-17}	10^{-8}–10^{-9}	• sensitive and offers structural information • interface between CE and MS complicated
indirect UV, fluorescence, amperometry	10–100 times less than direct method	—	• universal • lower sensitivity than direct methods

others:
Radioactivity, thermal lens, refractive index, circular dichroism, Raman

[a] assume 10 nl injection volume.

Samples are injected into the capillary tube at the opposite end from the detector using one of two methods, i.e. *hydrodynamic* or *electrokinetic*. These are illustrated in figure 4.47. The end of the tube is dipped into the sample solution and a very small volume (1 to 50 nl) introduced into the capillary either using gravity, positive pressure or a vacuum (hydrodynamic method) or

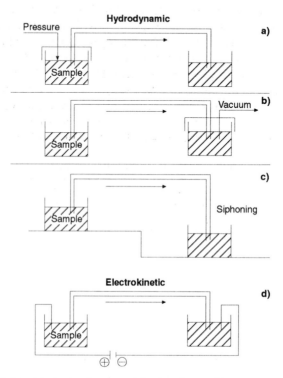

Figure 4.47.   Methods of sample injection for capillary electrophoresis

by applying a voltage across the capillary whilst the other end is immersed in the buffer solution causing electrophoretic migration into the tube (electro-kinetic method). Reproducible injections can be difficult to achieve, especially with electrokinetic injection, because of the number of parameters involved, e.g. time, pressure drop, solution viscosity and capillary dimensions. Typi-cally, the relative precision of injection is two or three per cent, although some systems now available are capable of better than one per cent. With the electrokinetic method, differences in electrophoretic mobility between sample components can result in discrimination effects whereby the composition of the injected sample may not be identical to that of the bulk sample.

Modes of hpce

Different separation mechanisms, which determine selectivity, can be exploited in hpce by appropriate choice of operating conditions. There are four principle *modes of operation* (table 4.22) and it should be noted that in only one, *micellar electrokinetic capillary chromatography (mecc)*, is it possible to separate neutral species from one another.

Capillary zone electrophoresis (cze), sometimes known as *free-solution capillary electrophoresis (fsce)*, is the simplest and curently the most widely used mode. The capillary is filled with a homogeneous buffer solution through which the various species migrate in discrete zones and at different velocities according to their electrophoretic mobilities. However, although species with either the same charge or opposite charges can be separated from one another, all neutral species have zero mobility and move as a single zone at the same velocity as the EOF. Selectivity is controlled and separations optimized mainly through the choice of buffer composition, pH and ionic

Table 4.22 Principal modes of hpce

MODE	SEPARATION MECHANISM	SEPARATION MEDIUM	TYPICAL SAMPLE TYPES
capillary zone electrophoresis (cze) or free-solution capillary electrophoresis (fsce)	differences in electrophoretic mobilities	electrolyte solution	relatively small charged species
micellar electrokinetic capillary chromatography (mecc or mekc)	differences in hydrophobic interactions with micelles (partitioning)	micellar electrolyte	relatively small neutral and charged species
capillary gel electrophoresis (cge)	molecular sieving and differences in electrophoretic mobilities	polyacrylamide or agarose gel	biopolymers
capillary isoelectric focusing (cief)	differences in isoelectric points	electrolyte/ ampholyte pH gradient	proteins

Table 4.23 Buffers commonly used in capillary electrophoresis

NAME	pK_a
phosphate	2.12 (pK_{a1})
citrate	3.06 (pK_{a1})
formate	3.75
succinate	4.19 (pK_{a1})
citrate	4.74 (pK_{a2})
acetate	4.75
citrate	5.40 (pK_{a3})
succinate	5.57 (pK_{a2})
MES	6.15
ADA	6.60
BIS-TRIS propane	6.80
PIPES	6.80
ACES	6.90
MOPSO	6.90
imidazole	7.00
MOPS	7.20
phosphate	7.21 (pK_{a2})
TES	7.50
HEPES	7.55
HEPPS	8.00
TRICINE	8.15
glycine amide, hydrochloride	8.20
glycylglycine	8.25
TRIS	8.30
BICINE	8.35
morpholine	8.49
borate	9.24
CHES	9.50
CHAPSO	9.60
CAPS	10.40
phosphate	12.32 (pK_{a3})

strength. Generally, low ionic strength and high pH favour fast migration and therefore the highest efficiencies and shortest analysis times. Effective buffering is essential because the EOF is a function of pH as is solute charge. Some commonly used buffers are listed in table 4.23. Each one should be used only over a pH range of $pK_a \pm 1$ (p. 48), polybasic acids such as phosphoric and citric being particularly versatile because each has three pK_a values. Buffers having large ions or which are zwitterionic, such as Tris, borate, histidine and CAPS, can be used at high concentrations because they generate relatively low currents, but the high UV absorbance of some is a disadvantage.

Additional substances (*buffer additives*) are often added to the buffer solution to alter selectivity and/or to improve efficiency, and the wall of the capillary may be treated to reduce adsorptive interactions with solute species. Organic solvents, surfactants, urea and *chiral selectors* are among the many additives that have been recommended (table 4.24). Many alter or even reverse the EOF by affecting the surface charge on the capillary wall, whilst some help to solubilize hydrophobic solutes, form ion-pairs, or minimize solute adsorption on the capillary wall. *Chiral selectors* enable racemic mixtures to be separated by differential interactions with the two enantiomers which affects their electrophoretic mobilities. Deactivation of the capillary wall to improve efficiency by minimizing interactions with solute species can be achieved by permanent *chemical modification* such as silylation or the

Table 4.24   Some additives used in capillary electrophoresis

ADDITIVE	EXAMPLE	USE
surfactants (anionic, cationic, or neutral)	SDS, CTAB, BRIJ, TWEEN	• EOF modification • solubilize hydrophobic solutes • ion pairing • MEKC above CMC
zwitterionic substances	MES, TRIS, CHAPS, CHAPSO	• increase ionic strength without increasing conductivity • affect selectivity of proteins
linear hydrophilic polymers	methyl cellulose, polyacrylamide, PEG, PVA	• reduce EOF • minimize sample adsorption at low concentrations • CGE at high concentrations
organic modifiers	methanol, acetonitrile, TFA	• alter EOF (generally reduce) • change selectivity in MEKC and chiral analyses
chiral selectors	cyclodextrins, crown ethers, bile salts	• chiral separations • solubilization of hydrophobic solutes
metal ions	K^+, Na^+, Cu^{2+}, Li^+	• alter selectivity in MEKC and CGE
hydrogen bonding/ solubilizing agents	urea	• melt double stranded DNA in CGE • solubilize proteins • alter selectivity in MEKC
complexing buffers	borate	• carbohydrate and catechol separations
quaternary amines	diaminopropane	• ion pairing • EOF reversal

adsorption of a polymeric coating. Alternatively, *dynamic deactivation* by buffer additives has the advantage of the modification being constantly renewed during electrophoresis so providing a more stable surface. Both approaches can eliminate or even reverse the EOF.

Cze is finding a wide range of applications, mostly in the biochemical, clinical and pharmaceutical fields where it is becoming a rival to hplc as the technique of choice. Amino acids, peptides and proteins, including glycoproteins have all been successfully separated, generally with higher efficiencies, better resolution, and more quickly than by hplc and at nanogram to picogram levels. Peptide mapping, or protein fingerprinting, where the identity of a protein is established from the peptide sequence after chemical or enzymatic cleavage into peptide fragments, the detection of drugs and their metabolites in biological fluids such as blood, plasma and urine, and the determination of inorganic cations and anions in aqueous samples are examples of the increasing use of this technique.

Micellar electrokinetic capillary chromatography (*mecc* or *mekc*) is a more versatile technique than cze due to its ability to separate neutral as well as ionic species. The term chromatography is used because a surfactant added to the buffer solution forms spherical aggregates of molecules or *micelles* that act as a second or pseudo-stationary phase with which solute species can interact. The interaction is hydrophobic and/or electrostatic and is analogous to a sorption mechanism in chromatography. The hydrophobic ends of the surfactant molecules are oriented inwards towards the centres of the micelles and the polar or ionic hydrophilic ends point outwards forming a charged surface which is in contact with the buffer solution. As the micelles are usually either cationic or anionic they migrate during electrophoresis, but if the buffer solution is neutral or basic the EOF carries all species towards the cathode and past the detector, as in cze. Neutral solutes interact with the micelles to varying degrees, not unlike partitioning in chromatography, and thus have different electrophoretic mobilities. The more hydrophobic a solute species is the more strongly it interacts with or partitions into the micelles. Consequently, for an anionic surfactant such as SDS (sodium dodecyl sulphate), which is widely used and which migrates against the EOF, the most hydrophobic neutral solutes have the longest 'retention times' migrating or 'eluting' with the same velocity as the micelles. Hydrophilic neutral solutes that do not interact with the micelles have the shortest retention times, eluting with the EOF. Selectivity can be varied widely in mecc by changing the surfactant and its concentration provided a minimum molar concentration known as the *critical micelle concentration* (CMC) is always exceeded. Examples of commonly used surfactants and their CMC values are given in table 4.25. Micelle formation and micelle-solute interactions can be dramatically affected by adding electrolytes and organic solvents such as methanol or acetonitrile, and

Table 4.25   Surfactants used in micellar electrokinetic capillary chromatography

SURFACTANT[a]	TYPE	CMC IN WATER (M)	AGGREGATION NO.
SDS	anionic	8.1×10^{-3}	62
CTAB	cationic	9.2×10^{-4}	61
Brij-35	nonionic	1.0×10^{-4}	40
sulfobetaine	zwitterionic	3.3×10^{-3}	55

[a] SDS, sodium dodecyl sulfate; CTAB, cetyltrimethylammonium bromide; Brij-35, polyoxyethylene-23-lauryl ether; sulfobetaine, N-dodecyl-N,N-dimethyl-ammonium-3-propane-1-sulfonic acid.

changing the buffer composition, pH and temperature. As in cze, these factors also influence the EOF by affecting the charge on the capillary wall.

Mecc is being applied in many areas where neutral solutes are to be separated. These include compounds of environmental interest, pharmaceuticals, drugs of abuse and nucleic acids. The high efficiency coupled with chromatographic partitioning afforded by the micellar pseudo-stationary phase gives it great versatility including chiral recognition through the use of chiral surfactants and chiral additives.

Capillary gel electrophoresis (*cge*) is a variation where selectivity based on molecular size and shape is introduced by filling the capillary tube with a polymeric material, usually a cross-linked polyacrylamide or agarose gel. A molecular sieving mechanism is therefore superimposed onto the basic electrophoretic process as is done in traditional gel slab or column separations. The larger the solute species, the slower its rate of migration, hence elution is in order of increasing relative molecular mass. A polymer-gel filled capillary provides additional advantages by minimizing band-spreading through solute-diffusion, preventing solute adsorption onto the capillary wall and eliminating electroosmosis. The latter results in the maximum resolution in the shortest possible distance. More recently, linear polymers capable of forming entangled polymer networks inside the capillary have been used. Very careful control of conditions both in gel formation and during electrophoresis is necessary to achieve acceptable reproducibility. The formation of bubbles or voids in the filled capillary can interrupt current flow, whilst excessive temperature fluctuations can degrade performance by causing movement of the polymer filling.

The main applications of cge are in separating protein fractions and oligonucleotides, and for DNA sequencing. Cyclodextrins have been incorporated into some polymers to introduce chiral selectivity.

Capillary isoelectric focusing (*cief*) separates amphoteric substances such as peptides and proteins on the basis of differences in their *isoelectric points*

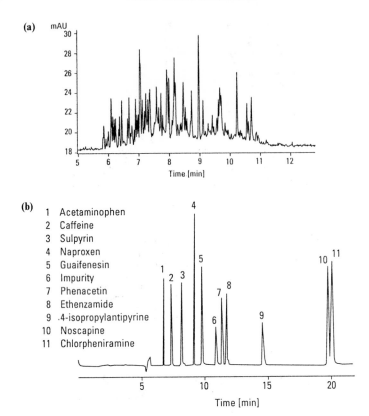

Figure 4.48. Some examples of separations by hpce. (a) CZE BSA peptide map. Conditions: 20 mM phosphate, pH 7, V = 25 kV, i = 16 μA, l = 50 cm, L = 57 cm, id = 50 μm with 3X extended pathlength detection cell, λ = 200 nm. (b) MECC separation of cold-relief medicine constituents. Conditions: 20 mM phosphate-borate, 100 mM SDS, pH 9, V = 20 kV, L = 65 cm, i.d. = 50 μm, λ = 210 nm. (c) CGE of 1 kbp ladder using minimally crosslinked polyacrylamide. Conditions: Bis-crosslinked polyacrylamide (3% T, 0.5% C), 100 mM Tris-borate, pH 8.3, E = 250 V/cm, i = 12.5 μA, l = 30 cm, L = 40 cm, i.d. = 75 μm, λ = 260 nm, polyacrylamide coated capillary. (d) CIEF of standard protein mixture. Polyacrylamide coated capillary

(*pI values*) defined as the pH at which the molecule is uncharged. A pH gradient is formed along the capillary using a mixture of *carrier ampholytes* having pI values spanning the desired pH range, typically 3 to 10. This is achieved by filling the capillary with a solution of the sample and ampholytes and applying a potential field. All charged species migrate along the capillary, cations and anions in opposite directions, until they become 'focused' at a point where they are uncharged, i.e. pI = pH. The zone occupied by each sample species is self-sharpening because diffusion away from the focal point

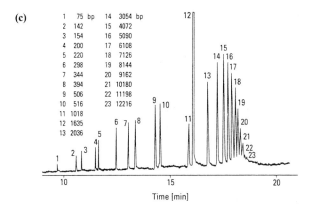

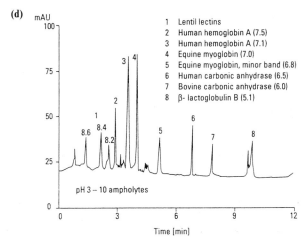

Figure 4.48—*Continued*

causes them to acquire a charge which results in a reversal in the direction of movement. The current flow virtually ceases when all species have become 'focused' at their respective isoelectric points after which they are passed sequentially through the detector by applying pressure at one end of the capillary or by adding a salt to one of the buffer reservoirs. Minimizing the EOF and adsorption effects by coating the walls of the capillary with a polymer is necessary to avoid disturbance of the separating species until focusing is complete. Larger samples can be analysed than in other modes of hpce, but this is limited as some proteins may precipitate at their isoelectric points.

Figure 4.48 illustrates some typical separations employing different modes of hpce.

References

1. CRAIG, L. C. and CRAIG, D., in *Techniques in Organic Chemistry*, A. Weissberger (Ed.), Vol. III, p. 248, Interscience, New York, 1956.
2. TSWETT, M., *Warsaw Soc. Nat. Sci.*, Biol. Sec., **14**, No. 6, 1903.
3. VAN DEEMTER, J., ZUIDERWEG, F. J. and KLINKENBERG, A., *Chem. Sci.*, **5**, 271, 1949.
4. GIDDINGS, J. C., *Dynamics of Chromatography*, Part 1, Marcel Dekker, New York, 1965.

Problems

1. Compound X has a distribution ratio of 2.7 for benzene/water. For an aqueous sample containing 4.5 mg of X per 50 cm^3, calculate:
   (a) the volume of benzene required to remove 99.0 % of X with a single extraction, and
   (b) the number of extractions with 50 cm^3 of benzene required to remove 99.0 % of X.

2. A 100 cm^3 volume of 6 M HCl contains 0.200 g of SbCl$_5$ for which the distribution ratio into diethyl ether is 4.25. If 25 cm^3 of ether is used for each extraction, how many are required to remove (a) 99.0 % and (b) 99.9 % of the SbCl$_5$ from the acid solution?

3. Acetylacetone (AcAc) is a chelating agent for many metals. 50 cm^3 of an aqueous solution of M^{2+} (5×10^{-3} M) is equilibrated with 20 cm^3 of ether containing an excess of AcAc. If 94 % of M is extracted into the ether, calculate the value of the distribution ratio given that M$^{2+} + 2AcAc^- \rightarrow M(AcAc)_2$ is the only reaction.

4. Use the following data to estimate the distribution coefficient, K_D, and the acid dissociation constant, K_a, for trifluoroacetylacetone distributed between water and chloroform:

$\log_{10}D$	0.3010	0.3010	0.3010	0.1461	-0.0458	-0.3979	-0.6989
pH	1.16	2.09	3.25	6.29	6.68	7.40	8.00

5. The following data relate to the distribution of 8-hydroxyquinoline (HOx) between chloroform and water:
$$K_{H_2Ox^+} = 8 \times 10^{-6} \qquad K_{HOx} = 1.4 \times 10^{-10} \qquad K_{DHOx} = 720$$
Calculate the distribution ratios and the concentrations of HOx remaining in the aqueous phase after equilibration of 0.1 % aqueous solutions at pH values of 1, 6 and 10 with equal volumes of chloroform.

6. Indium, cadmium and silver can be extracted into chloroform as their 8-hydroxyquinoline complexes, and the pH$_{\frac{1}{2}}$ values for these metals are 2.1, 6.3 and 8.8 respectively. Plot a graph of theoretical percent extraction against pH over the range 0 to 9 for each metal. Deduce the pH of incipient extraction (0.01 %) and complete extraction (99.99 %) for each metal, and comment on the feasibility of separating each from the other assuming that all the distribution coefficients are sufficiently high.

7. A gc column packed with a non-polar stationary phase gave the following retention times for a series of hydrocarbons:

Compound	t'_R (min)
methane and air	1.8
ethane	2.4
n-propane	3.6
propylene	4.3
isobutane	5.5
n-butane	7.5
isobutylene	8.6
trans-2-butene	10.6
cis-2-butene	12.3
isopentane	13.6
n-pentane	18.2

What conclusions can be drawn from a plot of log t_R versus carbon number?

8. Calculate the number of theoretical plates required to achieve baseline separation $(R_s = 1.5)$ for an $\alpha(= k'_1/k'_2)$ value of 1.10 in

(a) a packed column where $k'_2 = 50$
(b) a packed column where $k'_2 = 5$
(c) a capillary column where $k'_2 = 2$
(d) a capillary column where $k'_2 = 0.5$
(e) as (b) but for $\alpha = 1.2$.

9. The following values for plate height and gas velocity were obtained for n-hexane on a 2-metre Apiezon-L column:

H (cm)	0.635	0.510	0.423	0.465	0.552	0.632	0.692	0.749
$\bar{u}$(cm s^{-1})	0.91	1.51	3.0	4.2	5.55	7.0	8.0	9.0

Plot a graph of H versus $\bar{u}$ and calculate the optimum gas velocity, the corresponding minimum plate height and the maximum plate number.

10. The separation of two compounds on a packed and a capillary column gave the following data:

	packed column	capillary column
dead volume, cm^3	15	3.0
adjusted retention volume, cm^3		
component A	160	1.1
component B	170	1.2
number of theoretical plates	6 400	25 600

Which column is giving the better resolution and why?

11. A mixture of methyl esters of fatty acids was chromatographed on a Carbowax 20 M column giving the following peak areas and detector response factors:

ester	peak area, cm^2	response factor
methyl-n-butyrate	2.95	0.81
methyl-iso-valerate	0.86	0.88
methyl-n-octanoate	1.66	0.98
methyl-n-decanoate	4.52	1.00

Calculate the percentage composition of the mixture by internal normalization.

12. An hplc separation of a two-component pharmaceutical product yielded the following data:

compound	retention time (chart distance) mm	peak width at base mm	peak width at $\frac{1}{2}$-height mm
solvent	30	—	—
aspirin	75	6.5	3.0
caffeine	86	8.1	3.25

Calculate (a) the capacity factor, k', for each compound;
          (b) the plate number for each compound using base widths;
          (c) the plate number for each compound using widths at $\frac{1}{2}$-height;
          (d) the resolution of the two compounds, R_s.

13. Two compounds were separated by hplc with an R_s value of 0.75, the plate number for the second compound being 4 500. Calculate the number of plates required to obtain resolutions of (a) 1.0 and (b) 1.5.

14. The distances travelled by five compounds and the solvent front after a tlc separation on silica gel were as follows:

compound	distance travelled, cm
solvent	12.5
methyl stearate	9.1
cholesterol	1.5
α-tocopherol (vitamin E)	5.6
methyl oleate	9.1
squalene (hydrocarbon)	10.3

Calculate the R_f values of each compound and comment on the relative values.

Further Reading

ANDERSON, R., *Sample Pre-treatment and Separation*, Wiley, Chichester, 1987.
BRAITHWAITE, A. and SMITH, F. J., *Chromatographic Methods*, 4th Ed., Chapman and Hall, London, 1985.
DENNY, R. C., *A Dictionary of Chromatography*, 2nd Ed., Macmillan, London, 1982.
GILBERT, M. T., *High Performance Liquid Chromatography*, Wright, Bristol, 1987.
GROB, R. L., (Ed.), *Modern Practice of Gas Chromatography*, 2nd Ed., Wiley, New York, 1985.
GROSSMAN, P. D. and COLBURN, J. C., *Capillary Electrophoresis, Theory and Practice*, Academic Press, New York, 1992.
HAMILTON, R. J. and SEWELL, P. A., *Introduction to High Performance Liquid Chromatography*, 2nd Ed., Chapman and Hall, London, 1982.
KIRCHNER, J. G., (PERRY, E. S., (Ed.)), *Thin Layer Chromatography*, 2nd Ed., Wiley, New York, 1978.
KNOX, J. H., *High Performance Liquid Chromatography*, Edinburgh University Press, 1978.
LEE, M. L., YOUNG, F. J. and BARTLE, K. D., *Open Tubular Column Gas Chromatography, Theory and Practice*, Wiley, New York, 1984.

LINDSAY, S., *High Performance Liquid Chromatography*, Wiley, Chichester, 1987.
MELVIN, M., *Electrophoresis*, Wiley, Chichester, 1987.
PERRY, J. A., *Introduction to Analytical Gas Chromatography, History, Principles and Practice* (Chromatography Science Series, Vol. 14), Marcel Dekker, New York, 1981.
SEWELL, P. and CLARK, B., *Chromatographic Separations*, Wiley, Chichester, 1987.
WILLETT, J. E. *Gas Chromatography*, Wiley, Chichester, 1987.

Chapter 5

Titrimetry and Gravimetry

Titrimetric and gravimetric analyses, which are largely derived from the application of the principles of solution chemistry, represent some of the traditional aspects of chemical analysis. The methods remain widely used however because of their simple means of operation which enable comparatively unskilled operators to achieve precise results. The most significant modern development has been the use of multidentate organic chelating agents which has resulted in a substantial widening of the scope of these types of analysis. In this chapter, emphasis will be given to these more recent applications.

5.1 Titrimetry

SUMMARY

Principles

Fast solution reactions between analyte and a reagent; titration to stoichiometric point by volumetric or coulometric methods; end point detection by visual indicators, precipitation indicators or electrochemical means.

Apparatus

Burettes, pipettes, volumetric flasks, analytical quality chemical balance, indicator electrodes and coulometric generating electrodes.

Applications

Very widespread for precise routine and non-routine analysis in industrial and research laboratories. Typical uses: determination of acidic and basic impurities in finished products, control of reaction conditions in industrial processes, mineral and metallurgical analysis. Relative precision 0.1 to 1%.

Disadvantages

Storage of large volumes of solutions, instability of some reagent solutions, need for scrupulously clean glassware.

A titrimetric method involves the controlled reaction of a standard reagent in known amounts with a solution of the analyte, in order that the *stoichiometric* or *equivalence point* for the reaction between the reagent and the analyte may be located. If the details of the reaction are known and the stoichiometric point is located precisely, the amount of analyte present may be calculated from the known quantity of standard reagent consumed in the reaction. In most cases a standard reagent solution is prepared and added manually or automatically from a burette; an alternative procedure is coulometric generation of the reagent *in situ*. The stoichiometric point may be detected by use of a visual indicator or by an electrochemical method (chapter 6).

DEFINITIONS

Titration. The overall procedure for the determination of the stoichiometric or equivalence point.

Titrant. The solution added or reagent generated in a titration.

Titrand. The solution to which the titrant is added.

End Point. A point in the progress of the reaction which may be precisely located and which can be related to the stoichiometric or equivalence point of the reaction; ideally, the two should be coincident.

Indicator. A reagent or device used to indicate when the end point has been reached.

TITRIMETRIC REACTIONS

It is clear that reactions suitable for use in titrimetric procedures must be stoichiometric and must be fast if a titration is to be carried out smoothly and quickly. Generally speaking, ionic reactions do proceed rapidly and present few problems. On the other hand, reactions involving covalent bond formation or rupture are frequently much slower and a variety of practical procedures are used to overcome this difficulty. The most obvious ways of driving a reaction to completion quickly are to heat the solution, to use a catalyst, or to add an excess of the reagent. In the last case, a *back titration* of the excess reagent will be used to locate the stoichiometric point for the primary reaction. Reactions employed in titrimetry may be classified as acid–base; oxidation–reduction; complexation; substitution; precipitation.

End Point Detection

A prerequisite for a precise and accurate titration is the reproducible identification of an *end point* which either coincides with the stoichiometric point of the reaction or bears a fixed and measurable relation to it. An end point may be located either by monitoring a property of the titrand which is removed when the stoichiometric point is passed, or a property which can be readily observed when a small excess of the titrant has been added. The most common processes observed in end point detection are change of colour; change of electrical cell potential; change of electrical conductivity; precipitation or flocculation. (Electrochemical methods are discussed in chapter 6; precipitation indicators find only limited use.)

Visual Indicators

The use of a colour change to indicate the end point is common to a wide variety of titrimetric methods. Visual detection of end points is a major factor in maintaining the simplicity of titrimetry, hence the capability of the human eye to detect colour change plays an important role in these techniques.

In general terms a visual indicator is a compound which changes from one colour to another as its chemical form changes with its chemical environment

$$In_A = In_B + nX \qquad (5.1)$$

$$\text{colour 1} \qquad \text{colour 2}$$

where X may be H^+, M^{n+} or e^-, and the colour of the indicator is sensitive to the presence of H^+, M^{n+}, oxidants or reductants. An *indicator constant* is defined as

$$K_{In} = \frac{[In_B][X]^n}{[In_A]} \qquad (5.2)$$

whence

$$[X]^n = K_{In}([In_A]/[In_B]) \qquad (5.3)$$

and

$$npX = pK_{In} + \log_{10}([In_B]/[In_A]) \qquad (5.4)$$

If the indicator is present in an environment where a titration reaction generates or consumes the X species, the indicator will change with the concentration of X in the solution and the colour of the solution will be determined by the ratio $[In_B]/[In_A]$. As a general guide, the eye will register a complete change from one colour to the other when this ratio changes from 10:1 to 1:10. Substitution in equation (5.4) enables the concentration range of X over which the indicator will change colour to be calculated, i.e.

$$npX = pK_{In} \pm 1 \tag{5.5}$$

For example, the acid–base indicator methyl orange has a pK_{In} of 3.7 and will thus change colour over the pH range 2.7 to 4.7. The ultimate sharpness of the end point will further depend upon the rate at which pX is changing at the end point of the titration. The additional factors involved in determining this rate of change are examined later in the discussions of specific titration methods. Because the indicator competes with the analyte and reagent for X it is obvious that the indicator concentration must be kept as low as possible in order to minimize interference with the analyte–reagent equilibrium. It follows that the colours exhibited by an indicator must be of a high intensity.

Apparatus for Titrimetric Analysis
In general, the apparatus for titrimetric analysis is simple in construction and operation. A typical analysis procedure would involve measurement of the amount of sample either by mass or volume, and then addition of the titrant from a burette or micro-syringe. Apart from visual indication, the course of a titration may be followed by electrochemical or photometric means; in neither is the equipment required complex. A simple valve voltmeter, or conductivity bridge will suffice on the one hand, and a simple spectrophotometer or filter photometer with minor modifications on the other. Varying degrees of automation may be incorporated.

ACID–BASE TITRATIONS

Neutralization reactions between Lowry-Brønsted acids and bases are frequently employed in chemical analysis. Methods based on them are sometimes termed *acidimetric* or *alkalimetric*.

Visual Indicators for Acid–Base Titrations
Table 5.1 summarizes the details of some useful acid–base indicators. Exact agreement with the pH range expressed by equation (5.5) is by no means always observed. This is because some colour changes are easier to see than others and so the general approximation made in deriving equation

Table 5.1   A range of visual indicators for acid–base titrations

INDICATOR	pK_{In}	LOW pH COLOUR	HIGH pH COLOUR	EXPERIMENTAL COLOUR CHANGE RANGE/pH
cresol red	ca. 1	red	yellow	0.2–1.8
thymol blue	1.7	red	yellow	1.2–2.8
bromo-phenol blue	4.0	yellow	blue	2.8–4.6
methyl orange	3.7	red	yellow	3.1–4.4
methyl red	5.1	red	yellow	4.2–6.3
bromo-thymol blue	7.0	yellow	blue	6.0–7.6
phenol red	7.9	yellow	red	6.8–8.4
phenolphthalein	9.6	colourless	red	8.3–10.0
alizarin yellow R	ca. 11	yellow	orange	10.1–12.0
nitramine	ca. 12	colourless	orange	10.8–13.0

(5.5) is not uniformly close. Structurally, the indicators form three groups: phthaleins (e.g. phenolphthalein); sulphonephthaleins (e.g. phenol red); and azo compounds (e.g. methyl orange).

(a) phenolphthalein

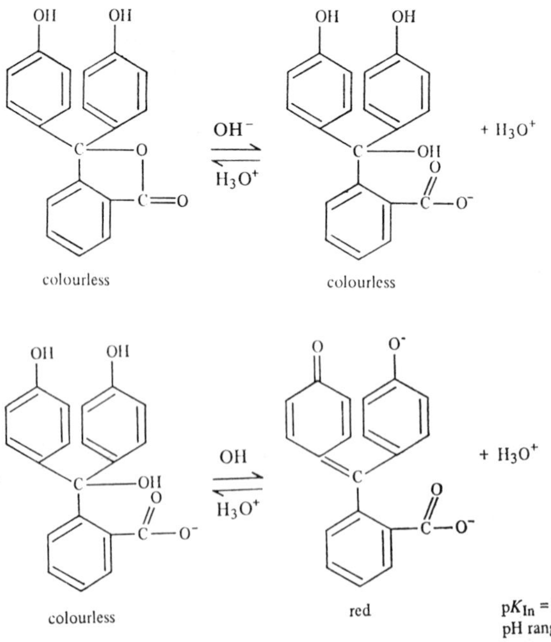

colourless

colourless

colourless   red

$pK_{In} = 9.6$
pH range 8.3 to 10.0

(b) phenol red (phenol sulphonephthalein)

Of the two colour changes undergone by this indicator only the one at pH 6.8 to 8.4 is commonly employed.

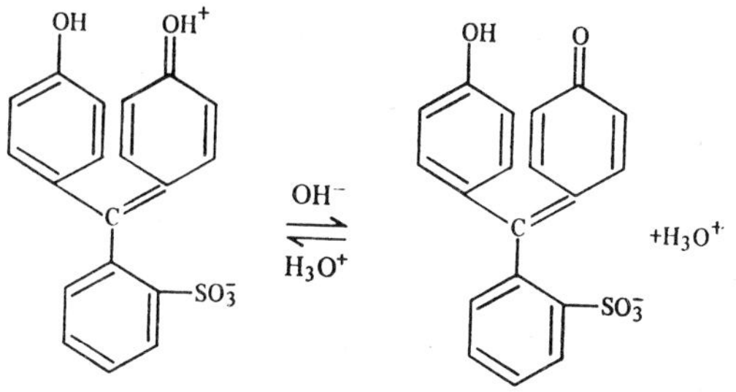

red

$pK_{In} = 1.5$
pH range 0.5 to 2.5

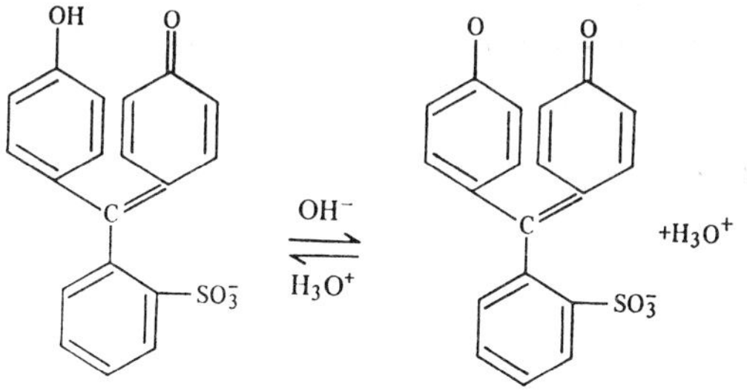

yellow

yellow

$pK_{In} = 7.9$
pH range 6.8 to 8.4

red

(c) methyl orange

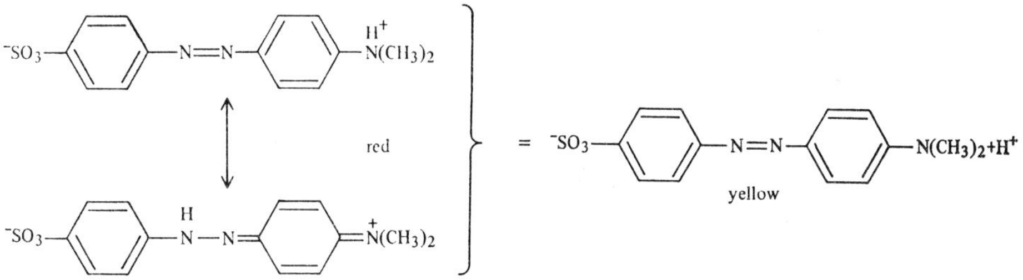

$pK_{In} = 3.7$
pH range 3.1 to 4.4

To select an indicator for an acid–base titration it is necessary to know the pH of the end point before using equation (5.5) or standard indicator tables. The end point pH may be calculated using equations (3.27), (3.29) or (3.30). Alternatively, an experimentally determined titration curve may be used (see next section). As an example, consider the titration of acetic acid (0.1 mol dm^{-3}), a weak acid, with sodium hydroxide (0.1 mol dm^{-3}), a strong base. At the end point, a solution of sodium acetate (0.05 mol dm^{-3}) is obtained. Equation (3.28) then yields

$$pH = 7.0 + 2.4 - 0.65 = 8.75$$

thus phenolphthalein is a suitable indicator (table 5.1).

Acid–Base Titration Curves

If the pH of the titrand is monitored throughout a titration, a graph of pH against amount of titrant added may be constructed. The characteristics of this curve are important in the selection of suitable titration conditions and indicators. Of particular importance are the position of the point of inflexion representing the neutralization point, the slope of the curve in the end point region, and the size of the end point 'break'. The influences of concentration and the strength of the acid or base are summarized in figures 5.1, 5.2 and 5.3.

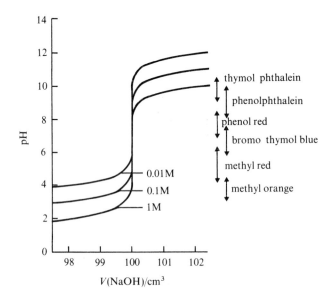

Figure 5.1.   The effect of concentration on titration curves and indicators for a strong acid (HCl) and a strong base (NaOH). 100 cm^3 of HCl is being titrated with NaOH of the same molarity in each case

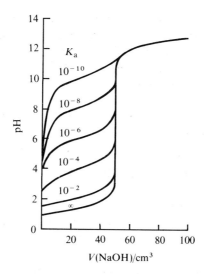

Figure 5.2.   The effect of K_a on the titration curves for weak acids with a strong base. 50 cm³ of 0.1 M acid is being titrated with 0.1 M NaOH in each case

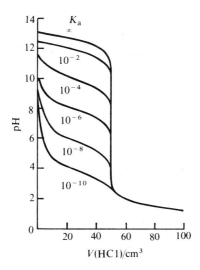

Figure 5.3.   The effect of K_b on the titration curves for weak bases with a strong acid. 50 cm³ of 0.1 M base is being titrated with 0.1 M HCl in each case

Where both the acid and the base are strong electrolytes, the neutralization point will be at pH $= 7$ and the end point break will be distinct unless the solutions are very dilute ($<10^{-3}$ mol dm^{-3}). The composition of the titrand at any point in the titration may be computed from the total amount of acid and base present. However, when one of the reactants is a weak acid or base the picture is less clear. The incomplete dissociation of the acid or base and the hydrolysis of the salt produced in the reaction must be taken into account when calculations of end points and solution composition are made. These points have been considered in chapter 3 and are used in the indicator selection procedure outlined in the preceding section of this chapter.

Acid–Base Titrations in Non-aqueous Solvents

Non-aqueous ionizing solvents can sometimes be used with advantage in acidimetry and alkalimetry. A basic solvent can encourage the loss of a proton from a very weak acid making it effectively much stronger. Similarly an acidic solvent can be used to amplify the basic nature of a solute. Manipulation of these principles enables the scope of acidimetry and alkalimetry to be extended to compounds that cannot be titrated in an aqueous medium. Standard reagents need to be selected with care as they must resist attack by the solvent as well as functioning as an acid or base in the appropriate solvent system. Tables 5.2 and 5.3 summarize data for non-aqueous titration media.

Table 5.2   Acidic solvents
Reagents: perchloric acid (acid) and potassium hydrogen phthalate (base)

SOLVENT	ANALYTE	INDICATOR
iso-propyl alcohol/ ethylene glycol	sodium carboxylates	phenol red
acetic acid	amines, heterocyclic bases, amides, urea	crystal violet neutral red
nitromethane/ acetic anhydride	very weak bases	methyl violet neutral red

APPLICATIONS OF ACID–BASE TITRATIONS

The importance of acids and bases in industrial processes is almost impossible to overstate. Correct pH conditions are often essential to the progress of manufacturing reactions. At the same time, it may be highly undesirable for the product to contain excessive acid or base. In the food and petroleum

Table 5.3 Basic solvents
Reagents: benzoic acid (acid) and tetra-*n*-butyl ammonium hydroxide (base)

SOLVENT	ANALYTE	INDICATOR
n-butylamine	carboxylic acids, phenols	thymol blue
ethylenediamine	amine hydrochlorides, phenols	azo violet
N,N-dimethyl formamide	phenols, salts and sulphonamides	azo violet
		thymol blue

industries, constant monitoring of acids and bases in both reaction mixtures and finished products is a common requirement which titrimetric methods frequently fulfil. At the process control stage, automatic methods are increasingly becoming the rule, and some details of such techniques are given in chapter 12. It is probable that the acidity of basicity present is derived from a mixture of acids or bases. The convention of *acid* or *base number* is used to overcome any ambiguity that might arise in the expression of acid or base content. Thus the acid or base content might be expressed as the number of milligrams of sulphuric acid or potassium hydroxide equivalent to 1 dm^3 or 1 kg of the sample, irrespective of the actual acids present. Acid–base titrations may be used also in analytical methods where the analyte reacts to produce a stoichiometric amount of an acid or base which can then be titrated. Probably the most well-known example is the Kjeldahl method for nitrogen determination in organic compounds. Subsequent to the destruction of carbonaceous material with sulphuric acid, the mixture is made strongly alkaline with sodium hydroxide. The ammonia produced from the original nitrogenous sample is then distilled off and absorbed in an excess of hydrochloric or boric acid. Finally the excess acid is back titrated to complete the analysis.

REDOX TITRATIONS

The concept of 'reduction potential' is introduced in chapter 6. When the reduction potentials of two species differ by 0.1 V or more, the resulting redox reaction will proceed rapidly and stoichiometrically so that it may be used as the basis for a titrimetric procedure. The end point of a redox titration may be observed by following the potential of the titrand with an indicator electrode or with a visual indicator. In two special cases, the reagent (potassium permanganate and iodine) is self-indicating (vide infra).

The properties and behaviour of some important redox reagents are summarized in table 5.4. When assessing these data three main points should be borne in mind.

Table 5.4   Some representative redox reagents

Oxidizing Agents

REAGENT AND FORMAL VALENCY	HALF REACTION	$E^\ominus$ VOLTS	CONDITIONS
manganese(VII) $KMnO_4$	$MnO_4^- + 8H^+ + 5e^- = Mn^{2+} + 4H_2O$	1.51	strong acid
	$MnO_4^- + 4H^+ + 3e^- = MnO_2 + 2H_2O$	1.69	weak acid/ neutral
	$MnO_4^- + e^- = MnO_4^{2-}$	0.56	strong base
cerium(IV) $Ce(SO_4)_2$	$Ce^{4+} + e^- = Ce^{3+}$	1.44	sulphuric acid solution
chromium(VI) $K_2Cr_2O_7$	$Cr_2O_7^{2-} + 14H^+ + 6e^- = 2Cr^{3+} + 7H_2O$	1.33	strong acid
iodine(V) KIO_3	$IO_3^- + 2Cl^- + 6H^+ + 4e^- = ICl_2^- + 3H_2O$	1.23	strong hydrochloric acid
bromine(V) $KBrO_3$ ($+KBr$)	$BrO_3^- + 5Br^- + 6H^+ = 3Br_2 + 3H_2O$	1.05	dilute acid
	$Br_2(aq) + 2e^- = 2Br^-$	1.09	dilute acid

Reducing Agents

REAGENT AND FORMAL VALENCY	HALF REACTION	$E^\ominus$ VOLTS	CONDITIONS
iron(II) $FeSO_4$	$Fe^{3+} + e^- = Fe^{2+}$	0.771	sulphuric acid
arsenic(III) H_3AsO_3	$H_3AsO_4 + 2H^+ + 2e^- = H_3AsO_3 + H_2O$	0.559	acid
titanium(III) $TiCl_3$	$TiO^{2+} + 2H^+ + e^- = Ti^{3+} + H_2O$	0.10	acid
sulphur(II) $Na_2S_2O_3$	$S_4O_6^{2-} + 2e^- = 2S_2O_3^{2-}$	0.08	neutral or dilute acid
carbon(III) $H_2C_2O_4$	$2CO_2(g) + 2e^- + 2H^+ = H_2C_2O_4$	-0.49	dilute sulphuric acid

(1) The division into oxidizing and reducing reagents has been made only on the basis of the common ways in which the reagents are employed. In principle it is possible to define oxidizing and reducing agents only for a specified reaction.

(2) Most half reactions involve large numbers of hydrogen ions and are therefore pH dependent.

(3) Reduction potentials are thermodynamic quantities and cannot be used to predict the rate at which a redox reaction will occur.

Visual Indicators for Redox Titrations

The behaviour of a reversible visual indicator in a redox titration may be represented by

$$In_{ox} + ne^- + nH^+ = In_{red}$$
$$colour\ 1 \qquad\qquad\qquad colour\ 2$$

(where In_{ox} and In_{red} represent the oxidized and reduced forms of the indicator). Nearly all indicator reactions involve hydrogen ions and the observed end points are therefore pH dependent. The potential for an indicator solution is given by substitution into the *Nernst equation* (chapter 6).

$$E = E_{In}^{\ominus} - \frac{0.059\ V}{n} \log_{10} \frac{[In_{red}]}{[In_{ox}][H^+]^n} \qquad (5.9)$$

Using the criteria for indicator colour change applied in developing equation (5.2), the potential range over which the indicator changes can be computed from an analogous equation

$$E = E_{In}^{\ominus} \pm \frac{0.059\ V}{n} - (0.059\ V) \log_{10} (1/[H^+]^n) \qquad (5.10)$$

If $n = 1$, the range of values of E over which a colour change is observed is 0.118 volt and is shifted by 0.059 volt for each unit change in pH. $E_{In}^{\ominus}$ is often known as the *transition potential* for the indicator. A representative list of redox indicators is given in table 5.5.

Table 5.5   Some common redox indicators

INDICATOR	OXIDIZED COLOUR	REDUCED COLOUR	TRANSITION POTENTIAL (VOLTS)	SOLUTION CONDITIONS
5-nitro-1,10-phenanthroline iron(II) complex	pale blue	red–violet	+1.25	H_2SO_4 (1 mol dm^{-3})
2,3'-diphenylamine dicarboxylic acid	blue–violet	colourless	+1.12	H_2SO_4 (7–10 mol dm^{-3})
1,10-phenanthroline iron(II) complex	pale blue	red	+1.11	H_2SO_4 (1 mol dm^{-3})
erioglaucin A	bluish red	yellow–green	+0.98	H_2SO_4 (0.5 mol dm^{-3})
diphenylamine sulphonic acid	red–violet	colourless	+0.85	dilute acid
diphenylamine	violet	colourless	+0.76	dilute acid
p-ethoxychrysodine	yellow	red	+0.76	dilute acid
methylene blue	blue	colourless	+0.53	acid (1 mol dm^{-3})
indigo tetra-sulphonate	blue	colourless	+0.36	acid (1 mol dm^{-3})
phenosafranine	red	colourless	+0.28	acid (1 mol dm^{-3})

One important group of colour indicators is derived from $1:10$ phenantholine (*ortho*-phenanthroline) which forms a $3:1$ complex with iron(II). The complex known as 'ferroin' undergoes a reversible redox reaction accompanied by a distinct colour change

$$E^{\Theta} = +1.06 \text{ V} \qquad (5.11)$$

blue   red

This indicator functions well giving a sharp colour change and is also resistant to oxidative decomposition. The transition potential may be modified by ring substitution with nitro or methyl groups (table 5.5).

A second important group includes the diphenylamine indicators. In the presence of a strong oxidizing agent, diphenylamine is irreversibly converted to diphenylbenzidine. This latter compound undergoes a reversible redox reaction accompanied by a colour change,

diphenylamine   diphenylbenzidine
(colourless)   (colourless)

$$E^{\Theta} = +0.76 \text{ V}$$

diphenyl benzidine violet
(violet) (5.12)

The use of a sulphonic acid derivative of diphenylamine overcomes the problem of low indicator solubility.

Selection of a Visual Indicator for a Redox Titration
Because of the relatively small number of indicators available and their pH dependence, selection is not as straightforward as in the case of acid–base titrations. For example, iron(II) may be titrated with cerium(IV) or chromium(VI) (table 5.4), whilst equation (5.9) in conjunction with table 5.5

will suggest suitable indicators. *p*-Ethoxychrysodine, diphenylamine and diphenylamine sulphonic acid will all suffice in dilute acid whilst both erioglaucin A and 1,10-phenanthroline-iron(II) have been used in stronger acid. A further problem arises, however, as a result of the complexing action of many anions. Where one oxidation state in a redox equilibrium is preferentially complexed the reaction will be displaced and the potential shifted accordingly. Notable in this complexing action is phosphoric acid, a reagent which can be used to advantage in some instances. The titration of iron(II) in a phosphoric acid medium is discussed in a later section. As a result of the uncertainties outlined above, selecting an indicator which will change at exactly the right point can be difficult, and large indicator blanks may have to be tolerated. Experimentally determined titration curves will generally be of considerable help in the selection of an indicator.

Self Indicating Reagents and Specific Indicators

Potassium permanganate and iodine, which are important redox reagents are both self indicating, i.e. the colour of the reagent in each case is intense and will impart a perceptible colour to a solution when present in very small excess. One drop of a solution of potassium permanganate ($0.02 \, mol \, dm^{-3}$) can be detected in a titrand solution of $100 \, cm^3$, and a similar amount of iodine by shaking the titrand with $5 \, cm^3$ of chloroform or carbon tetrachloride to produce an intense purple colour. Specific indicators react in a specific manner with one participant in the reaction. The best examples are starch, which produces an intense blue colour with iodine and potassium thiocyanate, which forms an intense red compound with iron(III).

APPLICATIONS OF REDOX TITRATIONS

Titrimetric methods based on the direct use of redox reactions are widely used. Their application to the determination of metals which have two well-defined oxidation states is well known. Analysis is often carried out either by first converting all the analyte metal ions to a higher oxidation state with oxidizing agents such as sodium peroxide and sodium bismuthate, or by reduction to a lower state using sulphur dioxide or sodium bisulphite. In all cases an excess of reagent is required which is then destroyed or removed before the sample is titrated. A more elegant means of quantitative reduction is to allow an acidified analyte solution to percolate through a column containing a metal *reductor*. The *silver reductor* contains powdered or granulated silver metal whilst the *Jones reductor* uses a zinc–mercury amalgam. The former is the milder reducing agent and is hence the more selective. An effluent from a column reductor, which is free from excess reducing agent, may be titrated directly with a suitable oxidizing agent. Table 5.6 compares the behaviour of these two reductors.

Where the reduction potentials of two analytes are sufficiently different a mixture may be analysed. Titanium(III), $E^{\ominus} = 0.10$ V, may be titrated with cerium(IV) in the presence of iron(II), $E^{\ominus} = 0.77$ V using methylene blue as indicator. Subsequently the total iron plus titanium may be determined using 'ferroin' as indicator. The determination of iron is illustrative of some practical problems which are encountered in direct titration procedures.

Table 5.6   A comparison of silver and Jones reductors

REDUCTION EFFECTED BY SILVER REDUCTOR	REDUCTION EFFECTED BY JONES REDUCTOR
$Fe^{3+} + e^- = Fe^{2+}$	$Fe^{3+} + e^- = Fe^{2+}$
$Cu^{2+} + e^- = Cu^+$	$Cu^{2+} + 2e^- = Cu$ (metal)
$H_2MoO_4 + 2H^+ + e^- = MoO_2^+ + 2H_2O$	$H_2MoO_4 + 6H^+ + 3e^- = Mo^{3+} + 4H_2O$
TiO^{2+} not reduced	$TiO^{2+} + 2H^+ + e^- = Ti^{3+} + H_2O$
Cr^{3+} not reduced	$Cr^{3+} + e^- = Cr^{2+}$

Dissolution of metallurgical or ore samples containing iron often requires the use of hydrochloric acid. Iron(II) is readily titrated with potassium permanganate ($E^{\ominus} = 1.51$ V) in strong acid solution. Unfortunately, oxidation of chloride ions ($E^{\ominus} = 1.36$ V) will also occur. This problem is overcome by using the 'Zimmerman–Rheinhardt' reagent which contains manganese(II) ions and phosphoric acid. The former displace the Mn(VII)–Mn(II) equilibrium and lower its reduction potential whilst the phosphoric acid preferentially complexes Fe(III), lowering the reduction potential for Fe(III)–Fe(II). Hence, the potassium permanganate, having its reduction potential lowered, stoichiometrically oxidizes Fe(II) without interacting with the chloride ions.

Among the most important indirect methods of analysis which employ redox reactions are the bromination procedures for the determination of aromatic amines, phenols, and other compounds which undergo stoichiometric bromine substitution or addition. Bromine may be liberated quantitatively by the acidification of a bromate–bromide solution mixed with the sample. The excess, unreacted bromine can then be determined by reaction with iodide ions to liberate iodine, followed by titration of the iodine with sodium thiosulphate. An interesting extension of the bromination method employs 8-hydroxyquinoline (oxine) to effect a separation of a metal by solvent extraction or precipitation. The metal–oxine complex can then be determined by bromine substitution.

COMPLEXOMETRIC TITRATIONS

Complex forming reactions find a wide utility in chemical analysis and have been used in titrimetric procedures for many years. Recently most attention has been concentrated on the use of ethylenediaminetetraaceticacid (EDTA) and consideration of this reagent in some detail is important. This study will also be useful in that it illustrates nearly all aspects of the use of inorganic and organic complexing agents in titrimetry.

ETHYLENEDIAMINETETRAACETICACID (EDTA)

In its unreacted form EDTA is a tetrabasic acid represented by

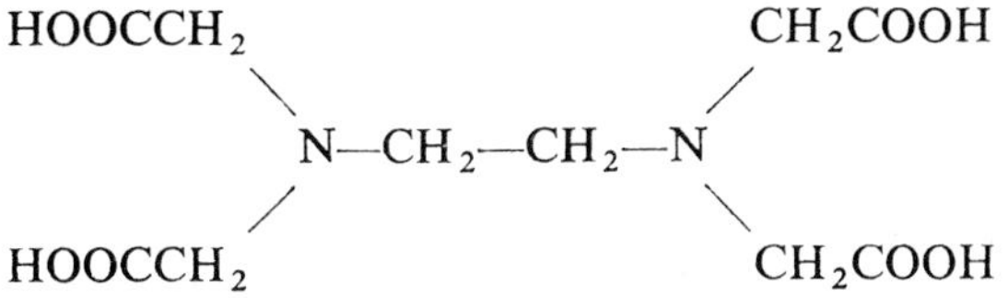

HOOCCH$_2$ CH$_2$COOH $pK_1 = 2.00$

$\diagdown \qquad \diagup$ $pK_2 = 2.67$

N—CH$_2$—CH$_2$—N $pK_3 = 6.16$

$\diagup \qquad \diagdown$ $pK_4 = 10.26$

HOOCCH$_2$ CH$_2$COOH

A useful abbreviation is H$_4$Y with H$_3$Y$^-$, H$_2$Y^{2-}, etc. referring to the various ions derived from the successive stages of dissociation. For practical purposes, the disodium salt Na$_2$H$_2$Y is preferred as a reagent. This salt has a distinctly higher solubility than the parent acid and avoids the high alkalinity produced by extensive hydrolysis in solutions of the tetrasodium salt. The four electron rich acetate groups together with the two nitrogen lone pairs constitute a sexadentate ligand which will form complexes with octahedral geometry (figure 5.4).

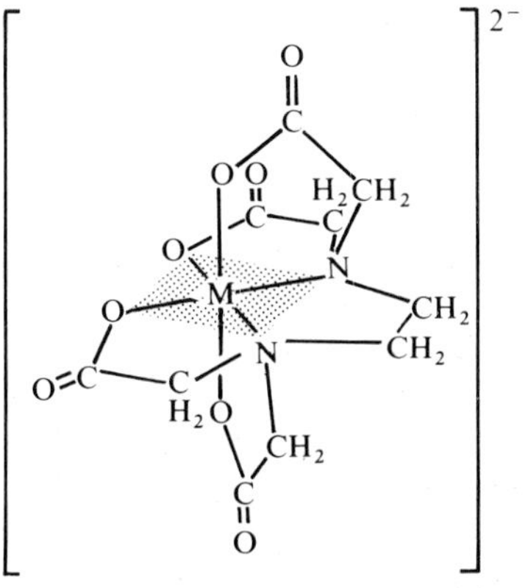

Figure 5.4.   A proposed structure for a metal–EDTA chelate showing its octahedral geometry

Although it is probable that only four bonds are formed in many complexes, the cage-like structures effectively prevent the formation of complexes other than those with 1:1 stoichiometry (chapter 3). This feature is of considerable analytical importance.

The Composition of Aqueous EDTA Solutions

Being a tetrabasic acid, EDTA dissociates in solution to give four different ionic species, H_3Y^-, H_2Y^{2-}, HY^{3-}, Y^{4-}, the relative amounts of which will depend upon the pH of the solution. The proportion of any species present is represented by its α-value, an idea introduced in chapter 3.

$$\alpha_4 = \frac{[Y^{4-}]}{C_L}, \qquad \alpha_3 = \frac{[HY^{3-}]}{C_L}, \qquad \alpha_2 = \ldots, \text{etc.} \qquad (5.13)$$

Where C_L is the total amount of uncomplexed EDTA, given by

$$C_L = [Y^{4-}] + [HY^{3-}] + [H_2Y^{2-}] + [H_3Y^-] + [H_4Y] \qquad (5.14)$$

It is convenient to evaluate α_4, which may be done by substituting into equation (5.14) from expressions for the dissociation constants K_1, K_2, K_3, K_4.

$$K_1 = \frac{[H_3Y^-][H^+]}{[H_4Y]}, \qquad \text{etc.} \qquad (5.15)$$

whence (5.13) becomes

$$\alpha_4 = \frac{K_1K_2K_3K_4}{[H^+]^4 + K_1[H^+]^3 + K_1K_2[H^+]^2 + K_1K_2K_3[H^+] + K_1K_2K_3K_4} \qquad (5.16)$$

and α_4 is seen to depend on the K_a values and the pH only.

Table 5.7   Values of α_4 for EDTA as a function of pH

pH	α_4	pH	α_4
2.0	3.7×10^{-14}	7.0	4.8×10^{-4}
3.0	2.5×10^{-11}	8.0	5.4×10^{-3}
4.0	3.6×10^{-9}	9.0	5.2×10^{-2}
5.0	3.5×10^{-7}	10.0	3.5×10^{-1}
6.0	2.2×10^{-5}	11.0	8.5×10^{-1}
		12.0	9.8×10^{-1}

The overall variations of solution composition are summarized in figure 5.5 and table 5.7, from which it will be seen that the major species present in the operating pH range of 2 to 10 are H_2y^{2-} and HY^{3-}.

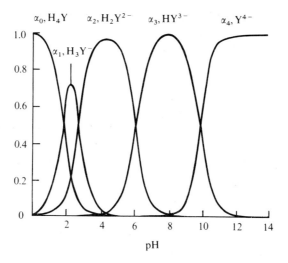

Figure 5.5.   The variation of α values with pH for EDTA species

The Formation and Stability of Metal–EDTA Chelates

EDTA forms stable complexes with a wide range of metal ions. The exceptional stability is conferred by the large number of donor groups and the subsequent isolation of the metal ion in the cage-like structure. It is clear from the foregoing section that complex formation will be pH dependent and may be represented by different equations depending on the pH and the ionic form of the EDTA. However, for purposes of comparison it is best to use the equilibrium

$$M^{n+} + Y^{4-} = MY^{(n-4)+} \tag{5.17}$$

and its associated formation constant

$$K_{MY} = \frac{[MY^{(n-4)+}]}{[M^{n+}][Y^{4-}]} \tag{5.18}$$

A representative list of formation constants defined on this basis is given in table 5.8.

These data may be used for the practical comparison of formation constants by use of the conditional constant principle (chapter 3), i.e. K'_{MY} is given by

$$K'_{MY} = K_{MY}\alpha_4 = \frac{[MY^{(n-4)+}]}{[M^{n+}]C_L} \tag{5.19}$$

which has been obtained by the combination of equations (5.13) and (5.18).

Table 5.8   Some selected metal–EDTA formation constants

CATION	K_{MY}	$\log_{10} K_{MY}$	CATION	K_{MY}	$\log_{10} K_{MY}$
Ag^+	2×10^7	7.3	Cu^{2+}	6.3×10^{18}	18.80
Mg^{2+}	4.9×10^8	8.69	Zn^{2+}	3.2×10^{16}	16.50
Ca^{2+}	5.0×10^{10}	10.70	Cd^{2+}	2.9×10^{16}	16.46
Sr^{2+}	4.3×10^8	8.63	Hg^{2+}	6.3×10^{21}	21.80
Ba^{2+}	5.8×10^7	7.76	Pb^{2+}	1.1×10^{18}	18.04
Mn^{2+}	6.2×10^{13}	13.79	Al^{3+}	1.3×10^{16}	16.13
Fe^{2+}	2.1×10^{14}	14.33	Fe^{3+}	1×10^{25}	25.1
Co^{2+}	2.0×10^{16}	16.31	V^{3+}	8×10^{25}	25.9
Ni^{2+}	4.2×10^{18}	18.62	Th^{4+}	2×10^{23}	23.2

From data in tables 5.7 and 5.8, the conditional constant may be calculated, e.g. for Mg^{2+}–EDTA at pH = 5

$$K'_{MgY} = (4.9 \times 10^8)(3.5 \times 10^{-7}) = 1.73 \times 10^2$$

and at pH = 10

$$K'_{MgY} = (4.9 \times 10^8)(3.5 \times 10^{-1}) = 1.73 \times 10^8$$

Hence, pH = 10 could be used for the titration of magnesium but not pH = 5 (stoichiometric reaction requires $K'_{MY} \geq 10^6$).

Secondary Complexing Agents

EDTA titration solutions frequently need to be buffered to a high pH to ensure stoichiometric formation of the complex. Unfortunately many metals will precipitate as hydroxides or hydrated oxides under these conditions. Use can be made of secondary or auxiliary complexing agents to retain the metal ion in solution. Typical reagents for this purpose are ammonium chloride, triethylamine and triethanolamine. Conveniently, the same reagents can provide the basis for a buffer system as well. These selected reagents will retain a range of metal ions in solution without seriously competing with the metal–EDTA equilibrium. The conditional constant concept again proves valuable in the assessment of practical situations. For example, when zinc is held in solution by an ammonia–ammonium chloride buffer at pH = 10, the fraction of zinc unreacted with EDTA which remains as Zn^{2+} ions can be expressed by

$$\beta = [Zn^{2+}]/C_M \qquad (5.20)$$

where C_M is the total zinc uncomplexed with EDTA, and is given by

$$C_M = [Zn^{2+}] + [Zn(NH_3)^{2+}] + [Zn(NH_3)_2^{2+}] + [Zn(NH_3)_3^{2+}] \\ + [Zn(NH_3)_4^{2+}] \qquad (5.21)$$

It is apparent that β will depend in part on the concentration of ammonia and in part on the formation constants K_1, K_2, K_3 and K_4 for the various ammonia complexes, where

$$K_1 = \frac{[Zn(NH_3)^{2+}]}{[Zn^{2+}][NH_3]} \qquad K_2 = \frac{[Zn(NH_3)_2^{2+}]}{[Zn^{2+}][NH_3]} \qquad \text{etc.}$$

Substitution into (5.20) gives

$$\beta = \frac{1}{1 + K_1[NH_3] + K_1K_2[NH_3]^2 + K_1K_2K_3[NH_3]^3 + K_1K_2K_3K_4[NH_3]^4} \tag{5.22}$$

whence $\beta = 8.0 \times 10^{-6}$ if the solution is 0.1 mol dm^{-3} with respect to ammonia. At pH $= 10$, the conditional constant for zinc–EDTA can then be computed (chapter 3).

$$K''_{ZnY} = K_{ZnY}\alpha_4\beta = \frac{[ZnY^{2-}]}{C_L C_M} \tag{5.23}$$

$$= (3.2 \times 10^{16})(3.5 \times 10^{-1})(8 \times 10^{-6})$$

$$= 8.9 \times 10^{10} \text{ mol}^{-1} \text{ dm}^3$$

It is thus concluded that the zinc–EDTA complex will be formed quantitatively at pH $= 10$ in a solution of ammonia (0.1 mol dm^{-3}).

The second reason for employing an auxiliary complexing agent is to mask the effect of an interfering ion. Both zinc and magnesium form stable EDTA complexes at pH $= 10$ and may be titrated in solutions buffered to that pH. If they are present during the titration of magnesium, zinc and many other heavy metals will interfere. However, heavy metals in general form stable cyanide complexes ($K_4 \approx 10^{20}$) and the addition of sodium or potassium cyanide to the titration mixture will reduce the conditional constants for the heavy metal–EDTA complexes so that they do not react to any significant extent. Meanwhile magnesium forms only a very weak cyanide complex and can be titrated without interference. For example, in the case of zinc, the stepwise formation constants for its cyanide complexes are $K_1 = 3 \times 10^5$, $K_2 = 1.3 \times 10^5$, $K_3 = 4.3 \times 10^4$, $K_4 = 3.5 \times 10^3$, whence from an expression similar to equation (5.22) $\beta \approx 10^{-19}$ when $[CN^-] = 1$ mol dm^{-3}. K''_{ZnY} then becomes 1.1×10^{-3}, indicating that the presence of cyanide effectively prevents the formation of the zinc–EDTA complex. Demasking at a subsequent stage in order that the zinc may be determined is effected by a formaldehyde–acetic acid mixture, i.e.

$$Zn(CN)_4^{2-} + 4H^+ + 4HCHO = Zn^{2+} + 4HOCH_2CN$$

Chloral hydrate (Cl$_3$CCHO) may also be used in demasking reactions.

End Point Detection for EDTA Titrations

The now familiar alternatives of visual and potentiometric detection are available. A number of organic dyes form coloured chelates with many metal ions. These coloured chelates are often discernible to the eye at concentrations of 10^{-6} to 10^{-7} mol dm^{-3} and can function as visual indicators. Most metal ion indicators will also undergo parallel reactions with protons bringing about similar colour changes. Hence, a careful consideration of pH is prudent when selecting an indicator. Some typical indicators appear in table 5.9. Of these, erichrome black T, which forms red complexes with over twenty metal ions, is amongst the most widely used. Its behaviour will serve as a general example of indicator function.

Firstly, an acid-base equilibrium is established in solution. When a metal cation is added, complexation equilibria are established concurrently.

$$\underset{\text{red}}{H_2In^-} \quad \overset{pK_a = 6.3}{\rightleftharpoons} \quad \underset{\substack{\text{blue} \\ \Updownarrow M^{2+} \\ MIn^- \\ \text{red}}}{HIn^{2-}} \quad \overset{pK_a = 11.5}{\rightleftharpoons} \quad \underset{\substack{\text{orange} \\ \Updownarrow M^{2+} \\ MIn^- \\ \text{red}}}{In^{3-}} \tag{5.24}$$

(note: H^+ ions are omitted for simplicity).

A pH in the range 8 to 10 conveniently maximizes the indicator form HIn^{2-} and enables the associated complexing reaction to go to effective completion by facilitating the removal of hydrogen ions. Thus, the pH dependent, conditional constant for the indicator is given by

$$K'_{MIn} = \frac{[M'][In']}{[MIn^-]} \tag{5.25}$$

for the reaction $MIn^- = M' + In'$ (see equation 5.1).

Here $[M']$ and $[In']$ represent the unreacted metal ion and indicator respectively. If the midpoint of the colour change occurs when

$$[MIn^-] = [In']$$

then equation (5.25) gives

$$pM' = pK'_{MIn} \tag{5.26}$$

which is the criterion for a coincident end point and equivalence point and an indicator must be selected to satisfy

$$pM' = pK'_{MIn} \pm 1 \quad \text{(see equation 5.5)}$$

In general, for the observation of a sharp end point the indicator complex must be reasonably strong $K_{MIn} > 10^4$ but less than $K_{MY} \times 10^{-1}$.

Table 5.9   Some metal ion indicators for EDTA titrations

INDICATOR	METAL DETERMINED IN DIRECT TITRATION	
eriochrome black T (used as its sodium salt)	Ba Cd Ca In Pb Mn rare earths widely used in back titrations and replacement titrations	Sc Sr Zn
pyrocatechol violet	Al Bi Cd Co Cu Ga Fe Pb	Mg Mn Ni Th Ti Zn
xylenol orange	Bi Cd Pb Sc Th Zn	
pyridylazonaphthol (PAN)	Cd Cu In Sc Zn widely used in back titrations and re-placement titrations	
murexide	Ca Co Cu Ni mainly of historical interest	

Potentiometric EDTA titrations are best carried out with a mercury pool electrode (figure 5.6) or a gold amalgam electrode. When this electrode dips into a solution containing the analyte together with a small amount of added Hg–EDTA complex, three interdependent reactions occur. For example, at pH = 8 the half cell reaction (a) which determines the electrode potential is related to the solution equilibrium by (b) and (c).

(a) $Hg^{2+} + 2e^- = Hg\,(1)$
(b) $Hg^{2+} + HY^{3-} = HgY^{2-} + H^+$
(c) $M^{n+} + HY^{3-} = MY^{(n-4)+} + H^+$

As the titration proceeds, the analyte ion, M^{n+} is increasingly complexed and eventually the equilibria involving Hg^{2+} are displaced and the potential of the electrode varies according to

$$E = E^{\ominus}_{Hg} + 0.0296 \text{ V} \log \frac{[M^{n+}]}{[MY^{(n-4)+}]} + \text{constant} \qquad (5.27)$$

In overall form this equation resembles that for the glass electrode (chapter 6) and a pM–EDTA curve resembles an acid-base titration curve. The mercury electrode is most usefully employed when coloured or turbid solutions are being titrated, or when dilute solutions and weak complexes lead to poor colour changes.

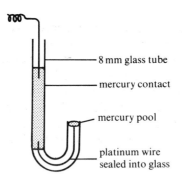

8 mm glass tube

mercury contact

mercury pool

platinum wire
sealed into glass

Figure 5.6.   A 'J-type' mercury pool electrode

EDTA Titration Procedures
If the analyte metal ion forms a stable EDTA complex rapidly, and an end point can be readily detected, a direct titration procedure may be employed. More than thirty metal ions may be so determined. Where the analyte is

partially precipitated under the reaction conditions thereby leading to a slow reaction, or where a suitable indicator cannot be found, back titration procedures are used. A measured excess of EDTA is added and the unreacted EDTA titrated with a standard magnesium or calcium solution. Provided the analyte complex is stronger than the Ca–EDTA or Mg–EDTA complex a satisfactory end point may be obtained with eriochrome black T as indicator. An alternative procedure, where end points are difficult to observe, is to use a displacement reaction. In this case, a measured excess of EDTA is added as its zinc or magnesium complex. Provided the analyte complex is the stronger, the analyte will displace the zinc or magnesium.

$$MgY^{2-} + M^{2+} = MY^{2-} + Mg^{2+} \tag{5.28}$$

The magnesium will be liberated quantitatively and may then be titrated with a standard EDTA solution. Where mixtures of metal ions are analysed, the masking procedures already discussed can be utilized or the pH effect exploited. A mixture containing bismuth, cadmium and calcium might be analysed by first titrating the bismuth at pH $= 1$–2 followed by the titration of cadmium at an adjusted pH $= 4$ and finally calcium at pH $= 8$. Titrations of this complexity would be most conveniently carried out potentiometrically using the mercury pool electrode.

APPLICATIONS OF EDTA TITRATIONS

Table 5.8 gives an indication of the range of elements that may be determined. Most procedures will require an analyte concentration of 10^{-3} mol dm^{-3} or more, although with special conditions, notably potentiometric end point detection, the sensitivity may be extended to 10^{-4} mol dm^{-3}. The analysis of mixtures of metal ions necessitates masking and demasking, pH adjustments and selective separation procedures. Areas of application are spread throughout the chemical field from water treatment and the analysis of refined food and petroleum products to the assay of minerals and alloys. Table 5.10 gives some selected examples.

TITRATIONS WITH COMPLEXING AGENTS OTHER THAN EDTA

A number of other reagents containing substituted amino and carboxylic acid groupings are used to a limited extent. Their general behaviour is very similar to that of EDTA, but the metal ion complexes formed may be stronger or weaker than parallel ones formed by EDTA. Stronger complexing agents (DETPA, DCTA) can extend the scope of complexometric titrations to

Table 5.10   Some examples of EDTA titration methods

ANALYTE AND MATRIX	OUTLINE PROCEDURE
Ca and Mg in water (water hardness)	*Ca determination* Add 2 cm^3 of NaOH solution (0.1 mol dm^{-3}) to 50 cm^3 of sample and titrate with EDTA using murexide indicator (table 5.8). *Mg determination* Destroy murexide colour with 1 cm^3 concentrated HCl add 3 cm^3 of NH$_3$—NH$_4$Cl buffer and titrate with EDTA using eriochrome black T.
Pb in minerals	Dissolve the sample in nitric acid. Separate Pb by extraction with CHCl$_3$ solution of sodium diethyl dithiocarbonate using alkaline cyanide solution to mask interferences. Titrate Pb in an ammonia-ammonium chloride medium at pH = 10 using eriochrome black T.
Zn in light alloys	Dissolve the sample in NaOH solution and precipitate ZnS with NaS solution. Dissolve ZnS in HCl solution, add ammonium citrate to mask Al and then titrate Zn at pH = 9 using eriochrome black T as indicator.
Ca, Mg and Zn in biological tissue	Ash the tissue and extract the residue with HCl solution. Separate Ca, Mg and Zn by retention on a cation exchange column followed by elution with HCl (5 mol dm^{-3}). Titrate Ca using murexide, Ca plus Mg using eriochrome black T, then add chloral hydrate to give a further endpoint with Zn.
Au in Au–Pt–Ag alloys	Dissolve sample in 'aqua regia' and filter off AgCl. Extract Au with diethyl ether, evaporate solvent, and react Au with K$_2$Ni(CN)$_4$. Buffer with NH$_3$—NH$_4$Cl, react liberated Ni with excess EDTA. Back titrate EDTA excess with Mn(II) using eriochrome black T.

elements not adequately complexed by EDTA and weaker ones (NTA) may show a better selectivity. Some reagents of these types are given in table 5.11. It is however beyond the scope of this text to discuss such reagents in detail. Inorganic ligands which have been used as complexometric titrants include the halide ions and pseudohalides CN$^-$ and SCN$^-$. Chloride ion may be determined by titration with mercury(II) nitrate solution, when stoichiometric formation of HgCl$_2$ is indicated by a consideration of the stepwise formation

Table 5.11   Some additional aminocarboxylic acid ligands

DCTA

1.2 diaminocyclohexaneN,N,N',N'tetraacetic acid

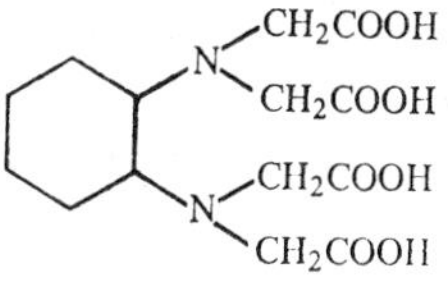

$$
\begin{array}{c}
\text{N}\diagup\text{CH}_2\text{COOH} \\
\qquad\diagdown\text{CH}_2\text{COOH} \\
\text{N}\diagup\text{CH}_2\text{COOH} \\
\qquad\diagdown\text{CH}_2\text{COOH}
\end{array}
$$

DETPA

diethylenetriamineN,N,N',N',N',pentaacetic acid

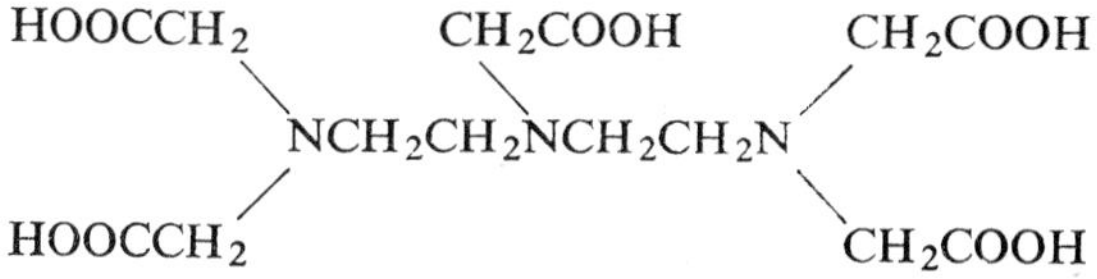

$$
\begin{array}{ccc}
\text{HOOCCH}_2 & \text{CH}_2\text{COOH} & \text{CH}_2\text{COOH} \\
\diagdown & \diagdown & \diagup \\
 & \text{NCH}_2\text{CH}_2\text{NCH}_2\text{CH}_2\text{N} & \\
\diagup & & \diagdown \\
\text{HOOCCH}_2 & & \text{CH}_2\text{COOH}
\end{array}
$$

NTA

nitrilotriacetic acid

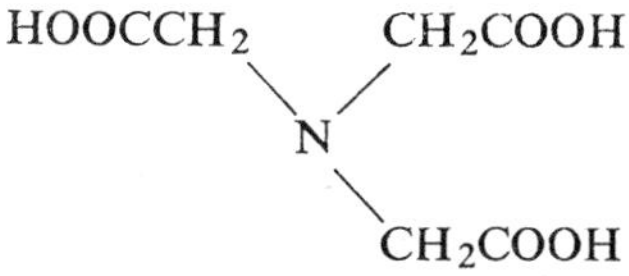

$$
\begin{array}{cc}
\text{HOOCCH}_2 & \text{CH}_2\text{COOH} \\
\diagdown & \diagup \\
 & \text{N} \\
 & \diagdown \\
 & \text{CH}_2\text{COOH}
\end{array}
$$

HMDTA

hexamethylenediaminetetraacetic acid

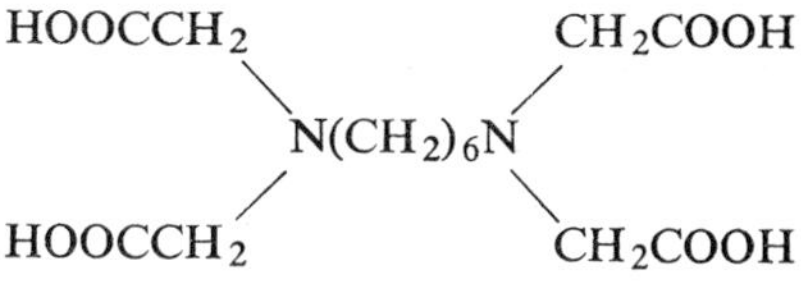

$$
\begin{array}{cc}
\text{HOOCCH}_2 & \text{CH}_2\text{COOH} \\
\diagdown & \diagup \\
 & \text{N(CH}_2)_6\text{N} \\
\diagup & \diagdown \\
\text{HOOCCH}_2 & \text{CH}_2\text{COOH}
\end{array}
$$

---

constants for $HgCl_4^{2-}$ ($K_1 = 5.5 \times 10^6$, $K_2 = 3.0 \times 10^6$, $K_3 = 7$, $K_4 = 10$) which reflect the low stability of $HgCl_3^-$ and $HgCl_4^{2-}$. The end point may be detected by the reaction of excess mercury ions with sodium nitroprusside or diphenylcarbazone, when a white precipitate is produced in each case. This titration can be used to determine chlorides in solution within the range 1–100 $\mu g\,cm^{-3}$ and has been widely employed for the analysis of water samples. Parallel reactions with Br^-, SCN^- and CN^- may also be used but iodides are more difficult to handle and quantitative analysis is not easy.

Cyanide reagents have been traditionally used in the metal finishing industry and the reaction,

$$Ag^+ + 2CN^- = Ag(CN)_2^- \qquad K_2 = 7 \times 10^{19} \qquad (5.29)$$

provides the basis for their direct titration. Excess silver ions react to produce AgCN

$$Ag^+ + Ag(CN)_2^- = 2AgCN \qquad (5.30)$$

which has a low solubility ($K_{sp} = 1.2 \times 10^{-16}$) showing a precipitate at the end point of the titration. The cyanide complexes $M(CN)_4^{2-}$ (M = Cu, Zn, Co, Ni) are significantly more stable than $Ag(CN)_2^-$ and may be determined in a back titration procedure. Excess cyanide is added to a slightly alkaline solution of the metal and the excess cyanide titrated with silver nitrate.

PRECIPITATION TITRATIONS

Titrimetric reactions in which the product is of low solubility have been used for very many years. Titrations employing such reactions are known as *precipitation titrations*. Although not of widespread importance, for a limited number of purposes such as the determination of halides these methods are invaluable. Silver nitrate is the most widely used reagent giving rise to silver salts which are characteristically sparingly soluble.

The end points of precipitation titrations can be variously detected. An indicator exhibiting a pronounced colour change with the first excess of the titrant may be used. The Mohr method, involving the formation of red silver chromate with the appearance of an excess of silver ions, is an important example of this procedure, whilst the Volhard method, which uses the ferric thiocyanate colour as an indication of the presence of excess thiocyanate ions, is another. A series of indicators known as *adsorption indicators* have also been utilized. These consist of organic dyes such as fluorescein which are used in silver nitrate titrations. When the equivalence point is passed the excess silver ions are adsorbed on the precipitate to give a positively charged surface which attracts and adsorbs fluoresceinate ions. This adsorption is accompanied by the appearance of a red colour on the precipitate surface.

Table 5.12   Substances determined by precipitation titrations with Ag^+

AsO_4^{3-}, Br^-, BH_4^-, CNO^-, CO_3^{2-}, CrO_4^{2-}, CN^-, Cl^-, $C_2O_4^{2-}$, epoxide, I^-, K^+, PO_4^{3-}, SCN^-, S^{2-}, SeO_3^{2-}, $V(OH)_4^+$, fatty acids, mercaptans

---

Finally, the electroanalytical methods described in chapter 6 may be used to scan the solution for metal ions. Table 5.12 includes some examples of substances determined by silver titrations and table 5.13 some miscellaneous precipitation methods. Other examples have already been mentioned under complexometric titrations.

Table 5.13   Miscellaneous precipitation titrations

ANALYTE	REAGENT	PRECIPITATE
Cl^-, Br^-	$Hg_2(NO_3)_2$	Hg_2Cl_2, Hg_2Br_2
SO_4^{2-}, MoO_4^{2-}	$Pb(NO_3)_2$	$PbSO_4$, $PbMoO_4$
Zn^{2+}	$K_4Fe(CN)_6$	$K_2Zn_3[Fe(CN)_6]_2$
PO_4^{3-}, $C_2O_4^{2-}$	$Pb(OAc)_2$	$Pb_3(PO_4)_2$, PbC_2O_4

5.2   Gravimetry

SUMMARY

Principles

Solution reaction between analytes and reagents to give sparingly soluble products; filtration, drying or ignition of precipitates; electrolytic deposition of metals; weighing.

Apparatus

Flasks, beakers, filter funnels, pipettes, filter crucibles, filter papers, oven, muffle furnace, chemical balance, desiccator.

Applications

Extensive numbers of inorganic ions are determined with excellent precision and accuracy; widely used in routine assays of metallurgical and mineralogical samples. Relative precision 0.1 to 1 %.

Disadvantages

Requires careful and time consuming procedures with scrupulously clean apparatus and very accurate weighings. Coprecipitation cannot always be avoided.

Gravimetry includes any analytical method in which the ultimate measurement is by weight. The simplest form may merely be the drying or heating of a sample in order to determine its volatile and non-volatile components, or possibly the sample might be distilled and the residue and fractions of distillate weighed. Metals may be deposited electrolytically and weighed (chapter 6). Of far greater scope and importance is the controlled precipitation of an analyte from solution, followed by the weighing of the precipitate. Subsequent attention will be restricted to these *precipitation methods*.

To be of gravimetric value, a precipitation process must fulfil certain conditions. The precipitate must be formed quantitatively and within a reasonable time. Its solubility should be low enough for a quantitative separation to be made. It must be readily filterable and, if possible, have a known and stable stoichiometric composition when dried so that its weight can be related to the amount of analyte present. Failing this, it must be possible to convert the precipitate to a stoichiometric weighable form (usually by ignition). In both cases the weighed form should be non-hygroscopic.

PRECIPITATION REACTIONS

If it is remembered that a chemical reaction can be displaced by changing the state of the products (chapter 3) then, provided the solubility product of a precipitate is small, quantitative reaction can be obtained. Practically, the precipitate may be formed directly on the addition of a reagent or on the

Table 5.14   Some inorganic precipitants

REAGENT	ANALYTE AND FORM PRECIPITATED	ANALYTE FORM WEIGHED
$NH_3(aq)$	Be hydrous oxide	BeO
	Al hydrous oxide	Al_2O_3
	Sc hydrous oxide	Sc_2O_3
	Fe hydrous oxide	Fe_2O_3
	In hydrous oxide	In_2O_3
	$(NH_4)_2U_2O_7$	U_3O_8
H_2S	ZnS	ZnO
	GeS	GeO_2
	As_2S_3	As_2O_3
$(NH_4)_2HPO_4$	$MgNH_4PO_4$	$Mg_2P_2O_7$
H_2SO_4	Sr, Cd, Pb and Ba sulphates	sulphates
HCl	AgCl	AgCl
	Si (silicic acid)	SiO_2
$AgNO_3$	AgCl	AgCl
	AgBr	AgBr
	AgI	AgI
$BaCl_2$	$BaSO_4$	$BaSO_4$

subsequent adjustment of solution conditions, e.g. pH. Precipitates may be of different chemical types, e.g. salts, complexes, hydroxides, hydrous oxides, and precipitating agents are conveniently divided into inorganic (table 5.14) and organic (table 5.15). The major areas of applications usually employ inorganic precipitants for inorganic analytes.

Table 5.15   Some organic precipitating agents

REAGENT	USES
8-hydroxyquinoline (oxine)	A non-specific reagent complexing with over twenty metals. pH control can be exploited to aid selectivity. Precipitates may be brominated to provide a volumetric finish (p. 149)
dimethylglyoxime	A highly specific reagent complexing with Ni(II) alone in alkaline media and Pd(II) alone in acid
sodium tetraphenylboron $(C_6H_5)_4B^-Na^+$	Forms 'salt like' precipitates with K^+ and NH_4^+. Hg^{2+}, Rb^+, Ca^+ interfere
benzidene	Forms a salt with SO_4^{2-} in acid solution $(C_{12}H_{12}N_2.H_2SO_4)$. The precipitate can be weighed, or titrated with NaOH or $KMnO_4$
cupferron	Precipitates a large number of heavy metals from dilute acid solution and some, e.g. Fe, Ti, Zr, V, U, Sn, Nb, Ta, from solutions as concentrated as 10% v/v HCl or H_2SO_4
α-benzoinoxime (Cupron)	Used as a specific precipitant for Cu from ammoniacal solution (Tartrate ions keep Fe, Al in solution), and for Mo from dilute acid. (Ni, Nb, Ta interfere)
tetraphenylarsonium chloride $(C_6H_5)_4AsCl$	Gives 'salt like' precipitates with a number of anionic species, e.g. ReO_4^-, MoO_4^-, WO_4^-, $HgCl_4^{2-}$, $SnCl_6^{2-}$, $CdCl_4^{2-}$, $ZnCl_4^{2-}$

The Solubility of Precipitates

The solubility product as a measure of solubility was introduced in chapter 3. For the solubility equilibrium

$$AB = A + B \qquad (5.31)$$

$$K_{sp} = [A][B] = \gamma_A C_A \gamma_B C_B \qquad (5.32)$$

where γ is the activity coefficient and C is the concentration of the species. The activity coefficient may be calculated from the Debye–Hückel equation for a temperature of 298 K,

$$-\log_{10} \gamma_A = 0.51 Z_A^2 \mu^{\frac{1}{2}} \qquad (5.33)$$

where Z is the charge on A and μ (mol kg^{-1}) is the ionic strength of its solution. Thus, γ_A, K_{sp} and in turn the solubility of AB will increase with the *ionic strength* of the solution environment. Figure 5.7 shows the effect of potassium nitrate on the solubility of barium sulphate.

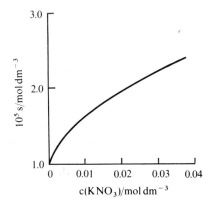

Figure 5.7.   The effect of ionic strength on the solubility s of BaSO$_4$

The *common ion effect* (chapter 3) is a further important factor affecting solubilities. Addition of A or B to the above system (equation 5.31) will shift the equilibrium to the left and reduce the solubility of AB. In practice, this situation would arise when an excess of a precipitating reagent has been added to an analyte solution. Such an excess leads to the possibility of complexation reactions occurring which will tend to increase the solubility of AB. For example, when aluminium or zinc is precipitated by hydroxyl ions, the following reactions with excess reagent can occur

$$Al(OH)_3 + OH^- = Al(OH)_4^- \qquad (5.34)$$

$$Zn(OH)_2 + 2OH^- = Zn(OH)_4^{2-} \qquad (5.35)$$

The net result of the three factors discussed above is that frequently an optimum (minimum) solubility is obtained when a small excess of the reagent is added. Figure 5.8 which is a solubility curve for AgCl illustrates this pattern.

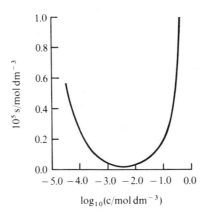

Figure 5.8. The solubility s of AgCl as a function of concentration c of KCl added to the solution

The effect of pH changes on precipitate solubilities merits special consideration. The dependence of solubility on pH may derive from the common ion effect where OH^- or H_3O^+ ions are generated by the dissolution of the precipitate, e.g.

$$Al(OH)_3 = Al^{3+} + 3OH^- \tag{5.36}$$

In other cases where solute ions react with OH^- or H_3O^+, a more complex pattern can arise, as in the example of calcium oxalate in which the solubility increases as the pH is lowered.

$$CaC_2O_4 = Ca^{2+} + C_2O_4^{2-} \tag{5.37}$$
$$\parallel H_3O^+$$
$$HC_2O_4^-$$
$$\parallel H_3O^+$$
$$H_2C_2O_4$$

Rates of Precipitate Formation and Particle Growth
The rate at which a precipitate can be produced in a filterable form varies widely and depends upon the solution environment. In the case of the oxalates of calcium and magnesium for instance, both compounds have fairly low solubility products (2.3×10^{-9} mol^2 dm^{-6} for calcium oxalate and

8.6×10^{-5} mol^2 dm^{-6} for magnesium oxalate). However, in the case of the calcium salt, precipitation is complete within a few minutes while the magnesium salt takes several hours. This difference in rates may even be used as a basis for the quantitative removal of calcium from solution without interference from magnesium. In order to expedite precipitate formation it is necessary first to look more closely at the processes by which precipitates are produced.

When a reagent is added to an analyte solution forming a sparingly soluble compound, the solubility product for that compound is immediately exceeded and the solution is said to be *supersaturated*. The *relative supersaturation* R_s is given by

$$R_s = (Q - S)/S \qquad (5.38)$$

where Q is the actual concentration of the solute and S is the equilibrium concentration. The rate at which Q reduces to S determines the rate of precipitation. Precipitate formation takes place first by the aggregation of small groups of ions or molecules, a process known as *nucleation*. It may occur either by the chance aggregation of molecules or ions in a *homogeneous nucleation* process, or by *heterogeneous nucleation* when aggregation is initiated by particulate impurities within the solution. Whilst the former process depends exponentially on the relative supersaturation of the solution, the latter is largely independent of it. After nucleation the precipitation continues by *particle growth*, with further ions or molecules adding to the aggregates. The rate of particle growth will also be dependent upon the relative supersaturation and on the surface area of the particles, but will not vary so dramatically as the rate of homogeneous nucleation. The relation between these rates and the relative supersaturation is summarized schematically in figure 5.9.

In a particular system, the nature of the precipitated particles will be determined by the relative rates of nucleation and particle growth. Where

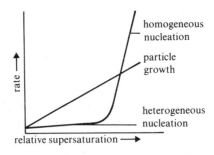

Figure 5.9.   Nucleation and particle growth rates related to relative supersaturation

nucleation predominates, small particles are produced and a colloid may result, whereas a predominance of particle growth will form a more tractable precipitate. From the practical standpoint, the difference $(Q - S)$ should be controlled and kept to a minimum. On mixing reagent and analyte solutions it is difficult to avoid a momentarily high $(Q - S)$ value (in the region where the two first make contact), especially if S is small. Furthermore, a low value of S is necessary for quantitative precipitation, so that a situation often arises in which the rate of homogeneous nucleation vastly exceeds that of particle growth. Consequently, the analyst frequently has to cope with colloidal precipitates or suspensions. Where S is larger, crystalline precipitates are more readily obtained. Colloidal suspensions occur as a result of electrical repulsions between particles which prevent agglomeration. These repulsive forces develop from ions adsorbed onto the surface of the precipitate which cause the formation of an electrical double layer. The adsorbed layer will contain an excess of the precipitate ion which predominates in solution whilst the diffuse counter ion layer will contain an excess of oppositely charged ions to maintain the overall electrical neutrality of the solution. In the familiar example where Cl^- has been precipitated as $AgCl$ by the addition of an excess of silver nitrate, the adsorbed layer will contain an excess of Ag^+, and the counter ion layer an excess of NO_3^- and OH^-.

Adsorption can be diminished by heating or by the addition of a highly charged strong electrolyte. This allows coagulation of the precipitate to proceed, but it must be remembered that if a filtered precipitate of this type is being washed the particles can be dispersed again as a colloid and pass through the filter. This effect, known as *peptization*, may be prevented by using hot washing solutions containing a suitable electrolyte. Precipitates formed by colloid agglomeration are amorphous and porous with very high surface areas. In almost all cases, precipitates are improved by heating them in contact with the solution before filtration. This is known as *digestion* and will promote the formation of larger particles with a reduced surface area, and more ordered arrangements within crystals. Thus both the surface adsorption and occlusion of impurities will be reduced.

Purity of precipitates
Steps are normally taken to prevent the simultaneous precipitation of materials other than the desired analyte species. Incorporation of impurities into the precipitate may however occur by *coprecipitation* or *post precipitation*. The former arises during the formation of the precipitate, and the latter after it has been formed. The various modes of coprecipitation are summarized in table 5.16.

Post precipitation involves the deposition of a sparingly soluble impurity of similar properties to the precipitate on the surface of that precipitate after

Table 5.16   The major types of coprecipitation and their relation to the precipitate type

TYPE OF COPRECIPITATION	MODE OF CONTAMINATION	PRECIPITATE TYPE MOST AFFECTED
isomorphic inclusion	substitution of the precipitate lattice with impurity ions of similar crystallinity	crystalline
non-isomorphic inclusion	solid solution of the impurity within the precipitate	crystalline
occlusion	physical trapping of impurities within precipitate particles	crystalline and colloidal
surface adsorption	chemisorption of impurities from the solution onto the precipitate surface	colloidal

it has been formed. It is particularly a problem where similar materials are being separated on the basis of their different rates of precipitation, e.g. calcium and magnesium oxalates or zinc and mercury sulphides. Coprecipitation can be reduced by the normal practical procedures of precipitation from hot dilute analyte solutions, followed by digestion of the precipitate. Inclusion phenomena when they occur are very difficult to overcome and it may be necessary to remove the impurity ion before precipitation. At the same time the degree of postprecipitation will be increased by digestion and where this form of contamination is paramount rapid filtration is essential. The extent to which coprecipitation may affect an analysis is well illustrated by the familiar precipitation of barium sulphate. Substances that may appear as impurities in this precipitate include sulphuric acid, alkali metal sulphates, ammonium sulphate, iron(III) sulphate, barium phosphate, barium carbonate, barium chloride, barium nitrate and barium chlorate. The excellent precision and accuracy often claimed for analyses based on this precipitate undoubtedly result from a compensation of errors.

PRACTICAL GRAVIMETRIC PROCEDURES

A typical gravimetric analysis procedure may be divided into five stages: sample pretreatment; precipitation; filtration; drying and ignition; weighing. The operations involved in the various stages are summarized by table 5.17. Two aspects, however, need slight amplification. Firstly, the possible generation of the precipitating agent within the solution in a *homogeneous precipitation* procedure (p. 210) should be considered. This method has the advantage of preventing local high concentrations of reagent and thus

Table 5.17 Practical gravimetric procedures

STAGE	PRACTICAL MANIPULATIONS	REMARKS
sample pre-treatment	dissolution of sample, separation or masking of interfering ions	prevents simultaneous precipitation and reduces inclusion of impurities
precipitation	from hot dilute solution, careful addition or homogeneous generation of precipitating agent in small excess, with stirring: digestion	promotes particle growth and reduces occlusion N.B. digestion can increase postprecipitation
filtration	cooled solution filtered, precipitate washed with dilute electrolyte solution	decreases solubility, reduces adsorbed impurities and prevents peptization
drying and ignition	careful drying at 110 to 140°C access of air during ignition	prevents sputtering losses, ensures complete oxidation to a stoichiometric product
weighing	weighings carried out to nearest 0.1 mg reheating and reweighing until constant weight is obtained samples stored in a desiccator	special care required for hygroscopic solids

promoting particle growth as well as minimizing the occlusion of impurities. Some homogeneous precipitation reactions are shown in table 5.18. Secondly, the filtration method must be selected to fit the treatment of the precipitate. Where the material is merely to be dried and weighed, a sintered glass crucible is generally the most satisfactory. If an ignition step is to be used, however, the precipitate may be collected on a filter paper and transferred to a silica or platinum crucible for ignition or filtered on an asbestos pad in a Gooch crucible. Sintered silica crucibles are also used for these precipitates.

Magnesium may be precipitated from solution as $MgNH_4PO_4 \cdot 6H_2O$, a compound which has been widely used as a basis for the gravimetric determination of the element over many years. An examination of its use will serve to illustrate some of the problems associated with inorganic precipitations. The initial precipitation is made from a solution at pH = 11 to 12 by the addition of ammonium phosphate in excess.

$$Mg^{2+} + NH_4^+ + PO_4^{3-} + 6H_2O = MgNH_4PO_4 \cdot 6H_2O$$

Table 5.18   Some homogeneous precipitation processes

PRECIPITATING AGENT	GENERATION REACTION
OH^-	$(NH_2)_2CO + 3H_2O = CO_2 + 2NH_4^+ + 2OH^-$ urea
PO_4^{3-}	$(CH_3O)_3PO + 3H_2O = 3CH_3OH + H_3PO_4$ trimethylphosphate
$C_2O_4^{2-}$	$(C_2H_5O)_2C_2O_4 + 2H_2O = 2C_2H_5OH + H_2C_2O_4$ ethyl oxalate
SO_4^{2-}	$(CH_3O)_2SO_2 + 2H_2O = 2CH_3OH + SO_4^{2-} + 2H^+$ dimethyl sulphate
CO_3^{2-}	$Cl_3CCOOH + 2OH^- = CHCl_3 + CO_3^{2-} + H_2O$ trichloroacetic acid
S^{2-}	$\overset{S}{\overset{\|}{CH_3CNH_2}} + H_2O = \overset{O}{\overset{\|}{CH_3CNH_2}} + H_2S$ thioacetamide
8-hydroxyquinoline	

8-acetoxyquinoline

Some of the problems encountered in the analysis are:

(a) *Precipitate stoichiometry*
The precipitate is not stoichiometric and it is necessary to convert it to the pyrophosphate by ignition (1100°C) to obtain the most precise results.

$$2MgNH_4PO_4 = Mg_2P_2O_7 + 2NH_3 + H_2O$$

(b) *Solubility*
For a precipitate used in gravimetric analysis magnesium ammonium phosphate has a rather high solubility ($0.1\,g\,dm^{-3}$ in water at 20°C). Hence solution volumes should be kept as low as possible and the precipitate must be filtered from cold solution. Furthermore, the solution composition must be carefully controlled to ensure the maintenance of conditions of minimum solubility. The presence of ammonium salts and hydrogen ions increases the solubility so that the precipitant $(NH_4)_3PO_4$ must be added in only small excess, and the pH must be maintained high (pH = 11 to 12). Supersaturation

is an additional problem, and quantitative precipitation is only achieved after the mixture has been allowed to stand for an extended time, e.g. overnight.

(c) *Precipitate purity*

H_3PO_4, $NH_4H_2PO_4$, $(NH_4)_2HPO_4$, $(NH_3)_3PO_4$, $Mg(H_2PO_4)_2$, $MgHPO_4$, $Mg_3(PO_4)_2$, basic Mg phosphates, $Mg(OH)_2$, $MgCl_2$ and NH_4Cl are all possible contaminants. Of these only NH_4Cl which is volatile and $MgHPO_4$ which ignites to $Mg_2P_2O_7$ are of no consequence. Amongst the rest are a number of compounds (including the precipitant itself) which have crystallinity similar to the precipitate, and are coprecipitated by isomorphic inclusion. The magnitude of the coprecipitation problem is such that to overcome it dissolution and reprecipitation of the initial precipitate is required. This is done by using the minimum amount of dilute hydrochloric acid and a very small amount of ammonium phosphate. Concentrated ammonia solution is then added to complete the reprecipitation.

The use of an organic precipitant may be exemplified by reference to the employment of 8-hydroxyquinoline (oxine), to determine aluminium. Aluminium oxinate, $Al(C_9H_6NO)_3$, can be quantitatively precipitated from aqueous solutions between pH = 4.2 and pH = 9.8. Due to the non-selective nature of oxine as a reagent many other metals are also precipitated within the same range. To achieve a separation of aluminium from these other elements pH control may be employed. Thus an acetic acid solution, buffered to pH = 5 by sodium acetate, will provide a medium for the separation of aluminium from troublesome contaminants such as Be, Mg, Ca, Sr and Ba. Ammoniacal solution (pH = 9) will enable a separation from As, B, F and P to be made, and if used in conjunction with H_2O_2 from Cr, Mo, Nb, Ta and V as well. Selectivity is further enhanced by the use of masking agents such as cyanide, tartrate or EDTA. By careful control of these parameters aluminium may be separated, and few interferences are observed if the precipitation is carried out from an ammoniacal–cyanide–EDTA solution. When large amounts of Ca or Mg are present the homogeneous precipitation procedure (table 5.18) is usefully employed at pH = 5. The precipitate is readily filtered and may be weighed after drying at 150°C as the anhydrous compound.

APPLICATIONS OF GRAVIMETRY

Gravimetric methods provide precise and accurate results and have found a wide utility in chemical analysis for many years. They are best suited to the determination of major constituents in samples because of the practical limitations in accurately weighing quantities of less than 0.1 g. The analysis of rocks, ores, soils, metallurgical and other inorganic samples for their major components has depended very much on gravimetric methods. Gravimetric procedures are, however, usually time consuming and demanding, with the result that there is a steady trend towards the use of quicker, instrumentally based methods. Notable for its impact in recent years on these traditional areas has been X-ray fluorescence analysis (chapter 8). Nevertheless gravimetric methods are still very much needed to calibrate these newer procedures. Tables 5.14 and 5.15 give a good indication of the scope of gravimetry.

Problems

1. Suggest suitable colour change indicators for the titration of (a) acetic acid (0.100 M) with sodium hydroxide (0.100 M) and (b) nitric acid (0.0100 M) with sodium hydroxide (0.0100 M). Give the reasons for your choice.

2. Sketch a titration curve (pH v. volume of titrant) for phosphoric acid titrated with sodium hydroxide. Comment upon the curve shape and the reasons for it.

3. On p. 198 the problem of titrating Fe(II) in the presence of chloride ions is discussed and the use of Zimmerman–Rheinhardt reagent suggested. An alternative is to use a different oxidizing agent. Suggest one, and give the reasons for your choice.

4. Hydroxylamine (NH_2OH) is oxidized by ferric iron in boiling sulphuric acid—an oxide of nitrogen being amongst the products. $25.00 \, cm^3$ of a solution of hydroxylamine ($2.00 \, g.dm^{-3}$) were boiled with an excess of ferric chloride in dilute sulphuric acid. $30.30 \, cm^3$ of potassium permanganate solution (0.0200 M) were required to reoxidize the ferrous ions produced. Deduce the identity of the oxide of nitrogen.

5. Use the data in tables 5.6 and 5.7 to deduce a pH suitable for the quantitative titration of Cu^{2+} in the presence of a small amount of Ag^+.

6. Calculate the conditional formation constant for the Zn-EDTA complex at pH 9.0 in a solution 0.100 M with respect to ammonia. Can zinc be titrated quantitatively with EDTA solution at this pH?

7. 0.150 g of an alloy containing some magnesium was dissolved in a suitable solvent. Magnesium oxinate was precipitated quantitatively, filtered, dried and weighed. The weight of precipitate was 0.250 g. What was the percentage of magnesium in the original sample?

Further Reading
COOPER, D. and DORAN, C., *Classical Methods Vol I*, Wiley, Chichester, 1987.
MENDHAM, J., DODD, D. and COOPER, D., *Classical Methods Vol II*, Wiley, Chichester, 1987.
SKOOG, D. A. and WEST, D. M., *Fundamentals of Analytical Chemistry*, 4th Ed., CBS College Publishing, New York, 1982.
JEFFEREY, G. H., BASSETT, J., MENDHAM, J. and DENNEY, R. C. (eds.). *Vogel's Textbook of Quantitative Inorganic Analysis*, 5th Ed., Longman, London, 1989.

Chapter 6

Electrochemical Techniques

If a solution forms part of an electrochemical cell, the potential of the cell, the current flowing through it and its resistance are all determined by the chemical composition of the solution. Quantitative and qualitative information can thus be obtained by measuring one or more of these electrical properties under controlled conditions. Direct measurements can be made in which sample solutions are compared with standards; alternatively, the changes in an electrical property during the course of a titration can be followed to enable the equivalence point to be detected. Before considering the individual electrochemical techniques, some fundamental aspects of electrochemistry will be summarized in this section.

DEFINITIONS

Electrochemical cell. A pair of electrodes (metallic or otherwise) in contact with an electrolyte solution.
Galvanic or Voltaic cell. An electrochemical cell which spontaneously produces current (or energy) when the electrodes are connected externally by a conducting wire.
Electrolysis or electrolytic cell. An electrochemical cell through which current is forced by a battery or some other external source of energy.
Half-cell reactions. Oxidation or reduction reaction occurring at an electrode (table 6.1).
Anode. The electrode at which oxidation occurs.
Cathode. The electrode at which reduction occurs.
Reversible cell. One in which the half-cell reactions are reversed by reversing the current flow; such a cell is said to be in thermodynamic equilibrium.
Standard Hydrogen Electrode (*SHE*). This consists of a platinum electrode coated with platinum black to catalyse the electrode reaction and over the

surface of which hydrogen at 760 mm of mercury is passed. The electrode is in contact with a solution of hydrogen ions at unit activity (1.228 M HCl at 20°C) and its potential is arbitrarily chosen to be zero at all temperatures.

Electrode potential E. The potential of an electrode measured relative to a standard, usually the SHE. It is a measure of the driving force of the electrode reaction and is temperature and activity dependent (p. 225). By convention, the half-cell reaction must be written as a reduction and the potential designated positive if the reduction proceeds spontaneously with respect to the SHE, otherwise it is negative. If the sign of the potential is reversed, it must be referred to as an oxidation potential.

Table 6.1   Typical half-cell reactions

REDUCTIONS	OXIDATIONS
deposition of metals	dissolution of metals
e.g. $Ag^+ + e^- = Ag$	e.g. $Cd = Cd^{2+} + 2e^-$
formation of hydrogen gas	formation of oxygen gas
$2H^+ + 2e^- = H_2$	$2H_2O = O_2 + 4H^+ + 4e^-$
	oxidation of halogens
	e.g. $2Cl^- = Cl_2 + 2e^-$
change of oxidation state	
e.g. $Fe^{3+} + e^- = Fe^{2+}$	$Sn^{2+} = Sn^{4+} + 2e^-$

Standard electrode potential $E^{\ominus}$. Electrode potential measured in solutions where all reactants and products are at unit activity (p. 225).

Theoretical cell potential. The algebraic sum of the individual electrode potentials of an electrochemical cell at zero current, i.e. $E_{cell} = E_{cathode} + E_{anode}$. In practice, when current flows in a cell or a liquid junction is present *(vide infra)*, and for certain electrode systems or reactions, the cell potential departs from the theoretical value.

Liquid-junction potential. A potential developed across a boundary between electrolytes differing in concentration or chemical composition. It is caused by different rates of migration of cations and anions across the boundary thereby leading to a charge separation. Its value is often several hundredths of a volt and variable, but it can be minimized by using a *salt bridge* connection, e.g. an agar gel saturated with KCl or NH_4NO_3 for which the potential is only 1 to 2 mV.

Ohmic drop IR. A potential developed when a current I flows in an electrochemical cell. It is a consequence of the cell resistance R and is given by the product IR. It is always subtracted from the theoretical cell potential and therefore reduces that of a galvanic cell and increases the potential required to operate an electrolysis cell.

Activation overpotential (overvoltage). The additional potential required to cause some electrode reactions to proceed at an appreciable rate. The result of an 'energy barrier' to the electrode reaction concerned, it is substantial for gas evolution and for electrodes made of soft metals, e.g. Hg, Pb, Sn and Zn. It increases with current density and decreases with increasing temperature, but its magnitude is variable and indeterminate. It is negligible for the deposition of metals and for changes in oxidation state.

Concentration overpotential or concentration polarization. The additional potential required to maintain a current flowing in a cell when the concentration of the electroactive species at the electrode surface is less than that in the bulk solution. In extreme cases, the cell current reaches a limiting value determined by the rate of transport of the electroactive species to the electrode surface from the bulk solution. The current is then independent of cell potential and the electrode or cell is said to be completely polarized. Concentration overpotential decreases with stirring and with increasing electrode area, temperature and ionic strength.

ACTIVITY DEPENDENCE OF ELECTRODE POTENTIALS—THE NERNST EQUATION

Electrode and therefore cell potentials are very important analytically as their magnitudes are determined by the activities of the reactants and products involved in the electrode reactions. The relation between such activities and the electrode potential is given by the *Nernst equation*. For a general half-cell reaction written as a reduction, i.e. $aA + bB + \cdots ne = xX + yY + \cdots$, the equation is of the form

$$E = E^{\ominus} - \frac{RT}{nF} \ln \frac{[X]^x[Y]^y \ldots}{[A]^a[B]^b \ldots} \tag{6.1}$$

where E is the electrode potential, $E^{\ominus}$ the standard potential, [ ] the activities of reactants and products at the electrode surface, R the gas constant, T the thermodynamic temperature, F the Faraday constant (96 487 C mol^{-1}), and n the number of electrons involved in the electrode reaction. At 298.15 K $RT/F \ln 10 = 0.059\ 158$ V thus

$$E = E^{\ominus} - (0.059 \text{ V}/n) \log_{10} \frac{[X]^x[Y]^y}{[A]^a[B]^b} \tag{6.2}$$

If the activities of all reactants and products are unity, $E = E^{\ominus}$. Theoretical cell potentials can be calculated using tabulated values of $E^{\ominus}$. For dilute solutions ($<10^{-1}$ M) concentrations can be used in place of activities, the error becoming insignificant below 10^{-3} M.

REFERENCE ELECTRODES

Electroanalytical measurements relating potential or current to concentration rely on the response of one electrode only, the other ideally being independent of solution composition and conditions. The latter is known as a reference electrode; two such electrodes having these properties and in common use are based on calomel and silver–silver chloride respectively.

Calomel Electrode
The electrode consists of two concentric glass tubes, the inner one of which contains mercury in contact with a paste of mercury, mercury(I) chloride (Calomel), and potassium chloride. This is in contact with a solution of potassium chloride in the outer tube which itself makes contact with the sample solution via a porous frit, fibre or ground-glass sleeve, figure 6.1.

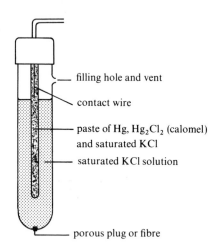

— filling hole and vent

— contact wire

— paste of Hg, Hg_2Cl_2 (calomel) and saturated KCl

— saturated KCl solution

— porous plug or fibre

Figure 6.1.   Saturated calomel reference electrode (SCE)

The half-cell and its associated electrode reaction may be represented as

$$Hg|KCl, Hg_2Cl_2(saturated)$$

$$Hg_2Cl_2 + 2e^- = 2Hg + 2Cl^-, E^\circ = 0.242 \text{ V}$$

The electrode potential is given by

$$E = 0.267\ 6 \text{ V} - (0.059/2 \text{ V}) \log_{10} [Cl^-]^2 \qquad (6.3)$$

and therefore depends on the chloride ion activity. The saturated Calomel electrode (SCE) is the easiest to prepare, although it is more temperature-sensitive than versions employing 1 M or 0.1 M potassium chloride.

Silver–Silver Chloride Electrode

This consists of a silver wire, coated with silver chloride and in contact with a solution of potassium chloride saturated with silver chloride. The solution is contained in a tube, the end of which is sealed with a porous plug or disc to facilitate contact with the sample solution. The half-cell and associated electrode reaction is represented by

$$Ag|KCl, AgCl(saturated)$$

$$AgCl + e^- = Ag + Cl^-, E^{\ominus} = 0.222 \text{ V}$$

The electrode potential is given by

$$E = 0.222 \text{ V} - 0.059 \text{ V} \log_{10}[Cl^-] \tag{6.4}$$

Like the Calomel electrode, the saturated KCl version of this electrode is the most convenient to prepare.

6.1 Potentiometry

SUMMARY

Principles

Measurement of the potential of a galvanic cell, usually at zero current; cell potential governed by the potential of an indicator electrode which responds to changes in the activity of the species of interest.

Instrumentation

Indicator and reference electrodes; potentiometer, pH meter or milli-voltmeter; electronics and recorder for automated systems.

Applications

Quantitative determination and monitoring of many species in solution over a wide range of concentrations (10^{-7} M to 1 M); relative precision 0.1 to 5%. Titrations are especially useful for coloured or turbid samples or for mixtures.

Disadvantages

Titrations are slow and time-consuming unless automated.

Potentiometry is the most widely used electroanalytical technique. It involves the measurement of the potential of a *galvanic cell*, usually under conditions of zero current, for which purpose potentiometers are used. Measurements may be 'direct' whereby the response of samples and standards are compared, or the change in cell potential during a titration can be monitored.

ELECTRODE SYSTEMS

The cell consists of an *indicator* and a *reference* electrode, the latter usually being the calomel or silver–silver chloride type. The potential of the indicator electrode is related to the activities of one or more of the components of the solution and it therefore determines the overall cell potential. Ideally, its response to changes of activity should be rapid, reversible and governed by the Nernst equation. There are two types of indicator electrode which possess the desired characteristics—metallic and membrane.

Metallic Indicator Electrodes

Metals such as silver, copper, mercury, lead and cadmium respond to variations in the activities of their own ions in a *Nernstian* and reproducible manner, e.g. for silver, the electrode reaction is $Ag^+ + e^- = Ag$, and the electrode potential is given by

$$E = E^{\ominus} - 0.059 \text{ V} \log_{10} \frac{1}{[Ag^+]} \tag{6.5}$$

Iron, nickel, cobalt, tungsten and chromium do not behave reproducibly due to crystal strain or oxide coatings. Metal electrodes which respond directly to solutions of their own ions are called 'Class I' or 'first order'.

Metals which form sparingly soluble salts will also respond to changes in the activity of the relevant anion provided the solution is saturated with the salt, e.g. for silver in contact with a saturated solution of silver chloride and containing solid silver chloride the electrode reaction is $AgCl + e^- = Ag + Cl^-$, and the electrode potential is given by:

$$E = E^{\ominus} - 0.059 \text{ V} \log_{10}[Cl^-] \tag{6.6}$$

Such electrodes are described as 'Class II' or 'second order'.

For titrations involving a change in oxidation state (redox systems) an inert electrode material such as platinum is used. The electrode potential is determined by the proportions of oxidized and reduced forms present, e.g.

$$Fe^{3+} + e^- = Fe^{2+}$$

and

$$E = E^{\ominus} - 0.059 \text{ V} \log_{10} \frac{[Fe^{2+}]}{[Fe^{3+}]} \tag{6.7}$$

Membrane or Ion-Selective Electrodes
These can be subdivided into

(a) glass electrodes
(b) solid-state electrodes
(c) liquid-membrane electrodes
(d) gas-sensing electrodes.

The construction and mechanism for the development of activity-dependent potentials is similar for all types, although gas-sensing electrodes are constructed slightly differently from the others and incorporate an internal pH-sensitive glass electrode. The first three types consist of a tube into one end of which is sealed an electrically-conducting membrane. The tube contains a solution or gel incorporating the ion to which the electrode is to respond, and another electrolyte, usually potassium or sodium chloride. The latter, together with a silver wire in contact with the solution or gel, constitutes an internal silver–silver chloride reference electrode. The cell is completed with a second or 'external' reference electrode. On immersion of both electrodes into a solution containing the ion to be monitored, a potential develops across the membrane the magnitude of which is related to the activities of the ion of interest in the internal and external (sample) solution. The response of many membranes is highly selective in that the *membrane potential* is a function of the activity of only one ion or a small number of ions.

In essence, the cell comprises two reference electrodes, whose potentials are constant, separated by the membrane whose potential governs the overall cell potential. Ideally, the response will be Nernstian, and at 298.15 K the cell potential is given by

$$E_{cell} = k - \frac{0.059 \text{ V}}{n} \log_{10} \frac{a_1}{a_2} \tag{6.8}$$

where k is a constant including the external and internal reference electrode potentials, and a_1 and a_2 are the activities of the ion to be measured in the external and internal solutions respectively. As a_2 is also constant, then

$$E_{cell} = k' - \frac{0.059 \text{ V}}{n} \log_{10} a_1 \tag{6.9}$$

where k' includes $\log_{10} a_2$.

Electrode Response and Selectivity
The term '*ion-selective*' is to be preferred to '*ion-specific*' in discussing membrane electrodes as most if not all are subject to the influence of ions other than the one to which they nominally respond. Interference from such ions and other species in solution may be either chemical or electrical in origin. As

electrode response is a function of the *activity* of the ion to be monitored rather than concentration, reactions resulting in its partial or complete complexation, precipitation or other chemical changes will reduce the activity of the ion and consequently the electrode response, i.e. the electrode responds only to the *'free'* ion in solution. These effects are illustrated for the fluoride ion as follows:

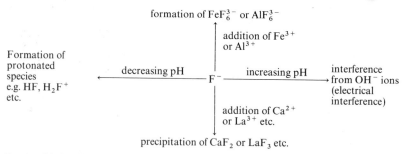

Ions which affect the membrane potential directly will produce an apparent increase in activity of the ion to which the electrode nominally responds. Under these circumstances the cell potential is more accurately given by the expression

$$E_{cell} = k' - \frac{0.059}{n} \log_{10}(a_1 + k_{1,2} a_2^{n/z})$$ (6.10)

where a_2 is the activity of the interfering ion of charge Z, and $k_{1,2}$ is the selectivity ratio for ion 1 over ion 2. If $k_{1,2}$ is zero, the electrode is truly specific for ion 1, whereas values greatly in excess of one indicate that the electrode would be subject to severe interference from the second ion. As selectivity ratios are influenced by overall solution composition they should be regarded as approximations only. Furthermore, confusion can arise because reciprocal values may be quoted, e.g. if $k_{1,2} = 0.005$ then $k_{1,2}^{-1} = 200$. Some examples of selectivity ratios are included in table 6.2.

(a) Glass Electrodes for pH Measurements

The membrane consists of a thin envelope of soft glass sealed into the end of a hard-glass tube. The tube is filled with a dilute solution of hydrochloric acid in which a silver wire is immersed thus forming a silver–silver chloride reference electrode, figure 6.2. The acid also provides a solution of hydrogen ions of constant activity a_2. The chemical composition and physical characteristics of the glass membrane are critical in determining the electrode response. Soda-glasses are highly hygroscopic, have a high electrical conductivity and show a good response whereas pyrex glass or quartz is virtually insensitive. Typically, the composition of a suitable soda-glass is 72% SiO_2, 22% Na_2O, and 6% CaO. The surface of the glass *must* be hydrated for the

Table 6.2   Characteristics of some ion selective electrodes

ION TO BE MONI- TORED	TYPE OF ELECTRODE	CONCENTRA- TION RANGE, M	OPTIMUM pH	APPROXIMATE VALUES OF SELECTIVITY CONSTANTS FOR INTERFERING IONS, $k_{1,2}$	
Na^+	glass	10^0-10^5	>7	H^+	10^2
				Cs^+, Li^+	0.002
				K^+	0.001
Br^-	solid-state	$10^0-5\times10^{-6}$	$2-12$	S^{2-}, I^-, CN^-	$\sim10^6$
Cl^-	solid-state	$10^6-5\times10^{-5}$	$2-11$	I^-, CN^-	$\sim10^9$
				S^{2-}	$\sim10^6$
				Br^-	$\sim10^5$
F^-	solid-state single crystal	10^0-10^6	$5-8$	OH^-	$\sim10^4$
Ca^{2+}	pvc-gel	$10^0-5\times10^{-7}$	$6-8$	Zn^{2+}	3.2
				Pb^{2+}	0.063
				Mg^{2+}	0.014
NO_3^-	pvc-gel	$10^0-7\times10^{-6}$	$3-10$	ClO_4^-	$\sim10^6$
				I^-	20
				Br^-	0.1
				NO_2^-	0.04
				Cl^-	0.004
CO_2	gas-sensing	$10^{-2}-10^{-4}$	—	volatile, weak acids interfere	
NH_3	gas-sensing	10^0-10^{-6}	—	volatile amines interfere	

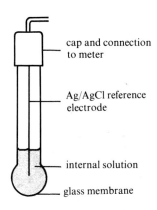

cap and connection to meter

Ag/AgCl reference electrode

internal solution

glass membrane

Figure 6.2.   Glass electrode

membrane to be pH sensitive. This is achieved by soaking the electrode in water or an aqueous buffer solution for several hours when the following ion-exchange reaction occurs

$$H^+(\text{solution}) + Na^+(\text{glass}) = Na^+(\text{solution}) + H^+(\text{glass})$$

The equilibrium is driven far to the right by this prolonged treatment, and results in the formation of hydrated gel layers at the inner and outer surfaces of the membrane. A diagrammatic representation of a cross-section of the membrane is shown in figure 6.3.

external solution $[H^+] = a_1$	hydrated gel $\leftarrow \sim 10^{-4}\,\text{mm} \rightarrow$ $H^+ + Na^+$	dry glass layer $\leftarrow \sim 10^{-1}\,\text{mm} \rightarrow$ Na^+ only	hydrated gel $\leftarrow \sim 10^{-4}\,\text{mm} \rightarrow$ $Na^+ + H^+$	internal solution $[H^+] = a_2$

Figure 6.3. Cross-section of a glass membrane

Although the surfaces of the gel layers contain only hydrogen ions, the ratio $[H^+]/[Na^+]$ decreases towards the dry glass layer which contains sodium ions only. Differences in hydrogen ion activity between the internal and external solutions result in a potential developing across the membrane by virtue of ion-exchange and a migration of charge. The charge is carried by sodium and hydrogen ions within the gel layers, by proton–proton exchange across the gel-solution interfaces and by the movement of sodium ions between vacant lattice sites (defects) in the dry glass layer. The potential developed is given by an equation similar to equation (6.9)

$$E_{\text{membrane}} = k'' - 0.059\ \text{V}\ \log_{10} a_{H^+} \tag{6.11}$$

where a_{H^+} is the activity of hydrogen ions in the external solution and k'' contains the logarithm of the activity of the internal solution and an *asymmetry potential* which is caused by strains in the curved membrane structure. The asymmetry potential changes with time primarily because of changes in the external gel surface with use or because of contamination. The electrode therefore requires frequent calibration when used for direct measurements.

Errors in the Use of a Glass Electrode

Because soda-glass membranes contain a high proportion of sodium ions, they exhibit a marked response to sodium ions in solution. The effect becomes increasingly significant as the hydrogen ion activity decreases, i.e. at high pH and it is sometimes referred to as the *alkaline error*. At pH 12, the error is about 0.3 of a pH unit if the solution is 0.1 M with respect to sodium ions, and 0.7 of a pH unit if the solution is 1 M in sodium ions. Other monovalent cations such as lithium and potassium have a similar but smaller effect. By replacing the sodium in the glass with lithium, and the calcium with strontium or barium, the error can be reduced by about one order of magnitude. So called 'wide range' glass electrodes which have alkaline errors as low as 0.1 of a pH unit in a 2 M sodium solution are commercially available. Below pH 1, the electrode is unlikely to have a Nernstian or a reproducible response, the effect being known as the *acid error*.

Glass Electrodes for the Determination of Cations other than Hydrogen

The alkaline error shown by a soda-glass membrane has been exploited in developing membranes with a high selectivity for sodium and other cations. For example, a study of glass composition has shown that the presence of Al_2O_3 or B_2O_3 can enhance the response to sodium ions relative to hydrogen ions. The membrane potential is then given by

$$E_{membrane} = k'' - 0.059 \text{ V} \log_{10}\left([H^+] + K\frac{\mu_{Na^+}}{\mu_{H^+}}[Na^+]\right) \quad (6.12)$$

where K is the equilibrium constant for the reaction

$$H^+(glass) + Na^+(solution) = Na^+(glass) + H^+(solution)$$

and is known as the *selectivity ratio*, μ_{Na^+} and μ_{H^+} being the mobilities of the sodium and hydrogen ions within the gel layers. For large values of K and at high pH, such an electrode exhibits a Nernstian response to sodium ions. Sodium and other cation-sensitive glass electrodes are available commercially; those with a high selectivity for H^+, Na^+, Ag^+ or Li^+ have proved to be the most satisfactory.

(b) Solid State Electrodes

These incorporate membranes fabricated from insoluble crystalline materials. They can be in the form of a single crystal, a compressed disc of microcrystalline material or an agglomerate of micro-crystals embedded in a silicone rubber or paraffin matrix which is moulded in the form of a thin disc. The materials used are highly insoluble salts such as lanthanum fluoride, barium sulphate, silver halides and metal sulphides. These types of membrane show a selective and Nernstian response to solutions containing either the cation or the anion of the salt used. Factors to be considered in the fabrication of a suitable membrane include solubility, mechanical strength, conductivity and resistance to abrasion or corrosion.

In all cases the mechanism by which the membrane responds to changes in the activity of the appropriate ion in the external solution is one of ionic conduction. Lattice defects in the crystals allow small ions with a low formal charge a degree of mobility within the membrane. It is this shift in charge centres which gives rise to the membrane potential, the process resembling the movement of sodium ions in the dry glass layer of a glass membrane.

A fluoride electrode, in which the membrane is a single crystal of lanthanum fluoride doped with europium to increase the conductivity, is one of the best ion-selective electrodes available. Conduction through the membrane is facilitated by the movement of F^- ions between anionic lattice sites which in turn is influenced by the F^- ion activities on each side of the membrane. If the electrode is filled with a standard solution of sodium fluoride, the membrane potential is a function of the fluoride activity in the sample solution only. Thus,

$$E_{cell} = k' - 0.059 \log_{10} a_{F^-} \tag{6.13}$$

where k' incorporates the logarithm of the activity of the internal fluoride solution. The electrode has a Nernstian response down to 10^{-5} M F^- (0.19 ppm) and selectivity ratios of less than 0.001 for all other anions except OH^-.

Electrodes responding to other halides, sulphide, cyanide, silver, lead, copper and cadmium are made using membranes fabricated from pure or doped silver sulphide (Ag_2S). The membrane potential is affected by the movement of Ag^+ ions between cationic lattice sites which in turn is determined by the activities of the Ag^+ ion in the internal and sample solutions. As the activity of the former is fixed, that of the latter alone influences the membrane potential. The electrode will also respond to the presence of S^{2-} ions because of their effect on the Ag^+ ion activity via the solubility product expression:

$$Ag_2S \text{ (membrane)} = 2Ag^+ + S^{2-} \text{ (solution)}$$

Silver halide and thiocyanate membranes would respond in a similar way to a silver sulphide membrane, Ag^+ ions being the mobile species, but by themselves make unsuitable membrane materials. A Nernstian response is, however, retained when they are incorporated into a Ag_2S matrix, the membrane behaving as if it were a pure halide or thiocyanate conductor, i.e.

$$AgX \text{ (membrane)} = Ag^+ + X^- \text{ (solution)}$$

where $X^- = Cl^-, Br^-, I^-$ or SCN^-.
As $K_{sp} = [Ag^+][X^-]$, the membrane potential is given by

$$E = k' - 0.059 \log_{10} \frac{K_{sp}}{a_{X^-}} \tag{6.14}$$

or

$$E_{cell} = k'' + 0.059 \log_{10} a_{X^-} \tag{6.15}$$

where k'' includes the logarithm of the activity of the internal silver solution and the solubility product of the silver salt incorporated into the membrane.

Similarly, doping Ag_2S with a divalent metal sulphide enables the electrode to respond to the corresponding metal (e.g. Pb, Cu or Cd). In all cases the electrode responds by virtue of the solubility product equilibria involved, the movement of Ag^+ ions within the membrane being the potential-governing factor. Selectivity ratios for some of these electrodes are given in table 6.2.

(c) *Liquid Membrane Electrodes*

Two types have been developed. One utilizes liquid ion-exchangers with a selective response to certain polyvalent cations or anions. The other involves the selective complexing ability of univalent cations by neutral macrocyclic antibiotics and cyclic polyethers. In earlier versions of both types, a water-immiscible liquid supported on an inert porous plastic disc formed the membrane but gradual leakage and dissolution of the membrane liquid necessitated the frequent replenishment of a reservoir inside the electrode body. Later versions employ PVC-gelled membranes incorporating the appropriate liquid and these behave more like a solid-state membrane, no replenishment of the liquid being required during use.

Electrodes with liquid ion-exchange membranes are typified by a calcium-sensitive electrode, figure 6.4. The membrane-liquid consists of the calcium

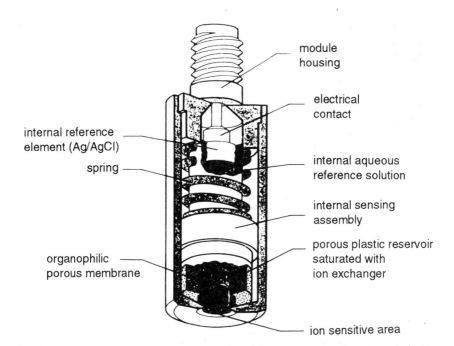

Figure 6.4.   Calcium sensitive ion-exchange membrane electrode

form of a di-alkyl phosphoric acid, $[(RO)_2POO^-]_2Ca^{2+}$, which is prepared by repeated treatment of the acid with a calcium salt. The internal solution is calcium chloride and the membrane potential, which is determined by the extent of ion-exchange reactions at the interfaces between the membrane and the internal and sample solutions, is given by

$$E = k' + \frac{0.059}{2} \log_{10} a_{Ca^{2+}} \qquad (6.16)$$

The electrode responds to Ca^{2+} down to 10^{-6} M (0.04 ppm) and is independent of pH in the range 6 to 10. Other electrodes of this type include $(Ca^{2+} + Mg^{2+})$, K^+, ClO_4^-, NO_3^- and BF_4^-. Some selectivity ratios are given in table 6.2.

Electrodes based on solutions of cyclic polyethers in hydrocarbons show a selective response to alkali metal cations. The cyclic structure and physical dimensions of these compounds enable them to surround and replace the hydration shell of the cations and carry them into the membrane phase.

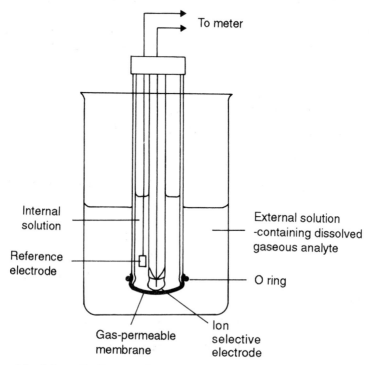

Figure 6.5. Schematic diagram of a gas-sensing electrode. Selective ion electrode is shown as a glass electrode. The reference electrode is an Ag–AgCl electrode. Other electrode combinations are possible

Conduction occurs by diffusion of these charged complexes, which constitute a 'space charge', within the membrane. Electrodes with a high selectivity for potassium over sodium ($> 1\ 000:1$) have been produced.

(d) Gas-Sensing Electrodes

The concentration of gases such as CO_2, NH_3, SO_2 and NO_2 in aqueous solutions can be measured with an electrode consisting of a glass electrode/reference electrode pair inside a plastic tube which is sealed with a thin gas-permeable membrane and containing an appropriate electrolyte solution (figure 6.5). The membrane consists of a microporous hydrophobic plastic film through which only gases can diffuse. On immersion of the electrode in a solution containing a dissolved gas (CO_2, NH_3, SO_2 or NO_2), rapid diffusion through the membrane occurs until an equilibrium is established in which the gas concentrations are the same in the internal electrolyte and the sample solution. Any change of the gas concentration in the internal electrolyte solution results in a change in its pH which in turn is sensed by the glass/reference pair. Such changes are directly proportional to the concentration of the gas in the sample solution.

The following equations summarize the equilibria involved for a CO_2-sensing electrode which utilizes a sodium hydrogen carbonate solution as the internal electrolyte:

CO_2 (sample) $\rightleftharpoons CO_2$ (membrane)

CO_2 (membrane) $\rightleftharpoons CO_2$ (internal electrolyte)

CO_2 (internal electrolyte) $+ 2H_2O \rightleftharpoons H_3O^+ + HCO_3^-$

The overall reaction may be represented as:

$$CO_2 \text{ (sample)} + 2H_2O \rightleftharpoons H_3O^+ + HCO_3^- \text{ (internal electrolyte)} \quad (6.17)$$

whence

$$K_e = \frac{[H_3O^+][HCO_3^-]}{[CO_2 \text{ (sample)}]}$$

By using a high concentration of HCO_3^- in the internal electrolyte, $[HCO_3^-]$ can be considered constant,

$$\therefore [CO_2 \text{ (sample)}] = k'[H_3O^+] = k'a_{H^+}$$

where k' includes K_e and the response of the electrode is given by

$$E_{membrane} = k'' - 0.059V \log_{10}[CO_2 \text{ (sample)}] \quad (6.18)$$

and k'' includes k' (cf. equation 6.11).

Thus, the pH response of the glass/reference pair is a function of the CO_2 content of the sample.

Similar equations for an ammonia-sensing electrode whose internal electrolyte is an ammonium salt lead to the relation

$$[NH_3 \text{ (sample)}] = k'[OH^-]$$

and the pH response of the glass/electrode pair is a function of the NH_3 concentration in the sample.

Gas-sensing electrodes differ from ion-selective electrodes in that no species in solution can interfere with the electrode response as only gases can diffuse through the membrane. However, it should be noted that any gas which causes a pH change in the internal electrolyte solution will affect electrode response.

DIRECT POTENTIOMETRIC MEASUREMENTS

Methods in which the cell potential for the sample solution is compared with that for one or more standards are rapid, simple and readily automated. The measurement of pH is the most common application of this type, one or more buffer solutions serving to calibrate the pH-meter (potentiometer). In all such measurements, calibration involves the evaluation of the constant k' in the equation

$$E_{\text{cell}} = k' + \frac{0.059 \text{ V}}{n} pX \qquad (6.19)$$

where pX is the negative decadic logarithm of the activity of species X (see equation 3.1). Thus, k' is determined for a solution of known pX and the value used for the measurement of solutions of unknown pX, i.e.

$$pX = \frac{E_{\text{cell}} - k'}{0.059 \text{ V}/n} \qquad (6.20)$$

In practice, the value of k' is never obtained as such, because the meter is adjusted so that the standard reads the correct value for its pX, the scale being Nernstian. As k' contains in addition to the reference electrode potentials, a liquid-junction potential and an asymmetry potential, frequent standardization of the system is necessary. The uncertainty in the value of the junction potential, even when a salt bridge is used, is of the order of 0.5 mV. Consequently the absolute uncertainty in the measurement of pX is always at least $0.001/(0.059/n)$ or 0.02 if $n = 1$, i.e. a relative precision of about 2% at best. For the most precise work a standard addition technique (p. 32) and close temperature control are desirable. All measurements should be made at constant ionic strength because of its effect on activities. Likewise, the presence of complexing agents or precipitants must be controlled as this can significantly decrease the activity of the analyte resulting in gross changes in electrode response at a given total concentration level.

POTENTIOMETRIC TITRATIONS

By monitoring the change of E_{cell} during the course of a titration, where the indicator electrode responds to one of the reactants or products, the stoichiometric or equivalence point can be located. A plot of E_{cell} against volume of titrant added gives a characteristic 'S-shaped' curve, figure 6.6(a), owing to the logarithmic relation between E_{cell} and activity. If the reactants are in a 1:1 mole ratio, the curve is symmetrical as shown and the equivalence point is the mid-point of the inflection. This is true for all acid-base, silver-halide and many other titrations. Where the mole ratio of reactants is not 1:1, an asymmetrical curve is produced and the equivalence point does not coincide with the inflection point. The error incurred in assuming the two points coincide is generally less than 1 % and can be eliminated by titrating standards. Figures 6.6(b) and (c) show the first and second derivative curves of a normal symmetrical titration curve. These may offer some advantage in locating the equivalence point. However, precise location of the inflection point is dependent on the collection of data in its immediate vicinity where the cell potential is likely to be unstable due to the low concentrations of electroactive species. Derivative curves should therefore be used with caution and are probably not worth the extra effort involved in their computation although some automated titration systems can produce a derivative curve directly.

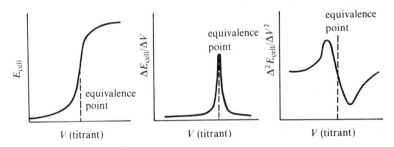

Figure 6.6.   Potentiometric titration curves. (a) Normal curve. (b) First derivative curve. (c) Second derivative curve

Potentiometric titrations yield more precise results than direct measurements, better than 0.2% if required. This is because the data collected effectively averages E_{cell} over a large number of readings. The principal types of potentiometric titration are summarized in the following sections.

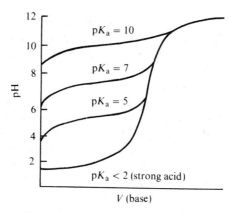

Figure 6.7.   Acid-base titration curves

Acid-base Titrations

The magnitude of the potential change or 'break' at the equivalence point depends on the strengths of the acid and base involved. Curves for the titration of acids with pK_a values ranging from 4 to 10 using a strong base as titrant are shown in figure 6.7. The size of the inflection diminishes as the acids become weaker (larger pK_a) until for a $pK_a > 10$ it is difficult to detect the equivalence point at all. A similar effect occurs for the titration of weak bases with a strong acid. The potentiometric method can be very useful for titrating mixtures of acids or a polyprotic acid where a visual indicator would show only one colour change. It can also be used to titrate the salt of a weak acid or base with a strong acid or base, e.g. in the standardization of a hydrobromic acid solution with sodium acetate for an epoxide determination. The titration data can be used to calculate the pK_a and equivalent weight of an unknown partly dissociated acid, HA. Thus, $HA = H^+ + A^-$ and at a point half-way through the titration,

$$[HA] = [A^-]$$

$$\therefore K_a = \frac{[H^+][A^-]}{[HA]} = [H^+] \tag{6.21}$$

or

$$pK_a = pH \tag{6.22}$$

Precipitation Titrations

Such titrations involve the progressive removal of a species from solution by precipitation. Ions frequently used for this purpose include Ag^+, Pb^{2+}, Cu^+, Zn^{2+}, CNS^-, SO_4^{2-}, S^{2-} and halides. The indicator electrode may be of the metallic or membrane type, e.g. Ag for a Ag^+–halide titration or F^- membrane for a La^{3+}–fluoride titration. Precipitation titration curves usually show departures from the theoretical curve due to adsorption of ions present in excess on the surface of the precipitate. Thus, if a solution of an iodide is titrated with silver nitrate, before the equivalence point the AgI precipitate adsorbs I^- ions thereby changing E_{cell}. After the equivalence point, the excess of Ag^+ ions is adsorbed on the precipitate producing a similar but opposite effect on E_{cell}. The net result is to make the inflection in the titration curve less sharp, figure 6.8(a). Mixtures can be titrated where the solubility products of the individual salts with the titrant ion differ significantly, e.g. mixtures of halides, whose silver salts have the following solubility products:

$$K_{AgI} = 8.5 \times 10^{-17} \, mol^2 \, dm^{-6}$$

$$K_{AgBr} = 3.3 \times 10^{-13} \, mol^2 \, dm^{-6}$$

$$K_{AgCl} = 1.72 \times 10^{-10} \, mol^2 \, dm^{-6}$$

Theoretical (dotted line) and experimental (continuous line) titration curves for such a mixture are shown in figure 6.8(b). The formation of mixed crystals and solid solutions limits the accuracy to 1 to 2% when the halides are present in similar concentrations.

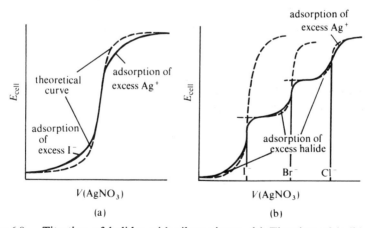

Figure 6.8.   Titration of halides with silver nitrate. (a) Titration of iodide with $AgNO_3$. (b) Titration of a mixture of chloride, bromide and iodide with $AgNO_3$

Redox Titrations
Platinum indicator electrodes respond to the ratio of oxidized to reduced forms for such oxidants as Ce^{4+}, MnO_4^-, $Cr_2O_7^{2-}$, I_2, $Fe(CN)_6^{3-}$, IO_3^-,

BrO_3^- and reductants such as Fe^{2+}, AsO_3^{3-}, Ti^{3+} and $S_2O_3^{2-}$. The titration curves are often unsymmetrical due to the stoichiometry of the reaction, but the error is small ($<1\%$) or can be eliminated by titrating standards. If the reduction potentials of two species differ by >0.2 V, separate inflections will be observed when titrating the mixture, e.g. the titration of Fe^{2+} and Ti^{3+} with potassium permanganate.

Some redox systems are slow to reach equilibrium at a platinum electrode after the addition of an increment of titrant, e.g.

$$Cr^{3+}/Cr_2O_7^{2-}, \quad Mn^{2+}/MnO_4^-, \quad S_2O_3^{2-}/S_4O_6^{2-}$$

For titrations involving those species, the response can be improved by forming a cell from a pair of identical platinum electrodes and passing a very small constant current (~ 5 μA) between them throughout the titration. As electrolysis is negligible, the solution composition is virtually unaffected by the passage of the current. The break in the titration curve is usually sharp and can be 100 to 200 mV or more. The use of a pair of platinum electrodes may be particularly advantageous if contamination of the sample solution is undesirable, or for non-aqueous titrations where a calomel or silver–silver chloride reference electrode may become blocked by precipitation of the salt at the liquid junction.

NULL-POINT POTENTIOMETRY

This is a method involving a two compartment cell with a salt bridge connection and having two identical indicator electrodes. The sample solution is placed in one compartment and a blank solution having the same total ionic strength in the other. Increments of a standard solution of the species to be determined are added to the blank compartment until the cell potential is zero. At this point, the activities of the species of interest in each compartment are equal and that of the sample solution can therefore be calculated. A concentrated standard solution should be used to minimize dilution errors. This method is particularly useful for the determination of trace amounts or where no suitable titrant can be found.

APPLICATIONS OF POTENTIOMETRY

The measurement of pH using a glass electrode is a necessary part of a large number of analytical procedures. In addition, the measurement and control of acidity is important in process streams and in such areas as soil science, water treatment technology and clinical diagnosis. A specific example of the latter is the measurement of CO_2 in blood and respiratory gases during operations using a gas-sensing electrode. In general, ion-selective electrodes

are finding increasing and widespread use for monitoring purposes in laboratories, in plant and for on-site analyses such as those related to oceanographic
and river studies. Because of their activity rather than concentration
dependence they are also valuable in fundamental studies of solubility and
complexation. One of the most successful and widely-used is the fluoride
electrode which is now the basis of routine methods for the determination of
fluoride in drinking water, toothpastes, soils, plant tissue and biological
fluids.

Potentiometric titrations are readily automated by using a motor-driven
syringe or an automatic burette coupled to a chart recorder or digital printout system. This is described in more detail in chapter 12. A microprocessor-
controlled titrator is discussed in chapter 13.

6.2   Polarography, Stripping Voltammetry and Amperometric Techniques

SUMMARY

Principles
> Measurement of the diffusion-controlled current flowing in an electro-
> lysis cell in which one electrode is polarizable. The current is directly
> proportional to the concentration of an electroactive species.

Instrumentation
> Micro-electrode (dropping mercury carbon or platinum), reference
> electrode, variable dc source, electronics and recorder.

Applications
> Quantitative and qualitative determination of metals and organic com-
> pounds at trace levels (10^{-4} M to 10^{-8} M); relative precision 2 to 3%.
> Amperometric titrations are more versatile and more precise than
> polarography.

Disadvantages
> Measurements very sensitive to solution composition, dissolved oxygen
> and capillary characteristics. Impurities in background electrolyte limit
> sensitivity.

The study of current-potential relations in an electrolysis cell where the
current is determined solely by the rate of diffusion of an electroactive
species is called *voltammetry*. To obtain diffusion-controlled currents, the
solution must be unstirred and the temperature of the cell thermostatically
controlled so as to eliminate mechanical and thermal convection. In addition,

a high concentration of an electrochemically inert *background* or *supporting* electrolyte is added to the solution to suppress the migration of electroactive species towards the electrodes by electrostatic attraction. Typically, the cell comprises a mercury or platinum micro-electrode, which is readily polarizable, and a calomel or mercury-pool reference electrode which is non-polarizable. By using a small polarizable electrode, conditions can readily be attained wherein the diffusion current is independent of applied potential and directly proportional to the concentration of electroactive species in the bulk solution. Measurement of such *limiting currents* forms the basis of quantitative analysis. The polarizable micro-electrode is usually made the cathode at which the electroactive species is reduced. The most widely used electrode is the dropping mercury electrode (DME) and the technique involving its use is known as *polarography*. A plot of current flowing in the cell as a function of the applied potential is called a *polarogram* or a *polarographic wave*, figure 6.9. At small applied potentials, only a *residual current*

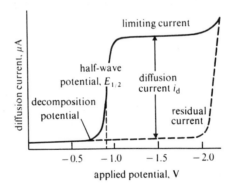

Figure 6.9.   Polarographic wave or polarogram

flows in the cell caused by the reduction of trace impurities in the sample solution and by charging of the mercury drops. The charging effect is analogous to the behaviour of a condenser. Above the *decomposition potential*, at which point reduction of an electroactive species is initiated, the current increases with applied potential, until it levels off at a limiting value. The difference between the limiting value and the residual current is known as the diffusion current, i_d. If, on increasing the applied potential further, other species in the solution are reduced, additional polarographic waves will be observed. Finally, the current will increase due to reduction of the supporting electrolyte or of the electrode material. In each polarographic wave, the potential at which the diffusion current reaches half the limiting value is known as the half-wave potential $E_{\frac{1}{2}}$ and is characteristic of the particular electroactive species involved. It is therefore useful for qualitative identification.

DIFFUSION CURRENTS

When the potential applied to a polarographic cell exceeds the decomposition potential of an electroactive species, its concentration at the surface of the mercury drop is immediately diminished. A concentration gradient is thereby established and more of that species diffuses from the bulk solution to the electrode surface (Fick's law of diffusion). The resulting current flow is proportional to the rate of diffusion which in turn is determined by the concentration gradient, i.e.

$$i = k(C - C_0) \qquad (6.23)$$

where C and C_0 are the concentrations of the electroactive species in the bulk solution and at the surface of the DME respectively. By progressively increasing the applied potential, reduction occurs more rapidly, C_0 eventually becomes virtually zero, and the concentration gradient reaches a maximum. At this point, the rate of diffusion and therefore the current flowing in the cell reaches a limiting value, i.e.

$$i_d = kC \qquad (6.24)$$

Further increases in the applied potential do not increase the current and the cell is said to be completely polarized or operating under conditions of high concentration overpotential (p. 225). The diffusion current i_d is hence directly proportional to the bulk concentration of the electroactive species.

HALF-WAVE POTENTIALS

Electroactive species are characterized by their $E_{\frac{1}{2}}$ values which are constants related to the standard electrode potentials and given by the equation

$$E_{\frac{1}{2}} = E^{\ominus} - \frac{0.059 \text{ V}}{n} \log \frac{k}{k_r} - E_{\text{reference}} \qquad (6.25)$$

where k and k_r are proportionality constants relating cell current to the rates of diffusion of oxidized and reduced forms of the electroactive species. Values are independent of the bulk concentration C but depend on the composition of the supporting electrolyte which can affect $E^{\ominus}$ if complexes are formed. The components of a mixture will give separate polarographic waves if the $E_{\frac{1}{2}}$ values differ by at least 0.1 V. Often, if two waves overlap, by careful choice of a complexing agent, the $E^{\ominus}$ and hence the $E_{\frac{1}{2}}$ value of one component can be changed by up to a volt or more. For example, in a potassium chloride solution Fe(III) and Cu(II) waves overlap whereas in fluoride medium, the value of $E_{\frac{1}{2}}$ for the FeF_6^{3-} anion is 0.5 V more negative, the Cu(II) wave being unaffected. Alkali metals which are reduced at potentials around -2.0 V can be determined using a tetraalkylammonium salt as the supporting electrolyte. Some other examples of the effect of complexing are given in table 6.3.

Table 6.3   The effect of complexation on half-wave potentials (volts)

ION	NO COMPLEXING AGENT $E_{\frac{1}{2}}/V$	KCN $E_{\frac{1}{2}}/V$	NH$_3$ $E_{\frac{1}{2}}/V$
Cu^{2+}	$+0.02$	*	-0.24 and -0.51
Zn^{2+}	-1.00	*	-1.35
Cd^{2+}	-0.59	-1.18	-0.81
Pb^{2+}	-0.40	-0.72	—

* not reducible

CHARACTERISTICS OF THE DME

A diagrammatic representation of a DME is shown in figure 6.10. Mercury from a reservoir is forced through a narrow-bore glass capillary (~ 0.06 mm

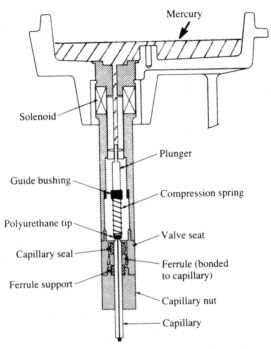

Figure 6.10.   Dropping mercury electrode

bore) by gravity. A succession of identical drops is formed which are detached at regular intervals (3 to 8 s), timing and reproducibility of drop dimensions being ensured by electrical control. This characteristic of the DME results in an oscillating cell current, the average of which is given by the Ilkovic equation

$$i_d = AnD^{\frac{1}{2}}m^{\frac{2}{3}}t^{\frac{1}{6}}C \tag{6.26}$$

where n denotes the number of Faradays, D is the diffusion coefficient for the electroactive species, m is the rate of flow of mercury, t is the drop time, C is the concentration in the bulk solution, and A is a constant. The equation is useful in comparing diffusion currents from electrodes with different capillary characteristics, i.e. different values of $m^{\frac{2}{3}}t^{\frac{1}{6}}$. The diffusion coefficient D is temperature sensitive to the extent of about 2.5% per kelvin; so that it is essential to control the cell temperature to ± 0.1 K if the highest precision is required.

In practice, a DME has certain advantages over other micro-electrodes:

1. The surface is continually and reproducibly renewed so that its past history is unimportant.
2. A reproducible average current is produced instantly on changing the applied potential.
3. Mercury has a large activation overpotential for hydrogen formation which facilitates the reduction of many species in acid solution.

Its use is restricted to the determination of reducible or easily oxidized species as at positive applied potentials greater than 0.4 V with respect to the SCE, mercury dissolves to give an anodic polarographic wave.

Current maxima

Polarographic waves often show a peak followed by a sharp fall to the limiting current plateau, the cause of which is related to streaming of the solution past the mercury drop. Known as a 'current maximum', it can be eliminated by adding a surfactant such as gelatin or methyl-red to the sample solution.

Oxygen waves

Two waves caused by the reduction of dissolved oxygen can interfere with the waves of other electroactive species unless the solution is purged with nitrogen prior to obtaining the polarogram

$$
\left.
\begin{array}{l}
O_2 + 2H^+ + 2e^- = H_2O_2 \\
O_2 + 2H_2O + 2e^- = H_2O_2 + 2OH^-
\end{array}
\right\}
\begin{array}{l}
E_{\frac{1}{2}} = -0.05 \text{ V} \\
\text{with respect to SCE}
\end{array}
$$

or

$$
\left.
\begin{array}{l}
H_2O_2 + 2H^+ + 2e^- = 2H_2O \\
H_2O_2 + 2e^- = 2OH^-
\end{array}
\right\}
\begin{array}{l}
E_{\frac{1}{2}} = -0.9 \text{ V} \\
\text{with respect to SCE}
\end{array}
$$

or

The reactions have been utilized for the determination of dissolved oxygen or hydrogen peroxide.

QUANTITATIVE ANALYSIS

Either calibration graphs prepared from standards or the method of standard addition (p. 32) can be used. For the former, the standards should be as similar as possible in overall chemical composition to that of the samples so as to minimize errors caused by the reduction of other species or by variation in diffusion rates. Often, the limiting factor for quantitative work is the level of impurities present in the reagents used.

MODES OF OPERATION USED IN POLAROGRAPHY

The earliest types of polarograph involved the manual changing of a dc applied potential stepwise and plotting it against corresponding current values. Later versions automated this process so that a chart recording could be obtained. More recently, *oscillographic* or *rapid scan* polarography, *pulse* polarography, *stripping voltammetry* and *ac* polarography were introduced primarily to increase sensitivity and to facilitate the resolution of closely-spaced polarographic waves. The first three of these techniques have found more application than the last.

Linear Sweep Oscillographic Polarography
A repetitive dc potential sweep of 0.5 to 1.0 V is synchronized with the growth of each mercury drop and the resulting current-potential curve displayed on the screen of a cathode-ray oscilloscope. Because the current increases with the surface area of the drop, the potential sweep is timed to occupy the last 2 s of a 7 s drop-life where the relative change of surface area is minimal. Exact synchronization is achieved mechanically by tapping the capillary every 7 s to detach the drop. Figure 6.11(a) shows the variation of the drop area and applied potential as a function of time, and figure 6.11(b) the current-potential curve as seen on the oscilloscope screen. The peak current i_p is not a polarographic maximum but is caused by the rapid potential sweep; it is however directly proportional to the concentration of the electroactive species. A twin-cell version of this type of instrument enables the polarogram of a blank solution to be automatically subtracted from that of the sample solution before display on the oscilloscope screen. Alternatively, by having the same sample solution in both cells and using sophisticated electronic circuitry, first and second derivative curves can be displayed. This enables components with $E_{\frac{1}{2}}$ values as little as 25 mV apart to be separated, but some

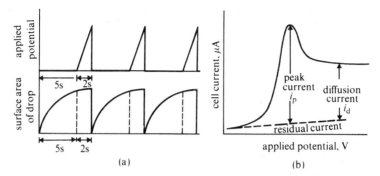

Figure 6.11. Oscillographic polarography. (a) Variation of drop area and applied potential with time. (b) Oscillographic display of a polarogram

sensitivity is sacrificed. Twin-cell operation suffers from the difficulty of maintaining a pair of exactly matched capillaries.

Differential Pulse Polarography
Pulse polarography seeks to minimize the contribution to measured cell currents arising from charging of the mercury drops (p. 244). This is achieved by superimposing a 10–100 mV pulse of short duration (40–60 ms) onto the

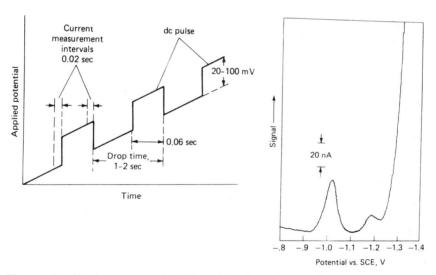

Figure 6.12. Timing sequence for differential pulse polarography (left) and resulting differential polarogram (right)

usual increasing dc potential. The pulse is applied during the last quarter of the growth of each drop and the diffusion current is measured just before the pulse and again during the last 20 ms. The difference between the two measured currents for each drop is plotted as a function of the dc potential. The timing sequence of operations and the resulting differential polarogram are shown in figure 6.12. Synchronization of the pulse with each drop is achieved in a similar manner to that used in *linear sweep oscillographic polarography*. The height of the peak maximum is directly proportional to the concentration of the electroactive species and it is possible to discriminate between species with half-wave potentials differing by only 0.05 V. Apart from suppression of charging current, sensitivity in pulse polarography is enhanced relative to conventional dc polarography because the application of a voltage pulse to each drop produces higher currents than are obtained when diffusion alone controls the electrochemical reaction.

Stripping Voltammetry

This is a modification of the polarographic technique whereby extremely low levels of certain metals are first pre-concentrated at a stationary electrode by cathodic deposition then determined by *anodic stripping*. The sample solution is electrolysed for an accurately measured period of time during which reducible species are deposited at a stationary mercury drop or mercury-coated solid electrode. The polarity of the cell is then reversed and a linear potential sweep, sometimes pulsed, is applied to produce anodic polarographic waves of the sample components concentrated in the mercury electrode. The initial preconcentration or pre-electrolysis step must take place under carefully controlled and reproducible conditions, as only a small fraction of the electroactive species is deposited at the microelectrode. Thus, variations in pre-electrolysis time, stirring rate, temperature, electrode potential and current density between samples and standards must be eliminated as far as possible. The ultimate sensitivity of the method is determined by the length of the pre-electrolysis time which can be up to an hour or more. However, by employing a differential pulse technique during the subsequent anodic stripping rather than a linear potential sweep, equivalent sensitivities can be obtained with pre-electrolysis times of a few minutes or less, and with better reproducibility. A diagram of an anodic stripping instrument and a typical stripping trace are shown in figure 6.13. Metals which form amalgams with mercury, e.g. Pb, Cd, Cu, Zn and Sn are readily determined by this technique but the ultimate sensitivity is limited by the level of impurities present in reagents and the background electrolyte solution. Nevertheless, concentrations in the range 10^{-8}–10^{-9} M (ppb) can be determined under favourable conditions and methods have been devised for Pb in blood and traces of Cu, Cd, Pb and Zn in drinking water.

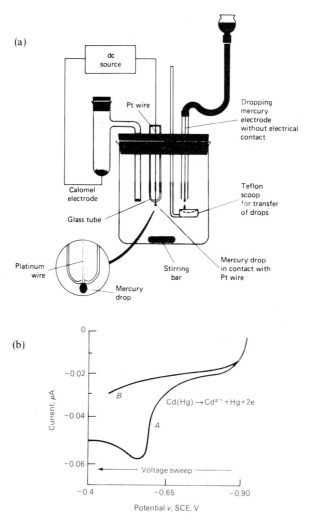

Figure 6.13. (a) Apparatus for stripping analysis. (b) Curve A: Current-voltage curve for anodic stripping of cadmium. Curve B: Residual current curve for blank. Reprinted with permission and adapted from R. D. DeMars and I. Shain, *Anal. Chem.* **29**, 1825 (1957). Copyright by the American Chemical Society

THE DISSOLVED OXYGEN ELECTRODE AND BIOCHEMICAL ENZYME SENSORS

There is a longstanding demand for a simple and portable instrument for the detection and measurement of oxygen dissolved in water. Suitable electrodes have been developed and more recently have been ingeniously used as the basis for a range of biochemical sensors.

Principles of Dissolved Oxygen Measurement
A typical commercial electrode is based on a galvanic cell comprising a lead anode and a silver cathode. In an alkaline electrolyte (e.g. 1M KOH) the following reaction will occur if a reducible species is present

$$Pb(s) + 4OH^-(aq) = PbO_2^{2-}(aq) + 2H_2O + 2e^- \qquad (6.27)$$

If the electrodes are surrounded by a semipermeable membrane through which only oxygen can diffuse from the external solution, reduction of oxygen will take place.

$$O_2(aq) + 4e^- + 2H_2O = 4OH^-(aq) \qquad (6.28)$$

The current flowing in the cell will depend upon the rate of diffusion of oxygen to the electrode surface and ultimately on the concentration of oxygen in the external solution. The construction of such an electrode is shown in figure 6.14.

Applications of the Dissolved Oxygen Electrode
Electrodes are capable of measurement at around the 1 ppm level. They are simple, robust and portable. As such there is considerable potential for their

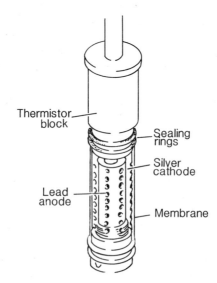

Figure 6.14.   Construction of an EIL temperature compensated oxygen electrode
(from EIL advertising material)

employment in ecological studies and they can provide constant monitoring of dissolved oxygen, e.g. in river waters. The potential for their employment in monitoring industrial liquors is clear, as is their applications in biological investigations of oxygen transport.

Principles of Enzyme Sensors

The biochemical oxidation of organic molecules is accomplished by enzymes known as oxidases. Specific oxidases exist for the oxidation of each compound. Oxidases require oxygen to react, and so decrease the amount of oxygen in their immediate environment. The degree of oxygen depletion is hence related to the enzyme activity, which is in turn determined by the amount of the oxidizable species present. Use of a dissolved oxygen electrode to measure the degree of oxygen depletion will thus provide a basis for the measurement of the amount of the oxidizable species which is present.

Specially designed electrodes in which the electrode tip is covered in a gel containing the appropriate oxidase are commercially available. A semi-permeable membrane retaining the gel, and permitting oxygen diffusion, completes the assembly.

Applications of Enzyme Sensors

The type of enzyme sensor described above is highly selective and can be sensitive in operation. There are obvious applications for the determination of small amounts of oxidizable organic compounds. However, it is perhaps too early to give a realistic assessment of the overall importance of enzyme sensors to analytical chemistry. This is especially so because of parallel developments in other biochemical sensors which may be based upon a quite different physical principle.

Commercially available kits allow for the determination of compounds such as ethyl alcohol, glucose, lactic acid and lactose.

AMPEROMETRIC TITRATIONS

If the limiting current flowing in a polarographic cell is measured during a titration in which any of the reactants or products are reducible at the micro electrode, the equivalence point can readily be detected. The choice of applied potential is not critical provided it corresponds to a point on the limiting current plateau. Values of i_d are simply plotted as a function of the volume of titrant added. Three types of titration curve may be observed, as shown in figures 6.15(a), (b) and (c). In each case, the equivalence-point is located at the intersection of two straight lines. Curvature in this region is due to partial dissociation of the products of the titration but as current readings can be taken well away from the equivalence-point, accuracy is not impaired (cf. potentiometric and visual indicator methods). Titrations which produce curves similar to that shown in figure 6.15(a) are sometimes referred to as 'dead-stop' methods because the current falls virtually to zero at the equivalence point. The most accurate determinations can be made if the curve is V-shaped, figure 6.15(c). Unless the titrant is at least twenty times as concentrated as the analyte solution, current readings must be corrected for volume changes by multiplying each one by the ratio $(V + v)/V$, where V is the solution volume before adding an increment of titrant of volume v.

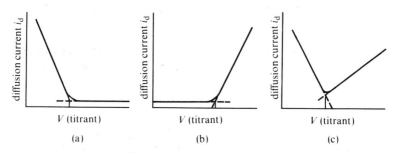

Figure 6.15. Amperometric titration curves. (a) Only substance titrated is reducible, e.g. Pb^{2+} against SO_4^{2-}, Ag^+ against halide. (b) Only titrant is reducible, e.g. SO_4^{2-} against Pb^{2+}, $Na_2S_2O_3$ against I_2. (c) Both reducible, e.g. Pb^{2+} against CrO_4^{2-}, Cu^{2+} against α-benzoin oxime

An alternative to the DME is the rotating platinum micro-electrode. Steady currents can be obtained by rotating the electrode at a constant speed of 600 min^{-1} or more. These are generally some twenty times larger than the currents derived from a DME because of mechanical convection created by the rotation. The rotating electrode is particularly useful for redox titrations involving such species as Br_2 and Fe(III) but the low hydrogen activation overpotential for platinum limits its applications to alkaline or weakly acid conditions. It is very sensitive to the presence of dissolved oxygen.

Bi-amperometric Titrations

Two identical stationary micro-electrodes (usually platinum) across which a potential of 0.01 to 0.1 V is applied can be used in place of either the DME or the rotating platinum micro-electrode. The equivalence point is marked by a sudden rise in current from zero, a decrease to zero, or a minimum at or near zero (figures 6.16(a), (b) and (c)). The shape of the curve depends on the reversibility of the redox reactions involved. The two platinum electrodes assume the roles of anode and cathode, and in all cases a current flows in the

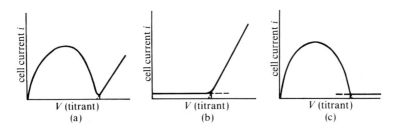

Figure 6.16.   Bi-amperometric titration curves. (a) Both electrode reactions reversible, e.g. Fe^{2+} against Ce^{4+}. (b) Only titrant reaction is reversible, e.g. AsO_3^{3-} against I_2. (c) Only substance titrated is reversible, e.g. I_2 against $Na_2S_2O_3$

cell only if there is a significant concentration of both the oxidized and reduced forms of one of the reactants. In general, two types of system can be envisaged:

1. Both Reactants Behave Reversibly

The curve is as shown in figure 6.16(a) and is exemplified by the titration of Fe(II) with Ce(IV) when a potential of 0.1 V is applied to the cell. At the outset, no current flows because only Fe(II) is present, the only electrode reaction possible being $Fe^{2+} \rightarrow Fe^{3+} + e^-$ at the anode. Upon the addition of the first increment of titrant and up to the equivalence point, the concentration of Fe(II) diminishes, whilst that of Fe(III) increases, and a current proportional to the smaller of the two concentrations flows in the cell, resulting in a maximum at the half-way stage. At the equivalence point only Fe(III) and Ce(III) are present and no current flows because neither electrode reaction Fe(III) $\rightarrow$ Fe(II) nor Ce(III) $\rightarrow$ Ce(IV) can proceed at the potential of 0.1 V. After the equivalence point, the current increases linearly with the rise in concentration of Ce(IV), that of Ce(III) being constant.

2. Only One Reactant Behaves Reversibly

If the titrant alone behaves reversibly, no current can flow until it is in excess, figure 6.16(b). This is the case if AsO_3^{3-} is titrated with I_2 at an applied potential of 0.1 V. After the equivalence point the current is linearly related to the concentration of excess I_2, that of I^- being constant. In cases where only the analyte forms a reversible couple, e.g. I_2 titrated with $Na_2S_2O_3$, the current before the equivalence point follows a similar path to that in the Fe(II)/Ce(IV) system, but afterwards remains at zero, figure 6.15(c).

A 'dead-stop' titration curve is produced if Ag^+ is titrated with a halide using a pair of identical silver electrodes. Only whilst both Ag^+ and Ag are present will a current flow in the cell, and this is linearly related to the Ag^+ concentration. Bi-amperometric titrations require only simple equipment but generally give poorer precision because the currents measured are not necessarily on the limiting current plateau.

Amperometric titrations are inherently more precise than polarography and are more generally applicable because the analyte need not itself be electroactive. Titrations involving the DME are not affected by changes in capillary characteristics as are conventional polarographic determinations whilst working at a pre-determined temperature is unnecessary provided that it remains reasonably constant throughout the titration.

APPLICATIONS OF POLAROGRAPHY AND AMPEROMETRIC TITRATIONS

Most metal ions are reducible at the DME, and multicomponent mixtures can often be analysed by selecting an appropriate supporting electrolyte and complexing agent. Polarography is used for the determination of trace metals in alloys, ultra-pure metals, minerals, foodstuffs, beverages and body-fluids although recently it has been largely superseded by atomic absorption spectrometry (chapter 8). It is ideally suited to the determination of metal impurities in AnalaR and other high-purity salts where the sample matrix can act as its own supporting electrolyte. Reducible anions such as BrO_3^-, IO_3^-, $Cr_2O_7^{2-}$ and NO_2^- can also be determined provided solutions

Table 6.4   Applications of polarography

ELEMENT OR COMPOUND DETERMINED	TYPE OF SAMPLE
Cu, Pb, Sn, Zn	foodstuffs
Ga, Zn, Cd, Ni	high purity aluminium
Cu, Pb, Ni, Co	steels
Mo, Ge, As, Sb	minerals and ores
Sn, Pb	beer and soft drinks
transition metals	high-purity salts
free sulphur	petroleum fractions
antioxidants	fuels
riboflavin	milk, pharmaceuticals
antibiotics, steroids	body fluids
vitamin C	fruit and vegetables
oxygen	seawater, gases

are well-buffered. Organic applications are restricted by an inability to distinguish between members of a homologous series and by the irreversibility of many reductions at the DME. Solutions must be well-buffered as many electrode reactions are pH dependent. The limited aqueous solubility of most organic compounds necessitates the use of polar solvents such as glycols, dioxane, formamide and methyl cellosolve with lithium or tetra-alkylammonium salts as the supporting electrolyte. Reducible organic functional groups include ketones, aldehydes, alkenes, nitriles, azo, peroxy, disulphide, nitro, nitroso and nitrite. Examples of the applications of polarography are given in table 6.4.

Amperometric titrations have an even wider range of application than polarography. Although the titrant may be added from a burette, in many applications it is electrically generated in a coulometric cell (p. 258). Such an arrangement lends itself to complete automation and is particularly valuable for the titration of very small quantities. For examples of coulometric titrations with amperometric equivalence point detection see table 6.5.

6.3   Electrogravimetry and Coulometry

Electrogravimetry, which is the oldest electroanalytical technique, involves the plating of a metal on to one electrode of an electrolysis cell and weighing the deposit. Conditions are controlled so as to produce a uniformly smooth and adherent deposit in as short a time as possible. In practice, solutions are usually stirred and heated and the metal is often complexed to improve the quality of the deposit. The simplest and most rapid procedures are those in which a fixed applied potential or a constant cell current is employed, but in both cases selectivity is poor and they are generally used when there are only one or two metals present. Selective deposition of metals from multi-component mixtures can be achieved by controlling the cathode potential automatically with a *potentiostat*. This device automatically monitors and maintains the cathode potential at a pre-determined value by means of a reference electrode and servo-driven potential-divider. The value chosen for the cathode potential is such that only the metal of interest is deposited and there are no gaseous products formed.

Separations of transition and non-transition metals by deposition of the former at a mercury-pool cathode can be used as a preliminary step in an analytical procedure. In such cases, the reduced metal cannot be weighed, although quantitative separations may be achieved.

COULOMETRY

SUMMARY

Principles
    Measurement of the quantity of electricity used in an electrochemical
    reaction at constant potential or constant current.

Instrumentation
    Working and counter-electrodes; source of constant current and timer,
    or a potentiostat and integrator; equivalence point detection system.

Applications
    Quantitative determination of many species in solution, especially at
    trace levels (micrograms or less); relative precision 0.2 to 5 %. Useful for
    unstable titrants and easy to automate.

Disadvantages
    Constant potential methods can be lengthy; constant current methods
    have similar disadvantages to other titrimetric techniques.

Coulometric methods of analysis involve measuring the quantity of
electricity required to effect a quantitative chemical or electrochemical
reaction and are based on Faraday's laws of electrolysis:

1. The quantity of a substance liberated at an electrode during electrolysis
is proportional to the quantity of electricity passing through the solution.

2. A given quantity of electricity liberates different substances in the
ratio of their molar masses divided by the number of electrons involved in
the corresponding electrode reactions.

The electrolysis cell consists of a *working-electrode* at which the species to
be determined is reduced or oxidized or at which a chemically reactive
species is formed, and a *counter-electrode*. In practice electrolysis may be at
constant potential, in which case the current diminishes to zero as the reaction
goes to completion, or at *constant current*. The quantity of electricity involved
in the former is measured by means of a chemical coulometer or by integrating
the area under the current–time curve. Constant current methods involve
the generation of a titrant for a measured length of time, the completion of the
reaction with the species to be determined being indicated by any of the

methods used in titrimetric analysis. For this reason, such procedures are described as *coulometric titrations*. In both constant potential and constant current coulometry, the current efficiency must be 100 %, i.e. all the electricity passing through the cell must be utilized in a reaction involving the species to be determined, either directly or indirectly.

COULOMETRY AT CONSTANT POTENTIAL

The technique is similar to electrogravimetry at constant cathode potential (p. 257), but is much more versatile because the substance to be determined need not form a stable, adherent deposit on the electrode. The quantity of electricity consumed in the reaction is determined by connecting a chemical coulometer in series with the cell or by using an electronic or electro-mechanical integrator. Chemical coulometers take the form of electrolysis cells where the total amount of product(s) liberated at the electrode(s) can readily be measured. The simplest to use are gas coulometers, in which the amount of hydrogen and oxygen (or nitrogen) liberated from a suitable electrolyte solution (e.g. sodium sulphate) is measured in a gas burette. During the analysis close control of the cathode (or anode) potential with a potentiostat (p. 257) is essential to avoid electrode reactions other than that related to the species of interest. A convenient working-electrode is a mercury cathode at which numerous metals and organic compounds can be reduced. Examples include the determination of lead in cadmium, nickel in cobalt and trichloroacetic acid in the presence of the mono- and dichloro-derivatives. Precisions of better than 1 % are easily attained, but analyses can take up to an hour or more.

COULOMETRIC TITRATIONS

The titrant is generated at a working electrode by the passage of a constant current until the equivalence point is indicated by potentiometric, ampero-metric or, less commonly, visual or photometric means. An accurate timer is required to enable the total quantity of electricity used to be calculated. A schematic diagram of a coulometric titrator is shown in figure 6.17. Constant-current sources, often just a battery, should be capable of delivering currents of 10 to 100 mA with an accuracy of 0.5 % or better. Electric timers with solenoid-operated brakes or electronic timers should be used to eliminate cumulative errors arising from starting and stopping the motor, a sequence which may be repeated many times as the equivalence point is approached.

The titrant is usually generated directly in the solution whence it can immediately react with the species to be determined. The electrodes are typically made of platinum coils or sheets. The counter-electrode is normally

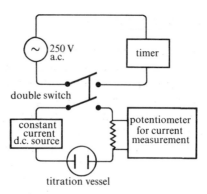

Figure 6.17.   Circuit diagram for coulometric titrations

isolated from the bulk solution by enclosing it in a tube sealed with a sintered-glass disc to prevent contamination of the solution with undesirable electrolysis products. An alternative arrangement, especially useful for acid-base titrations, is to generate the reagent externally before adding it to the titration vessel. An apparatus for generating and separating hydrogen and hydroxyl ions using an electrolyte solution such as sodium sulphate is shown in figure 6.18. To eliminate errors due to impurities in reagents used in the preparation of the sample, a 'pre-titration' method is sometimes used. A blank solution is first titrated, the sample is added to the titrated solution and the solution titrated again. This procedure is particularly advantageous for determinations at very low levels.

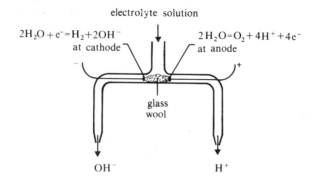

Figure 6.18.   Device for external generation of H^+ and OH^-

APPLICATIONS OF COULOMETRIC TITRATIONS

The coulometric generation of titrants is widely applicable to redox, precipitation, acid-base and complexing reactions. Of particular value is the determination of many organic compounds with bromine and of mercaptans and halides with the silver ion. Amperometric equivalence point detection is the most common. An attractive feature of the technique is that the need to store standard and possibly unstable reagent solutions is obviated. In fact many applications involve the use of electrogenerated reagents such as halogens and chromium(II) which are difficult or impossible to store. The technique is especially useful for the determination of very small amounts, i.e. in the microgram or nanogram ranges. Relative precisions of 0.2 to 5% can be attained at trace levels, which is better than most other titrimetric procedures. Some examples of titrations with electrogenerated titrants are given in table 6.5.

Table 6.5   Applications of coulometric titrations at constant current

ELEMENT OR COMPOUND DETERMINED	TITRANT GENERATED	MEANS OF EQUIVALENCE POINT DETECTION	EXAMPLE OF APPLICATION
water	I_3^- (Karl-Fischer reagent)	amperometric	organic solvents, petroleum products
Cr, V	Fe^{2+}	amperometric	steels, oils, asphalt
Ag	I^-	amperometric	lubricating oils
phenols, olefins	Br_2	amperometric	petroleum products
thymol	Br_2	amperometric	herbs
mercaptans, chloride	Ag^+	amperometric	fuels
aromatic amines	H^+	potentiometric	organic chemicals
oxygen	Cr^{2+}	amperometric	seawater, gases

6.4   Conductometric Titrations

SUMMARY

Principles

Measurement of the conductance of an electrolyte solution using an ac source. Rate of change of conductance as a function of added titrant used to determine the equivalence point.

Instrumentation

Platinum electrodes; low-potential ac source; conductance bridge.

Applications

Acid-base titrations, especially at trace levels. Relative precision better than 1 % at all levels.

Disadvantages

Conductance is a non-specific property; high concentrations of other electrolytes can be troublesome.

The electrical conductance of a solution is a measure of its current-carrying capacity and is therefore determined by the total ionic strength. It is a non-specific property and for this reason direct conductance measurements are of little use unless the solution contains only the electrolyte to be determined or the concentrations of other ionic species in the solution are known. Conductometric titrations, in which the species of interest are converted to non-ionic forms by neutralization, precipitation, etc. are of more value. The equivalence point may be located graphically by plotting the change in conductance as a function of the volume of titrant added.

A conductance cell consists of two platinum electrodes of large surface area across which an alternating low-voltage potential is applied. Generally, 5 to 10 V at 50 to 10 000 Hz is employed. A dc potential cannot be used as the current flow would lead to electrolysis and hence changes in solution composition. The cell is incorporated into one arm of a Wheatstone-bridge type of circuit and the conductance measured by adjustment of a calibrated resistor to balance the bridge. Conductance measurements made in this way have a precision of 0.1 % or better but a limiting factor is the temperature-sensitivity of ionic conductances. The cell temperature should be held constant to within ± 0.1 K throughout a series of measurements if a precision of better than 0.5 % is required.

IONIC CONDUCTANCES

Defined as the reciprocal of resistance (siemens, Ω^{-1}) conductance is a measure of ionic mobility in solution when the ions are subjected to a potential gradient. The *equivalent conductance* λ of an ion is defined as the conductance of a solution of unspecified volume containing one gram-equivalent and measured between electrodes 1 cm apart. Due to interionic effects, λ is concentration dependent, and the value, λ_0, at infinite dilution is used for comparison purposes. The magnitude of λ_0 is determined by the charge, size and degree of hydration of the ion; values for a number of cations and anions at 298.15 K are given in table 6.6. It should be noted that H_3O^+ and

Table 6.6   Ionic conductances

CATIONS	λ_0^+/Ω^{-1} $(m^2\ mol^{-1})$	ANIONS	λ_0^-/Ω^{-1} $(m^2\ mol^{-1})$
H_3O^+	350	OH^-	198
K^+	73.5	$\frac{1}{2}SO_4^{2-}$	79.8
NH_4^+	73.4	Br^-	78.4
		Cl^-	76.3
$\frac{1}{3}Fe^{3+}$	68	NO_3^-	71.4
Ag^+	62		
$\frac{1}{2}Mg^{2+}$	53	CH_3COO^-	40.9
Na^+	50	$\frac{1}{2}C_2O_4^{2-}$	24
Li^+	38.7		

OH^- have by far the largest equivalent conductances. For this reason and because H_2O has a very low conductivity, acid-base titrations yield the most clearly defined equivalence-points. Some specific examples of titration curves are shown in figure 6.19. In all cases, the equivalence points are located at the intersection of lines of differing slope. Curvature in these regions is due to partial dissociation of the products of the titration reaction. Conductance readings must be corrected for volume changes unless the titrant is at least twenty times as concentrated as the solution being titrated (cf. amperometric titrations). The shape of each curve can be explained in terms of the λ_0 values given in table 6.6.

Titration of Strong Acids or Bases, figure 6.19(a)
Before the equivalence-point, H_3O^+ is progressively replaced by Na^+ which has a much smaller equivalent conductance. After the equivalence point, the conductance increases linearly with increasing concentration of OH^- and Na^+. Using LiOH in place of NaOH would give an even sharper change of slope.

Titration of Weak Acids or Bases, figures 6.19(b) and (c)
Curve (b) shows the titration of acetic acid, $K_a = 1.75 \times 10^{-5}$ mol dm^{-3} at two different concentrations. The initial additions of OH^- establish a buffer solution in which the H_3O^+ concentration is only slowly reduced. The resulting fall in conductance is increasingly counteracted by the addition of Na^+ and the formation of CH_3COO^- thus leading to a minimum in the curve. After the equivalence-point, there is a more rapid increase due to the addition of excess OH^- and Na^+. Concentrated solutions of weak acids give more pronounced changes of slope at the equivalence point than dilute solutions. The change in slope is well defined for very weak acids, curve (c), e.g. boric acid, $K_a = 6 \times 10^{-10}$ mol dm^{-3} but for those with values greater

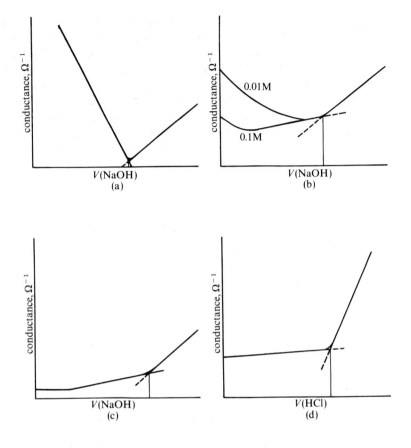

Figure 6.19.   Conductometric titration curves (a) Titration of strong acid with
NaOH. (b) Titration of acetic acid, $K_a = 1.75 \times 10^{-5}$ mol dm^{-3} with NaOH.
(c) Titration of boric acid $K_a = 6 \times 10^{-10}$ mol dm^{-3} with NaOH. (d) Titration
of sodium acetate with HCl

than 10^{-5} mol dm^{-3}, curvature in the buffer region renders detection of the
equivalence point impossible.

Titration of Salts of Weak Acids or Bases, figure 6.19(d)
The curve shown is for the titration of sodium acetate with hydrochloric
acid. Before the equivalence point, there is a slight increase in conductance
as Cl$^-$ replaces CH$_3$COO$^-$ which is converted to undissociated acetic acid.
After the equivalence point, the conductance increases linearly with the
addition of H$^+$ and Cl$^-$.

The course of precipitation and complexometric titrations can also be followed conductometrically, but changes of slope are generally less pronounced. Redox reactions are difficult to follow because of the high background concentration of H_3O^+ and other electrolytes usually present. Conductometric titrations are especially useful for very dilute solutions as the percentage change in conductance is independent of concentration. Unlike potentiometric and visual indicator methods, measurements need not be made close to the equivalence point so that there is an advantage in conductometric methods when the chemical reactions involved are relatively incomplete.

Problems

1. Calculate the theoretical cell potentials for the following systems:

(a) $Pb|Pb^{2+}(0.100\,M)||Cd^{2+}(0.001\,M)|Cd$

(b) $Ag|AgBr$ (saturated), $Br^-(0.01\,M)$, $H^+(10\,M)|H_2$ (1 atmosphere) Pt

$(E^{\ominus}_{Pb^{2+}/Pb} = +0.126; E^{\ominus}_{Cd^{2+}/Cd} = -0.403; E^{\ominus}_{Ag^+/Ag} = +0.799; K_{AgBr} = 7.7 \times 10^{-13})$

2. (a) Calculate K_{sp} for MX_2 in the system:

$M|MX_2$ (saturated), $X^-(0.400\,M)|H_2$ (1 atmosphere) Pt

$(E^{\ominus}_{M^{2+}/M} = +0.0118; E_{cell} = +0.204$, M as anode)

(b) Calculate $E^{\ominus}$ for the $\frac{1}{2}$-cell reaction:

$$MX_2 + 2e = M + 2X^-$$

3. A 0.2 g sample of toothpaste containing fluoride was treated with 50 cm³ of a suitable buffer solution and diluted to 100 cm³. Using a fluoride ion-selective electrode, a 25.00 cm³ aliquot of this solution gave cell potentials of -155.3 mV before and -176.2 mV after spiking with 0.1 cm³ of a 0.5 mg cm⁻³ fluoride standard. Calculate the pF⁻ corresponding to each cell potential and the percentage by weight of fluoride in the toothpaste.

4. Plot the titration curve in the vicinity of the end-point and its first and second derivatives from the following data and compare the end-point values:

Burette reading cm³	Cell potential mV
47.60	372
47.70	384
47.80	401
47.90	512
48.00	732
48.10	748
48.20	756

5. An organic acid (0.6079 g) was dissolved in 45.67 cm³ of an NaOH solution, and the excess base titrated with 3.25 cm³ of 0.12 M HCl. In a standardizing titration, 39.33 cm³ of the NaOH solution was equivalent to 31.69 cm³ of the 0.12 M HCl. Calculate the molecular weight of the unknown acid.

6. The iron in a 0.100 g sample was converted quantitatively to Fe(III) and titrated coulometrically with electrogenerated Ti(III). A current of 1.567 mA for 123.0 seconds was required to reach the end-point. Calculate the percentage of iron in the sample. (1 faraday = 96 485 coulombs mol⁻¹)

7. Sketch the shape of the titration curve for the following conductometric titrations:

(a) silver acetate v. lithium chloride
(b) ammonium chloride v. sodium hydroxide
(c) sulphuric acid in glacial acetic acid v. sodium hydroxide
(d) a bromide solution v. silver nitrate

(Use the data in table 6.6).

8. A 50 cm³ aliquot of a lead nitrate solution in 0.1 M KNO_3 was titrated amperometrically with 0.05 M potassium dichromate. The burette readings and corresponding values of cell current were as follows:

Volume of $K_2Cr_2O_7$, cm³	Cell current, μA
0	84.0
1.00	66.2
2.00	48.3
3.00	31.7
4.00	15.3
5.30	12.0
5.50	21.9
6.00	43.9
6.50	67.0

Plot the titration curve and determine the end-point volume and concentration of the original lead nitrate solution.

Further Reading

BAILEY, P. L., *Analysis with Ion-Selective Electrodes*, 2nd Ed., Heyden, London, 1980.
EVANS, A., *Potentiometry and Ion Selective Electrodes*, Wiley, Chichester, 1987.
GREEF, R., PEAT, R., PETER, L. M., PLETCHER, D. and ROBINSON, J., *Instrumental Methods in Electrochemistry*, Ellis Horwood (Wiley), Chichester, 1985.
RILEY, T. and TOMLINSON, C., *Principles of Electroanalytical Methods*, Wiley, Chichester, 1987.
RILEY, T. and WATSON, A., *Polarography and Other Voltammetric Methods*, Wiley, Chichester, 1987.
SVEHLA, G., *Automatic Potentiometric Titrations*, Pergamon, Oxford, 1977.

Chapter 7

An Introduction to Analytical Spectrometry

Spectrometric techniques form the largest and most important single group of techniques used in analytical chemistry, and provide a wide range of quantitative and qualitative information. All spectrometric techniques depend on the emission or absorption of *electromagnetic radiation* characteristic of certain energy changes within an atomic or molecular system. The energy changes are associated with a complex series of *discrete* or *quantized energy levels* in which atoms and molecules are considered to exist. To understand how studies of such transitions between energy levels can yield information requires some knowledge of the properties of electromagnetic radiation and of the nature of atomic and molecular energy.

ELECTROMAGNETIC RADIATION

Electromagnetic radiation has its origins in atomic and molecular processes. Experiments demonstrating reflection, refraction, diffraction and interference phenomena show that the radiation has wave-like characteristics, while its emission and absorption are better explained in terms of a particulate or quantum nature. Although its properties and behaviour can be expressed mathematically, the exact nature of the radiation remains unknown.

The wave theory supposes that radiation emanating from a source consists of an electromagnetic field varying periodically and in a direction perpendicular to the direction of propagation. The field may be represented as electric and magnetic vectors oscillating in mutually perpendicular planes. If the periodic variation is simple harmonic in nature, the radiation may also be described in terms of a sinusoidal wave, figure 7.1. The wave can be characterized by its frequency v, its length λ, and its amplitude A. Frequency is defined as the number of complete cycles or oscillations per second, wavelength as the linear distance between any two equivalent points in successive cycles and amplitude

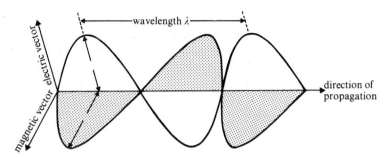

Figure 7.1.   Wave representation of electromagnetic radiation

as the maximum value reached by the vectors in a cycle. The amplitude and frequency of the wave are determined by the intensity of the radiation and the energy associated with it respectively. Thus,

$$I \propto A^2$$

and

$$E \propto v$$

where I is the intensity and E the energy. Frequency and wavelength are simply related by the expression

$$v = \frac{c}{\lambda} \tag{7.1}$$

where c is the velocity of propagation of the wave in a vacuum (2.998×10^8 m s^{-1}) and is independent of frequency. It is important to remember that energy and frequency are *directly* proportional but that energy and wavelength are *inversely* related. A unit used to characterize radiation in some spectrometric techniques is the wavenumber σ defined as the number of waves per centimetre and related to frequency and wavelength by the expression

$$\sigma = \frac{1}{\lambda} = \frac{v}{c} \tag{7.2}$$

The various spectrometric techniques have developed independently, and at different times. They differ as to whether frequency, wavelength or wavenumber is measured, with the result that comparisons between techniques is sometimes difficult—equation (7.2) is a useful aid in this respect.

Quantum theory considers radiation as a stream of 'energy packets'—*photons* or *quanta*—travelling through space at a constant velocity (c when in a vacuum). The energy of a photon is related to the frequency of the radiation, as defined in wave theory, by the expression

$$E = hv \tag{7.3}$$

where h is Planck's constant (6.6×10^{-34} J s).

The energy associated with electromagnetic radiation covers a wide range of magnitude as do the corresponding frequencies and wavelengths—the whole being termed the *electromagnetic spectrum*. Figure 7.2 shows the range of energy and the corresponding frequencies, wavelengths and wavenumbers. Atomic and molecular processes involving the emission or absorption of radiation are shown opposite the appropriate energy ranges. The electromagnetic spectrum is divided into a number of regions; these are artificial

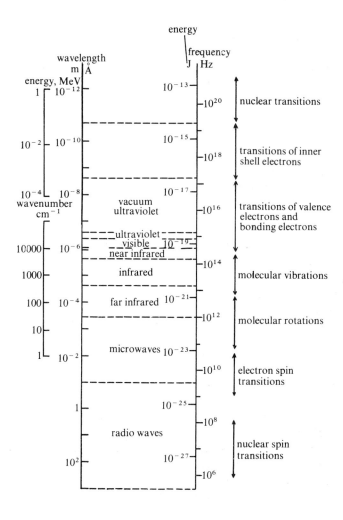

Figure 7.2.   The electromagnetic spectrum

divisions in the sense that they have been defined solely as a result of differences in the instrumentation required for producing and detecting radiation of a given frequency range. It must be emphasized that the radiation itself is fundamentally the same over the entire electromagnetic spectrum and that even the boundaries between the instrumentally based divisions are diffuse.

ATOMIC AND MOLECULAR ENERGY

The total energy of an atom or molecule includes contributions from several sources. These originate from within the nucleus, from interactions between electrons and nuclei (*electronic energy*), from electron and nuclear spin, from the vibrational and rotational motion of molecules and from translation of atoms or molecules through space. Except for translational motion, all these forms of energy are considered to be discontinuous or *quantized*. According to quantum theory, for each form of energy, an atom or molecule can exist in certain discrete energy states or quantized levels defined by a set of quantum numbers and in accordance with a set of mathematical rules. The magnitudes of the various forms of energy and the differences between adjacent levels vary considerably. Electronic energy is comparatively large, and it will be seen from figure 7.3 that the levels are much more widely spaced than vibrational

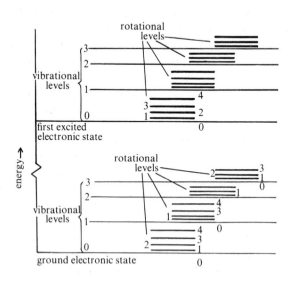

Figure 7.3.   Electronic, vibrational and rotational energy levels, not to scale (only the first two electronic levels are shown)

levels, which in turn are more widely spaced than rotational levels. Electron and nuclear spin energies are small and the levels lie very close together. For simplicity, the latter and nuclear energy levels, which are even more widely spaced than electronic levels, are not shown on the diagram. Exactly which set of energy levels an atom or molecule occupies under ambient conditions is determined by the rules of quantum theory and by the Maxwell–Boltzmann equation (*vide infra*). They are designated *ground state* levels, higher ones being termed *excited state* levels.

THE ABSORPTION AND EMISSION OF ELECTROMAGNETIC RADIATION

If a substance is irradiated with electromagnetic radiation, the energy of the incident photons may be transferred to the atoms or molecules raising them from the ground state to an excited state. This process, known as *absorption*, is accompanied by attenuation of the incident radiation at a particular frequency and can occur only when the energy difference between the two levels is exactly matched by the energy of the photons. The frequency of the radiation is given by

$$E_2 - E_1 = \Delta E = h\nu \qquad (7.4)$$

where E_1 and E_2 are the energies of the two levels and ΔE is the difference between them. The absorbed energy is rapidly lost to the surroundings by collisions allowing the system to revert or *relax* to the ground state. Sometimes the energy is not dissipated in this way but is re-emitted a few milliseconds later—a process known as *fluorescence*.

By heating substances to high temperatures in a flame or in an electric discharge some of the kinetic energy imparted to the atoms is utilized in exciting electrons into higher energy levels, from which they relax to the ground state with the *spontaneous emission* of radiation. The frequency of the emitted radiation corresponds to the difference in energy between the excited and ground states, equation (7.4).

Even for the simplest compounds, the number of energy levels and hence the number of possible transitions is large. It has already been seen that these transitions are related to a variety of atomic and molecular processes involving energy changes of quite different magnitudes. Radiation can thus be absorbed or emitted over the entire range of the electromagnetic spectrum from low energy radiowaves to high energy γ-rays. Depending on the nature of the analytical information required, and for instrumental reasons, only transitions related to a particular atomic or molecular process and confined to a narrow region of the electromagnetic spectrum are studied at any one time. The information is presented in the form of a *spectrum*, a graphical representation of the degree of absorption or the intensity of emission of electromagnetic radiation as a function of frequency, wavelength or wavenumber.

THE COMPLEXITY OF SPECTRA AND THE INTENSITY OF SPECTRAL LINES

The number and intensities of lines which may appear in a spectrum are determined by three factors,

1. the *populations* of the energy levels from which the transitions originate
2. the values of individual *transition moments* or *transition probabilities*
3. quantum mechanical *selection rules*.

In brief, the statistical probability for the occurrence of a transition can be calculated and lies between zero and one. It can be shown that weak spectral lines have small transition moments while strong lines have values approaching one. One factor which determines the value of a transition moment is the magnitude of the change in dipole moment associated with the transition. Further mention of this will be made in later sections. In addition, the selection rules may show that a particular transition is *forbidden*. As a consequence of these conditions, spectra often have fewer lines or bands than might be expected, a fact which facilitates interpretation. A detailed discussion of transition moments and selection rules, which are mathematical concepts, is beyond the scope of this book and the known selection rules will therefore be assumed. Spectral line widths also affect the appearance of the spectrum. Lines arising from transitions in atoms and gaseous molecules are characteristically narrow as are those due to electron and nuclear spin transitions which involve only small energy changes. Molecular absorption spectra in the infrared, visible and ultraviolet regions consist of sets of very closely spaced lines which are not normally resolved by the instrumentation used. They are broadened by collisions between solute and solvent molecules, so that the overall appearance is of a number of broad overlapping band envelopes.

The relative populations of energy levels, that is the proportions of the analyte species occupying them, have a direct bearing on line intensities and are determined by the spacings of the levels and the thermodynamic temperature. The relation is expressed in the *Maxwell–Boltzmann* equation,

$$\frac{n_2}{n_1} = \frac{g_2}{g_1} \exp\left(-\frac{\Delta E}{kT}\right) \tag{7.5}$$

where n_1 and n_2 are the numbers of species in energy states E_1 and E_2 separated by ΔE, g_1 and g_2 are statistical weighting factors, k is the Boltzmann constant (1.38×10^{-23} J K^{-1}) and T is the thermodynamic temperature. Calculations show that at room temperature and when ΔE exceeds 10^3 J mol^{-1}, only the lowest level of a set will be populated to a significant extent. Thus, absorption spectra recorded in the infrared, visible, ultraviolet regions and beyond arise from transitions from the ground state only. Furthermore for atomic emission spectra to be observed, a considerable increase in temperature (to more than 1 500 K) is required to give appreciable population of the higher energy levels. Indeed the practical importance of the Maxwell–Boltzmann equation lies in demonstrating the effect of changes in thermal

Table 7.1
Analytical spectrometric techniques

NAME OF TECHNIQUE	PRINCIPLE	MAJOR APPLICATIONS
arc/spark spectrometry or spectrography plasma emission spectrometry flame photometry	atomic emission	qualitative and quantitative determination of metals, largely as minor or trace constituents quantitative determination of metals as minor or trace constituents
X-ray fluorescence spectrometry	atomic fluorescence emission	qualitative and quantitative determination of elements heavier than nitrogen as trace to major constituents
atomic fluorescence spectrometry		quantitative determination of metals at minor or trace constituents
atomic absorption spectrometry	atomic absorption	quantitative determination of metals as minor or trace constituents
γ-spectrometry	nuclear emission	qualitative and quantitative determination of elements at trace levels
ultraviolet spectrometry	molecular absorption	quantitative determination of elements and compounds, mainly at trace levels
visible spectrometry		quantitative determination of elements and compounds, mainly as trace and minor constituents
infrared spectrometry		identification and structural analysis of organic compounds
nuclear magnetic resonance spectrometry	nuclear absorption	identification and structural analysis of organic compounds
mass spectrometry	structural fragmentation or ionization of atoms	identification and structural analysis of organic compounds identification and determinations of elements and isotopes at trace levels

excitation conditions on the intensity of atomic emission spectra. Therefore it is possible to have some measure of control over the sensitivity of atomic emission techniques but little or no such control over molecular absorption.

ANALYTICAL SPECTROMETRY

The set of energy levels associated with a particular substance is a unique characteristic of that substance and determines the frequencies at which electromagnetic radiation can be absorbed or emitted. Qualitative information regarding the composition and structure of a sample is obtained through a study of the positions and relative intensities of spectral lines or bands. Quantitative analysis is possible because of the direct proportionality between the intensity of a particular line or band and the number of atoms or molecules undergoing the transition. The various spectrometric techniques commonly used for analytical purposes and the type of information they provide are given in table 7.1.

INSTRUMENTATION

The detailed design and construction of instruments used in the various branches of spectrometry are very different and at first sight there may seem to be little in common between an X-ray fluorescence analyser and a flame photometer. On closer examination, however, certain common features emerge and parallels between the functions of the various parts of the different instruments are observed. The basic functions of any spectrometer are threefold (figure 7.4).

    (a) the production of radiation of frequencies appropriate to energy changes within the sample
    (b) spectral examination of this radiation to facilitate qualitative analysis of the sample
    (c) measurements of the intensity of radiation at frequencies selected from the information obtained in (b) to facilitate quantitative analysis of the sample.

Depending on the technique, the sample may itself fulfil the function of (a) or it may be positioned between (a) and (b) or between (b) and (c).

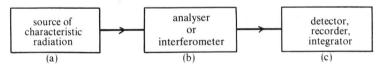

Figure 7.4.   Schematic layout of a spectrometer

Many instruments utilize a *double beam* principle in that radiation absorbed or emitted by the sample is automatically compared with that associated with a blank or standard. This facilitates the recording of data and corrects for matrix effects and instrumental noise and drift. Instrumentation for the generation of radiation is varied and often peculiar to one particular technique. It will be discussed separately in the relevant sections. Components (b) and (c), however, are broadly similar for most techniques and will be discussed more fully below.

The Analyser

The function of this subunit is to present so-called *monochromatic* radiation to the detector, i.e. to separate or disperse the radiation so that selected frequencies corresponding to particular energy transitions within the sample may be individually examined. For instruments designed to operate in the ultraviolet, visible and infrared regions of the spectrum, there are two approaches to this problem.

A filter may be employed which selectively absorbs all except the required range of frequencies—a technique known as *filter photometry*. The absorption characteristics of some standard filters suitable for use in the visible region are given in figure 7.5. This is a simple but inflexible approach with a poor selectivity as it is difficult to make filters with narrow range transmission characteristics. It follows also that a very large number of different filters is

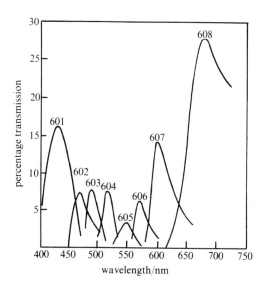

Figure 7.5. Characteristics of standard Ilford filters for the visible region

needed. As a result, filters are used only in unsophisticated instruments where their rather poor selectivity can be tolerated. An alternative approach is to separate or *disperse* the radiation spatially by deflecting it through different angles according to frequency using a *monochromator*. This device consists of a prism or diffraction grating, focusing mirrors or lenses and a slit system. Polychromatic radiation entering the monochromator is directed on to a narrow entrance slit and thence to the prism or grating. By rotating the prism or grating, radiation of different frequencies and consisting of sharp images of the entrance slit can be made to pass successively through a fixed exit slit and on to the detector. If the output from the detector is fed to a chart recorder, a continuous trace of the spectrum is obtained. Alternatively, a photographic film or plate or a series of exit slits and detectors positioned along the focal plane of the monochromator enables radiation of many frequencies to be detected simultaneously without rotating the prism or grating.

Prism Dispersion

In many spectrometric instruments, the monochromator is based on prism optics (figure 7.6). This is particularly true of cheaper instruments. When a beam of polychromatic radiation passes through a prism, the light is refracted from its original path. The higher the frequency of the radiation, the greater

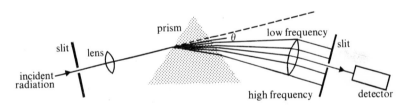

Figure 7.6.   Dispersion of radiation by a prism monochromator

will be the angle of refraction θ. It should be noted that the angular dispersion, $d\theta/d\nu$ does not vary linearly with ν. This results in the frequencies of a recorded spectrum appearing relatively more widely spaced at the high-frequency end than at the low-frequency end.

Diffraction grating dispersion

For most instruments operating in the ultraviolet, visible and infrared regions dispersion of radiation is produced by a diffraction grating. This consists of a polished piece of glass into the surface of which a large number of very closely spaced parallel grooves are cut. The surface is coated with aluminium to render it highly reflective. The grooves must be cut extremely accurately with a spacing that is of the same order of magnitude as the wavelength of the radiation which is to be dispersed. Holographic gratings, which are produced by a rapid photographic process, are superior to those which have mechanically cut grooves. Dispersed radiation of a given wavelength contains less 'stray light', i.e. radiation of wavelengths other than the nominal one, which leads to a wider range of linear response in measurements of absorbance as a function of concentration (p. 355 *et seq.*) The principles of dispersion by a grating are shown in figure 7.7.

A beam of monochromatic radiation of wavelength λ falls on a grating with parallel grooves d apart. Consider two incident rays striking the grating at equivalent points A and B and at an angle φ to the normal. The rays are

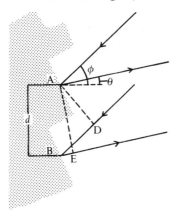

Figure 7.7.   Operating principle of a diffraction grating. φ is the angle of incidence; d is the groove spacing; θ is the angle of reflection

diffracted in all directions, one such direction making an angle θ to the normal. The ray reflected at B travels an additional distance given by

$$BD + BE = d(\sin \varphi + \sin \theta) \qquad (7.6)$$

If BD + BE represents an exact number of wavelengths, the rays reflected at A and B will constructively interfere. At all other values of θ, the net reflected intensity will be zero. Thus, for a polychromatic beam, the angle of reflection θ will vary with wavelength according to the equation

$$d(\sin \varphi + \sin \theta) = n\lambda \qquad (7.7)$$

where n is an integer.

The angular dispersion of the grating is given by

$$\frac{d\theta}{d\lambda} = \frac{1}{d \cos \theta} \qquad (7.8)$$

By using the grating so that the value of θ remains within a narrow range $\cos \theta$ and hence the angular dispersion can be considered constant. This offers a decided advantage over prism dispersion when interpolations are to be made from standard line spectra.

Monochromators for dispersing X-radiation utilize single crystals which behave like a diffraction grating. The spacing of the crystal lattice determines the angles at which radiation is reflected and generally two or more different crystals are required to cover the X-ray region of the spectrum.

In the microwave and radiowave regions, virtually monochromatic incident radiation is generated and the need for a monochromator is thus obviated.

Interferometers

Infrared spectrometers employing an *interferometer* and having no monochromator are now predominant. These non-dispersive instruments, known as Fourier transform (FT) spectrometers, have increased sensitivity and can record spectra much more rapidly than the dispersive type. This is because instead of scanning a spectrum over a given wavenumber range, a process that takes a dispersive instrument from 1 to 4 minutes, the interferometer enables all the data to be collected virtually simultaneously in the form of an *interferogram* then mathematically transformed (using Fourier integrals) by computer into a conventional spectrum.

A diagram of a typical interferometer (Michelson type) is shown in figure 7.8. It consists of fixed and moving front-surface plane mirrors (A and B) and a beamsplitter. Collimated infrared radiation from the source incident on the beamsplitter is divided into two beams of equal intensity that pass to the fixed and moving mirrors respectively. Each is reflected back on itself, recombining at the beamsplitter from where they are directed through the sample compartment and on to the detector. Small movements of mirror B along the direction of the radiation beam incident upon it causes constructive or destructive interference between the two halves of the recombined beam as the optical path difference between them is altered. Continuous movement of B backwards and forwards over a short distance generates a dynamic interference pattern or interferogram that can be monitored by the detector. The interferogram is sampled several thousand times by the microcomputer during one cycle of the mirror movement, which takes about 0.1 second, and a digitized representation of it stored in memory. The digitized interferogram is mathematically related to the spectral output of the source, i.e. its spectrum, which can be derived from it by performing a point-by-point fast Fourier transformation (FFT), a process accomplished by the computer in less than 1 second. If the radiation is then passed through an absorbing sample, a modified interferogram can be recorded. Fourier transformation of this second interferogram produces a spectrum of the source modified by the characteristic absorptions of the sample. Subtraction of the first spectrum from the second then yields the

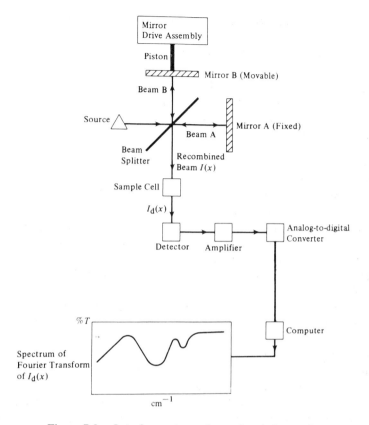

Figure 7.8.   Interferometer and associated electronics

spectrum of the sample which is the equivalent of a spectrum recorded by a double beam dispersive spectrophotometer.

A full-range ir spectrum can be recorded in about 1 second and sensitivity can be enhanced by accumulating multiple scans which are then computer-processed to increase the signal-to-noise ratio. One of the major uses for FT-ir spectrometers is in coupling them to gas chromatographs in what is called a 'coupled' or 'hyphenated' technique, i.e. gc-ir (p. 111). This enables the spectra of eluting peaks to be recorded without the need to trap the component or stop the gas flow. Computer enhancement and manipulations of the recorded spectra are additional features of both stand-alone FT-ir spectrometers and gc-ir systems.

Detectors
The detector in a spectrometer must produce a signal related to the intensity

of the radiation falling on it. For instruments operating in the visible region a *photovoltaic* or *barrier-layer* cell is the simplest and cheapest available. Current produced when radiation falls on a layer of a semiconductor material, e.g. selenium, sandwiched between two metallic electrodes, is proportional to the power of the incident radiation and can be monitored by a galvanometer. Barrier layer cells are robust and are often used in portable instruments but they are not very sensitive and tend to be unstable during extended use.

A more sensitive device is the *photoelectric cell* or *vacuum phototube*. Electrons dislodged when radiation strikes the surface of a sensitive photocathode are collected by an anode and constitute a current which is proportional to the radiation intensity. After amplification, the signal can be fed to a galvanometer, digital voltmeter or potentiometric chart recorder. Much higher sensitivity is obtained from a *photomultiplier tube* which contains up to sixteen electrodes (*dynodes*) successive ones being stabilized at increasingly positive potentials. Each primary electron ejected from the photocathode by incident radiation and collected by the first dynode causes two or more secondary electrons to be ejected by the latter. These are collected by the second dynode where the same process occurs. An *electron cascade* develops which results in an amplification factor of at least 10^6 by the time the electrons reach the anode. Photomultiplier tubes are very widely used but are more expensive than phototubes due to the need for stabilizing circuitry for each dynode to ensure long-term operational stability of the detector.

Diode array detectors consist of silicon *integrated circuit* (*ic*) *chips* incorporating up to one or two hundred pairs of photodiodes and capacitors. Each photodiode measures about 0.05×0.5 mm and is sensitive to radiation covering a wide spectral range, e.g. 200–800 nm. An array of several hundred photodiodes on one or two chips, positioned to receive dispersed radiation from a monochromator, enables a wide range of wavelengths to be monitored continuously without the need for mechanical scanning. The current generated by each photodiode is proportional to the intensity of radiation it receives and it can be measured sequentially many times a second under the control of a microprocessor. The digitized signals can be stored by a micro- or minicomputer for immediate electronic processing and visual display on a cathode ray screen (VDU) in the form of a spectrum. Diode arrays are also utilized in *vidicon tube* detectors which are similar in construction to a small television tube. Such tubes can also be used in non-scanning spectrometers but generally resolution is limited if detection over a wide wavelength range is required. More complex instruments incorporate an integration facility enabling the total signal accumulated over a period of time to be measured. This facility is commonly used in emission spectrometry. In the infrared region, *thermal* or *photon* detectors are used. *Thermal detectors* depend on the heating effect of infrared radiation producing a temperature sensitive response. They include *thermocouples* (voltage changes), *thermistors* (resistance changes), *pyroelectric* (electric polarization changes) and *Golay cells* (changes in the pressure of an enclosed gas). *Photon detectors* are more sensitive and have a more rapid response. They are semiconductor devices made of such materials as *mercury cadmium telluride* (*MCT*) which at liquid nitrogen or helium temperatures generate a variable voltage according to the energies of incident photons of infrared radiation.

Radiations outside the ultraviolet, visible and infrared regions cannot be detected by conventional photoelectric devices. X-rays and γ-rays are detected by gas-ionization, solid-state ionization, or scintillation effects in crystals. *Non-dispersive* scintillation or solid-state detectors combine the functions of monochromator and detector by generating signals which are proportional in size to the energy of the incident radiation. These signals are converted into electrical pulses of directly proportional sizes and thence processed to produce a spectrum. For radiowaves and microwaves, the radiation is essentially monochromatic, and detection is by a radio receiver tuned to the source frequency or by a crystal detector.

Further Reading
DENNEY, R. C., *A Dictionary of Spectroscopy*, 2nd Ed., Macmillan, London, 1982.

Chapter 8

Atomic Spectrometry

Electronic energy levels are widely spaced, and spectral transitions between them are therefore observed towards the high energy end of the electromagnetic spectrum, i.e. in the visible, ultraviolet and X-ray regions. The spectra are characterized by narrow lines and may be simple or complex depending on the number of excited states involved in the excitation process. The wavelengths of the observed absorption and emission lines are characteristic of a particular element, and the intensity of a given spectral line is proportional to the number of atoms undergoing the corresponding transition. *Atomic spectrometric* techniques thus provide means both for the qualitative identification of elements and for their quantitative determination. Because of the large dipole changes associated with electronic transitions, the sensitivity of the techniques is high and they are used primarily in the field of trace and minor component elemental analysis. A summary of those used in analysis is given in table 8.1. To understand these techniques, it is helpful to review certain aspects of the theory of atomic structure and electronic transitions.

ATOMIC STRUCTURE AND SPECTRA

According to quantum theory, the electrons of an atom occupy quantized energy levels or orbitals defined by a set of four quantum numbers whose permitted values (table 8.2) are determined by mathematical rules. The natural tendency is for electrons to occupy orbitals of the lowest possible energy consistent with the *Pauli exclusion principle* which states that no two electrons in one atom may be defined by the same set of values for the four quantum numbers n, l, m_l and s. The shapes and energies of the orbitals are determined by the values of the quantum numbers and by complex interelectronic effects. For most elements, therefore, orbital energies cannot be calculated exactly although an energy level diagram showing relative values

282

Table 8.1 Techniques based on atomic spectrometry

SPECTROMETRIC TECHNIQUE	RADIATION PROCESS	MEANS OF EXCITATION	SAMPLE CONTAINER/MEDIUM	MONOCHROMATOR	DETECTOR
arc/spark emission	emission	electric arc/spark	carbon or graphite electrode	diffraction grating or prism	photomultiplier or photographic plate
laser microprobe emission	emission	laser	original sample material	diffraction grating	photomultiplier
glow discharge emission	emission	glow discharge lamp	machined sample disc	diffraction grating	photomultiplier
plasma emission	emission	electromagnetic induction	gas plasma	diffraction grating	photomultiplier
atomic absorption	absorption	uv/visible radiation	flame or heated rod/furnace	diffraction grating or prism	photomultiplier
atomic fluorescence	fluorescence emission	uv/visible radiation	flame or heated rod/furnace	diffraction grating or prism filter	photomultiplier
flame photometry (flame atomic emission)	emission	flame	flame	diffraction grating, prism or filter	photomultiplier
X-ray fluorescence	fluorescence emission	X-radiation	cell	single crystal diffractor	gas ionization, crystal scintillation or semiconductor
plasma mass spectrometry	n/a	electromagnetic induction	gas plasma	magnetic field	electron multiplier

Table 8.2   Quantum classification of electronic energies

QUANTUM NUMBER	SYMBOL	PARAMETER SPECIFIED	PERMITTED VALUES
principal	n	radial distance	integral from 1 to ∞
secondary	l	angle	integral from 0 to $(n-1)$
magnetic	m_l	angle	integral from 0 to $\pm l$
spin	s	spin	$\pm \frac{1}{2}$

can be drawn. For example, part of the energy level diagram for the outer orbitals of sodium is shown in figure 8.1. In its ground state, the single valence electron of sodium occupies the orbital defined by $n = 3$, $l = 0$, $m = 0$, $s = -\frac{1}{2}$. By suitable excitation, the electron can be promoted to an orbital of higher energy from which it can return to the ground state with

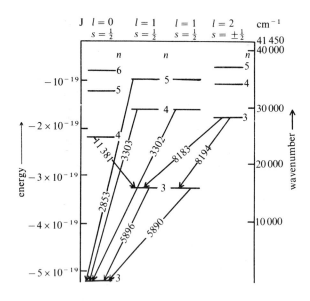

Figure 8.1. Energy level diagram of the sodium atom. The energy levels are denoted by the values for the principal quantum number n, the orbital quantum number l, and the spin quantum number s. Levels with $l = 0$ are not split; for $l = 0$ two separate levels are drawn ($s = +\frac{1}{2}$, $s = -\frac{1}{2}$); for $l > 1$ the splitting is too small to be shown in the figure. Wavelengths of a few spectral transitions are given in nanometers

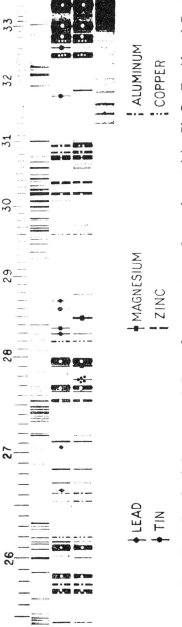

Figure 8.2.   Examples of emission spectra. 1. Iron reference spectrum; 2. sample containing Pb, Sn, Zn, Al, and Cu; 3. sample containing Mg, Zn, Al, Cu and a trace of Pb; 4. Blank

emission of radiation at discrete wavelengths in the visible and ultraviolet regions. The number of allowed transitions is limited by the selection rule $\Delta l = \pm 1$ which results in a rather simple emission spectrum consisting of a series of closely spaced doublets at the wavelengths indicated in the diagram. The doublets arise because of the slightly different energies associated with the two values of the spin quantum number s. The spectra of elements with more than one valence electron are more complex and may contain hundreds of lines (see figure 8.2). The complexity of the spectra also depends upon the means of excitation. Thermal excitation by flames (temperatures in the range 1 500 to 3 500 K) results in relatively simple spectra from valence electrons and involves only a small number of excited states. Higher temperatures (up to 10 000 K) produced by electrical discharges and radio frequency heated plasmas involve excitation to many more levels and consequently result in more complex spectra, including contributions from ionized species. Conversely, excitation by absorption of ultraviolet or visible radiation, the principle upon which atomic absorption and atomic fluorescence spectrometry depends, usually involves only resonance transitions (*vide infra*) and therefore results in very simple spectra. Finally, X-radiation excites electrons occupying inner orbitals and results in fluorescence emission in the X-ray region, the spectra again being relatively simple. The various means of excitation will be discussed more fully in later sections.

Intensity of Spectral Lines

The intensities of atomic spectral lines are determined by their transition probabilities, which are largest for transitions to and from the ground state giving rise to the so-called *resonance lines*. Weak lines, arising from nominally 'forbidden' transitions, may also appear if the selection rules are relaxed by perturbation effects which alter orbital symmetries. Line intensities are also governed by the time atoms spend in the excitation region, i.e. the flame, the electrical discharge or the beam of electromagnetic radiation. This *residence time* can be very short for volatile analytes and where thermal excitation is used because strong vertical thermal currents tend to sweep the sample rapidly out of the excitation zone. Clearly, the longer the residence time, the more sensitive the measurement process becomes and this is an important parameter that needs to be optimized in developing an analytical procedure.

INSTRUMENTATION

The basic instrumentation used for spectrometric measurements has already been described in the previous chapter (p. 274). Methods of excitation, monochromators and detectors used in atomic emission and absorption techniques are included in table 8.1. Sources of radiation physically separated from the sample are required for atomic absorption, atomic fluorescence and X-ray fluorescence spectrometry (cf. molecular absorption spectrometry), whereas in flame photometry, arc/spark and plasma emission techniques, the sample is excited directly by thermal means. Diffraction gratings or prism monochromators are used for dispersion in all the techniques including X-ray fluorescence where a single crystal of appropriate lattice dimensions acts as a grating. Atomic fluorescence spectra are sufficiently simple to allow the use of an interference filter in many instances. Photomultiplier detectors are used in every technique except X-ray fluorescence where proportional counting or scintillation devices are employed. Photographic recording of a complete spectrum facilitates qualitative analysis by optical emission spectrometry.

QUALITATIVE AND QUANTITATIVE ANALYSIS

Only arc/spark and plasma emission and X-ray emission spectrometry are suitable techniques for qualitative analysis as in each case the relevant spectral ranges can be scanned and studied simply and quickly. Quantitative methods based on the emission of electromagnetic radiation rely on the direct proportionality between emitted intensity and the concentration of the analyte. The exact nature of the relation is complex and varies with the technique; it will be discussed more fully in the appropriate sections. Quantitative measurements by atomic absorption spectrometry depend upon a relation which closely resembles the Beer–Lambert law relating to molecular absorption in solution (p. 353).

8.1 Arc/Spark Atomic (Optical) Emission Spectrometry

SUMMARY

Principles
> Emission of electromagnetic radiation in the visible and ultraviolet regions of the spectrum by atoms and ions after electronic excitation in electrical discharges.

Instrumentation
> Emission spectrometer incorporating sample and counter electrodes, means of excitation, prism or grating monochromator, photomultiplier detection system, microprocessors or computers for data processing, interference correction and data display.

Applications
> Used for qualitative detection of metals and some non-metals

particularly at trace levels. Quantitative determination of metals mainly in solid samples such as minerals, ores and alloys. Relative precision 3 % to 10 %.

Disadvantages

Instability of electrical discharges and matrix effects lead to only moderate precision. Photographic recording is time consuming.

An electric arc or spark will cause volatilization and dissociation of a solid or liquid sample and electronic excitation of atomic species present. At the high temperatures (4 000 to 10 000 K) attained within such discharges, collisions between atoms and electrons of the discharge result in a high degree of excitation and the spectra produced are thus comparatively rich in lines. Elements with low ionization potentials may become completely ionized whilst others may undergo thermal excitation largely as unionized atoms. Emission lines originating from the high temperature 'core' of an arc tend to be characteristic of ionic species whereas those coming from the cooler outer 'mantle' will be more characteristic of atoms. Each element is characterized by the wavelengths at which radiation is emitted and the intensity of spectral lines is directly related to the concentration of that element in the sample. The technique therefore provides a means of qualitative identification and quantitative analysis.

INSTRUMENTATION

A diagram of an emission spectrometer is shown in figure 8.3. The electric discharge occurs between a pair of electrodes (preferably enclosed in a special housing for reasons of safety) one of which contains the sample. Emitted radiation is dispersed by a prism or grating monochromator and detected by photomultipliers.

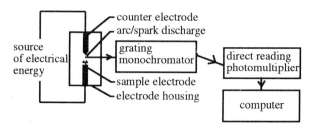

Figure 8.3.   Schematic diagram of an emission spectrometer

Excitation by Electric Discharge

During the many years that atomic emission spectrometry has been employed for chemical analysis a variety of types of excitation sources have been used. In earlier times electric discharges, dc-arcs and ac-sparks, found considerable favour. The inherent instability of the discharges has meant that as more stable alternatives have been developed they have been progressively replaced by them. Where electrical excitation is still employed it is achieved by an electrically controlled spark with far greater stability and much improved precision for the analysis.

Excitation by Laser

A rather specialized emission source, which is applicable to the study of small samples or localized areas on a larger one, is the *laser microprobe*. A pulsed ruby laser beam is focussed onto the surface of the sample to produce a signal from a localized area ca. 50 μm in diameter. The spectrum produced is similar to that produced by arc/spark sources and is processed by similar optical systems.

SAMPLE PREPARATION

For the simplest analyses of metals and alloys, electrodes made of the sample material itself may be used, a disc or cylinder being cut, cast or pressed from powder. Homogeneity of the electrode is of prime importance. A pointed graphite rod or tungsten is used as a counter electrode. For non-metals, high purity graphite electrodes are used, the lower electrode, or anode, having a cup machined in the end to contain the powdered sample. Intimate mixing of the sample with graphite powder and a spectroscopic buffer or carrier (*vide infra*) is necessary to optimize and control excitation conditions. A laser is focused directly on to the sample and will provide a signal from a very small part of it. Particularly valuable is the potential of the laser for the analysis of very small specimens.

QUALITATIVE AND QUANTITATIVE ANALYSIS

Compared to flame excitation, random fluctuations in the intensity of emitted radiation from samples excited by arc and spark discharges are considerable. For this reason instantaneous measurements are not sufficiently reliable for analytical purposes and it is necessary to measure integrated intensities over periods of up to several minutes. Modern instruments will be computer controlled and fitted with VDUs. Computer-based data handling will enable qualitative analysis by sequential examination of the spectrum for elemental

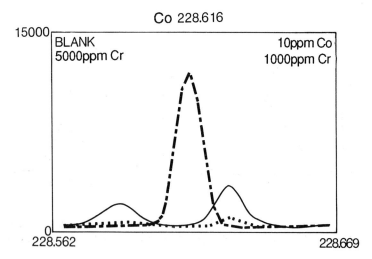

Figure 8.4.   Wavelength characterization scan (from ARL laboratories). Key: ——— 5000 ppm Cr; – – – 1000 ppm Cr; – · – · – 10 ppm Co.

lines. Peak integration may be used for quantitative analysis and peak overlay routines for comparisons with standard spectra, detection of interferences and their correction (figure 8.4). Alternatively an instrument fitted with a poly-chromator and which has a number of fixed channels (ca 30) enables simul-taneous measurements to be made. Such instruments are called *direct reading spectrometers*.

INTERFERENCES AND ERRORS ASSOCIATED WITH THE EXCITATION PROCESS

Problems may arise from the different volatility of elements in the sample which will in turn be modified by the matrix. Fractional distillation can occur and a serious error result unless the sample is completely burned because the emission intensity for each element will vary independently with time, figure 8.5(a). On the other hand, this effect may well be exploited as in the case of volatile impurities in a non-volatile matrix. By recording the spectrum only in the early part of a burn, they may be detected with little interference from the matrix. The prespark routine discussed earlier (p. 275) may also be used to improve precision.

Another aspect of the volatility problem concerns the sensitivity of the method—the greater the volatility of a material the shorter its residence time in the discharge. It follows that volatility and sensitivity will be related and that the lowest detection limits will be associated with low volatility. Excessively rapid volatilization will introduce a serious error by *self-*

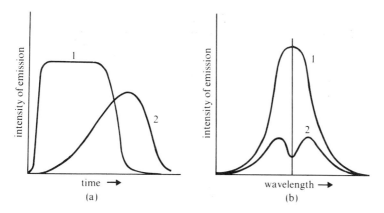

Figure 8.5.   (a) Variations of line intensity during burning of sample; element 1
is more volatile than 2. (b) energy-wavelength profile and self-absorption

absorption. The high concentration of the element in the cooler part of the
discharge will absorb emitted radiation from the core, thus depressing the
intensity of the line. In exceptional cases, there may even be *self-reversal* of
the line giving a minimum in the intensity-wavelength profile, figure 8.5(b).
Self-reversal is also observed with intense lines from major constituents of a
sample. Volatility differences present a problem as they arise not only from
variations in the properties of the analytes, but also from chemical reactions
taking place between analyte and matrix or the carbon of the electrodes. A
further problem is the formation of cyanogen radicals, $CN\cdot$, which provide
interference from molecular spectral bands. In earlier years much ingenuity
was employed in minimizing complex interference effects but they remain a
problem leading to relatively poor precision and limits of detection.
Dissatisfaction with arc/spark analysis on these grounds is the primary
reason for the displacement of arc/spark emission by more recently
developed techniques such as atomic absorption and ICP–AES, leaving
spark methods pre-eminent only in fields such as metallurgical analysis
where matrix effects are less important and where speed of analysis and
simplicity of operation are of the greatest importance.
    The intensity I of an atomic spectral line is related to the number n of atoms
(not ions) present in the discharge and to the temperature T of the discharge
by the equation

$$I = \frac{Ag_2hvb(1 - \alpha)n}{B} \exp\left(-\frac{\Delta E}{kT}\right) \qquad (8.1)$$

where A is the transition probability, g_2 is a statistical weighting factor
associated with the Maxwell–Boltzmann equation (p. 272), h is Planck's

constant $(6.6 \times 10^{-34}$ J s), v is the frequency of the emitted radiation, b is the diameter of the discharge in the direction of the detector, α is the degree of ionization, ΔE is the difference in energy between the two levels associated with the transition and B is a constant at a given temperature. Under constant excitation conditions, n is proportional to C, the concentration of the analyte and therefore

$$I = k'C \qquad (8.2)$$

where k' is a constant.

Equation (8.2) also holds for an ion line under constant excitation conditions. In practice, an internal standard should always be used because of the complex nature of the relation between I and C. The internal standard line and the analyte line should have similar wavelengths and intensities so as to be affected in a similar manner by changes in excitation conditions. A pair of lines fulfilling these requirements is known as a *homologous pair*. Spectra of a series of standards each containing a fixed amount of the internal standard are recorded. A major constituent of the sample sometimes fulfils the role of an internal standard as its concentration can be considered to be constant. The ratios of readings for the homologous pair are then plotted against the amount of analyte in the corresponding standard to produce a calibration or *working curve*. Figure 8.6 shows such a curve for the determination of magnesium using molybdenum as an internal standard. The homologous pair are Mg at 279.8 nm and Mo at 281.6 nm. The corresponding ratios are calculated for samples and the concentration of magnesium determined from the working curve. Precision is in the range 3 to 10%, direct reading spectrometers and spark or plasma sources giving the most reliable data.

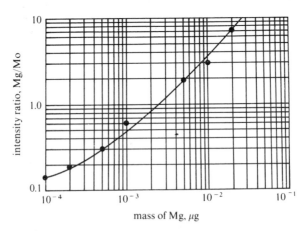

Figure 8.6.   Working curve for the analysis of magnesium with molybdenum as internal standard

Arc/spark emission methods have been widely used for the determination of metals and some non-metals particularly as minor and trace constituents. In recent years, however, the technique has been extensively displaced by atomic absorption spectrometry, and plasma emission methods. Detection limits for many elements are of the order of 1 to 10 ppm (table 8.3) and as over sixty can be determined simultaneously the technique has few rivals when a rapid qualitative survey is required. It is ideally suited to the analysis of solid samples and finds widespread application in the metallurgical industry for the analysis of alloys, in the semi-conductor and nuclear industries for the detection of undesirable impurities, and in geochemical prospecting. Quantitatively, precision is only moderate which is often not acceptable for the determination of minor and major constituents, but because of extremely rapid processing of results by computer-aided direct reading spectrometers, the technique is an invaluable aid in production control. For example, a complete analysis on a steel sample just drawn from a furnace can be completed in two or three minutes and the composition of the bulk melt adjusted, if required, before casting. In the field of geochemistry, an important application is in the charting of areas of land with respect to the composition of underlying rock strata. The rapid screening of hundreds of samples taken at pre-determined positions enables the geochemist to

Table 8.3   Detection limits for a 1 mg sample in a dc arc using graphite electrodes

ELEMENT	LINE MEASURED, nm	DETECTION LIMIT, ppm
Ag	328.0	0.5
Al	394.4	1
B	249.8	1
Be	234.8	2
Cd	326.1	50
Co	340.5	5
Cu	327.3	0.1
Fe	302.0	10
Mg	285.2	0.01
Mn	279.4	2
Mo	313.3	5
Pb	283.3	10
Sn	284.0	5
Ti	337.3	5
V	318.4	10
W	294.4	100
Zn	334.5	100

produce a detailed map of the distribution of valuable mineral and ore deposits. However, the development of plasma emission and atomic absorption spectrometry has steadily displaced arc/spark methods from many applications. Notwithstanding the need to take samples into solution, the former techniques are often preferred because of greater sensitivity, better precision and relative freedom from interferences. Nevertheless, arc/ spark emission remains important especially in metallurgy where speed of sample preparation and analysis may outweigh other considerations.

8.2   Glow Discharge Atomic Emission Spectrometry

SUMMARY

Principles
> Emission spectrum generated by bombardment of sample by argon ions in a "sputtering process".

Instrumentation
> Glow discharge lamp (analogous to hollow cathode lamp) in which the sample acts as the cathode. Attached to a standard atomic emission spectrometer.

Applications
> Limited to samples that are electrically conducting. Gives high stability signals. Valuable in metallurgy.

Disadvantages
> Limited scope for sample types. Gives analysis of surface only.

A recent development in emission spectroscopy has been the *glow discharge lamp (GDL)* as an excitation source. This source overcomes the reliability variations discussed in section 8.1 which lead to errors and poor precision. For electrically conducting samples such as metals and alloys it shows an impressive performance. Apart from the modified source, the overall instrumentation remains similar to that discussed in section 8.1. A silent electric discharge through an argon atmosphere will lead to the formation of positive argon ions. Under the influence of a potential gradient these ions will impact on the surface of the cathode leading to vaporization and excitation of atoms from its surface. This process, known as *sputtering*, is also the basis of the hollow cathode lamp used in atomic absorption spectrometry (*vide infra*).

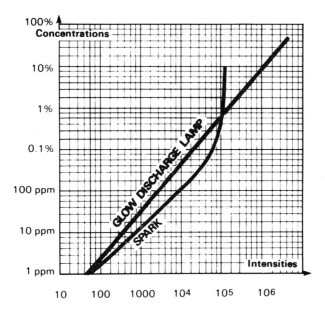

Figure 8.7.   Typical GDL and spark emission calibration curves contrasting range
and linearity (with permission from Jobin-Yvon)

If the cathode comprises an electrically conducting sample, as in metallurgical analysis, its surface atoms will be volatilized and excited, producing on relaxation an emission spectrum characteristic of the composition of the sample surface. Furthermore, as volatilization depends only on the sputtering process all elements will be volatilized at more or less the same rate. Hence many of the complex matrix problems experienced by arc/spark methods will be avoided. As a result, calibration curves will be linear over a wider range with much better precision for the analysis (figure 8.7). Additionally, as the sputtering process steadily erodes the surface ($2-5\,\mu m\,min^{-1}$) changes in composition in the surface layers may also be studied.

INSTRUMENTATION

The overall instrumentation with the exception of the GDL resembles that for other forms of atomic optical emission spectrometry. The construction of the GDL is shown schematically in figure 8.8. Initially the lamp is flushed with argon, evacuated and then filled with argon to low pressure for the sputtering and glow discharge to operate. After analysis the system is flushed with argon at a high flow rate, the sample replaced and subsequent analysis carried out by the same sequence. The surface of the sample needs to be smooth.

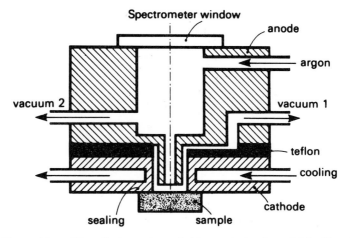

Figure 8.8.   Glow discharge lamp (with permission from Jobin–Yvon)

The GDL can be used for the analysis of trace, minor and major constituents in electrically conducting samples, especially metallurgical specimens. Similar detection limits to arc/spark methods are observed but with greater freedom from interferences and much improved precision (figure 8.9).

8.3   Plasma Emission Spectrometry

SUMMARY

Principles
   Emission of electromagnetic radiation in the visible and ultraviolet regions of the spectrum by atoms and ions after electronic excitation in a high temperature gas plasma.

Instrumentation
   Emission spectrometer incorporating a sample nebulizer, grating mono-chromator, photomultiplier detection system and microprocessor con-troller. Excitation by dc-arc plasma jet, or inductively coupled plasma.

Applications
   Very widespread for qualitative and quantitative analysis of metals and some non-metals, particularly at trace levels. Relative precision 0.5 to 2%.

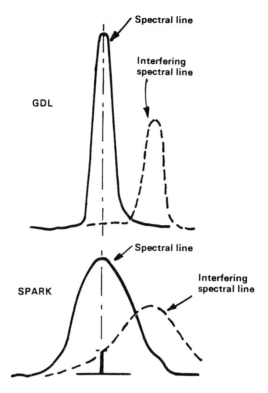

Figure 8.9.   Comparison of spectral characteristics of GDL and spark emission
signals (with permission from Jobin-Yvon)

Disadvantages

Samples require dissolution before analysis. Instrumentation is complex
and expensive to purchase and operate.

At high temperatures ($\sim 6\,000$ K) a gas such as argon will contain a high
proportion of ions and free electrons constituting a *plasma*. Additional energy
may be supplied to the electrons in the plasma by the application of an
external electromagnetic field. By collisions between the electrons and other
species in the plasma this additional energy is uniformly distributed. If the gas
pressure is high the mean free path of the electrons decreases and the
probability of collision increases making the energy transfer more efficient
and leading to a substantial temperature enhancement. Small plasmas suitable for emission spectrometry sources with power densities of $1\,\mathrm{kW\,cm^{-3}}$
and temperatures of $8\,000$–$10\,000$ K operating at ambient pressures can be

produced in this way. Plasma excitation offers an attractive alternative to that of arc/spark because of the high temperatures attainable and the homogeneity of the source. The plasma has a substantially better signal stability and thus analytical precision. Two types of plasma sources have reached the stage of commercial manufacture. These are based on the use of a dc-arc discharge in one case and on radio frequency induction heating in the other.

INSTRUMENTATION

The outline of the construction of a plasma emission spectrometer may be inferred from figure 8.3 with a plasma source replacing the electrode system. Instruments are computerized and make use of automatic sample handling.

The Direct-Current Plasma Source
In a *direct-current plasma source* (*DCP*) initial heating of an inert gas, usually argon, is produced by a dc-arc. Experimentally it is arranged for the plasma to be established in a high velocity gas stream. When the edges of the plasma are cooled with an inert gas vortex the cooler outer parts have much reduced ionization relative to the rest of the plasma. Consequently, they are unable to carry the arc current, so that the current density in the rest of the plasma is increased. As a result the temperature increases from the 6 000 K typical of an arc, to 9 000 or 10 000 K. A common version of the DCP (figure 8.10) uses an inverted V-shaped electrode assembly with the excitation region established at the intersection of two argon streams. A

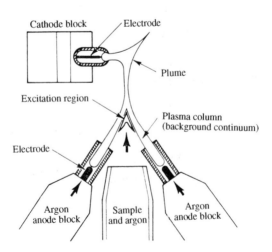

Figure 8.10.   dc plasma jet

nebulizer is used to inject the sample at this intersection point. Superior stability and sensitivity are claimed for this arrangement together with a much improved signal to noise ratio. A particular shortcoming of the dc source is the dependence of the signal on the surface of the carbon electrodes. As the plasma operates, the electrodes are eroded so that after 2–3 hours replacement or reshaping is needed. On the other hand, a distinct advantage is the stability of the DCP with varying sample matrices, a characteristic which is particularly valuable where the solvent contains a high concentration of dissolved solids.

The Inductively Coupled Plasma Torch

A second plasma emission source which has become the most popular is the inductively coupled plasma torch (ICP). Figure 8.11 illustrates a typical source construction, which consists of three concentric quartz tubes. An aerosol of the sample solution in injected into the plasma through the central tube in a stream of argon flowing at about $1 \, dm^3 \, min^{-1}$. A higher flow of argon ($15 \, dm^3 \, min^{-1}$) is injected between the second and outer tubes. The plasma is ignited by a high voltage discharge from a Tesla coil and sustained by the induction heating from the rf coil which operates at a power level of about $2 \, kW$. A third current of argon ($0.5 \, dm^3 \, min^{-1}$) entering between the inner

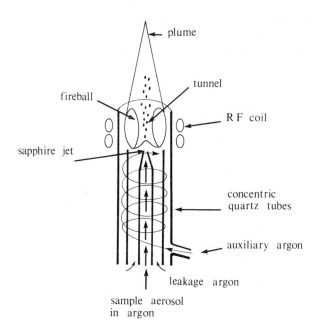

Figure 8.11. ICP torch

and outer tubes has the effect of lifting the plasma clear of the quartz tube, minimizing damage to the orifice by the high temperature. In this device a sample aerosol may be injected into a plasma at a temperature of 10 000 K. In this way the sample argon stream drives a tunnel through the centre of the fireball and a plasma with toroidal geometry results. Thus, the sample passes through the high temperature region at the centre and is heated with virtually no contact with the plasma itself. It follows that sample material cannot be trapped within the plasma itself and any consequent memory effects are avoided. Having emerged from the fireball, the sample stream forms a tail or plume containing the sample elements free of molecular associations. On cooling, the ionized atoms will relax to their ground states emitting characteristic radiation with little background or continuum. The exact point at which relaxation occurs will depend upon the temperature and the ionization energy of the element in the sample. Thus the optimum alignment for the optical channel of the instrument will vary from element to element.

It can be seen from the above that the sample stream emerging from the plasma will be rich in free ions and atoms of the elements from the sample. Thus, the ICP could provide an attractive source for analytical methods other than those based upon straightforward emission. Instruments using the ICP source as a basis for atomic fluorescence have been developed. More importantly, however, the ICP source has been applied to provide a source of ions for mass spectrometry (*vide infra*).

SAMPLE INTRODUCTION FOR PLASMA SOURCES

For the majority of applications, the sample is taken into solution and introduced into the plasma as an aerosol in the argon stream. The sample solution is pumped by a peristaltic pump at a fixed rate and converted into an aerosol by a nebulizer (see atomic absorption spectrometry). Various designs of nebulizer are in use each having strengths and weaknesses. The reader is directed to the more specialist texts for a detailed consideration of nebulizers. There is an obvious attraction in being able to handle a solid directly, and sample volatilization methods using electric spark ablation, laser ablation and electrothermal volatilization have also been developed.

ANALYTICAL MEASUREMENTS

Qualitative analysis may be made by searching the emission spectrum for characteristic elemental lines. Photographic recording is not normally used, but with modern high resolution optics and computer control, the emission spectrum may be examined for the characteristic lines of a wide range of elements (figure 8.4). Quantitative measurements are made on the basis of line intensities which are related to the various factors expressed in equation (8.1). Under constant excitation conditions the simple proportionality expressed in equation (8.2) applies. Straightforward calibration with standards is thus possible.

Over seventy elements may be detected and measured by ICP–OES and in principle could be determined simultaneously by using an instrument fitted with a polychromator having the appropriate number of channels. Apart from the high cost of providing so many channels, physical limitations in fitting additional channels become severe after 35 or 40, and most simultaneous instruments are limited to this number. Many laboratories will require the flexibility to deal with smaller and varying groups of elements. The alternative approach is to make measurements on a *sequential* basis with instrumental parameters being optimized in turn by a preprogrammed microcomputer. Using this system, a group of 10–12 elements can be determined in 2–3 minutes. Notwithstanding this flexibility, in laboratories where large numbers of samples are routinely analysed for large numbers of elements, simultaneous instruments capable of analysing for a portfolio of 30–35 elements may well be employed. Recent technological developments are making use of diode array detectors (chapter 7) which can respond at a number of different wavelengths simultaneously. Furthermore, they may be re-programmed quickly and easily. Thus the different merits of sequential and simultaneous operation may be effectively combined.

Plasma sources are characterized by the great stability of the discharge and

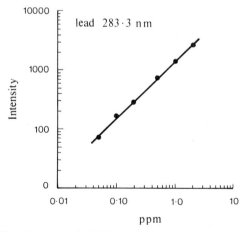

Figure 8.12. Calibration curve for ICP determination of lead in sea water showing the wide linear range

Table 8.4   Atomic spectroscopy detection limits (micrograms/litre) (From Perkin Elmer, *Guide to Techniques and Applications of Atomic Spectroscopy*, 1988.)

ELEMENT	FLAME AA	Hg/ HYDRIDE	GFAA	ICP EMISSION	ICP-MS
Ag	0.9		0.005	1	0.04
Al	30		0.04	4	0.1
As	100	0.02	0.2	20	0.05
Au	6		0.1	4	0.1
B	700		20	2	0.1
Ba	8		0.1	0.1	0.02
Be	1		0.01	0.06	0.1
Bi	20	0.02	0.1	20	0.04
Br					1
C				50	50
Ca	1		0.05	0.08	5
Cd	0.5		0.003	1	0.02
Ce				10	0.01
Cl					10
Co	6		0.01	2	0.02
Cr	2		0.01	2	0.02
Cs	8		0.05		0.02
Cu	1		0.02	0.9	0.03
Dy	50				0.04
Er	40				0.02
Eu	20				0.02
F					100
Fe	3		0.02	1	1
Ga	50		0.1	10	0.08
Gd	1200				0.04
Ge	200		0.2	10	0.08
Hf	200				0.03
Hg	200	0.008	1	20	0.03
Ho	40				0.01
I					0.02
In	20		0.05	30	0.02
Ir	600		2	20	0.06
K	2		0.02	50	10
La	2000			1	0.01
Li	0.5		0.05	0.9	0.1
Lu	700				0.01
Mg	0.1		0.004	0.08	0.1
Mn	1		0.01	0.4	0.04
Mo	30		0.04	5	0.08
Na	0.2		0.05	4	0.06
Nb	1000			3	0.02
Nd	1000				0.02
Ni	4		0.1	4	0.03

Table 8.4 *cont'd.*

ELEMENT	FLAME AA	Hg/ HYDRIDE	GFAA	ICP EMISSION	ICP-MS
Os	80				0.02
P	50000		30	30	20
Pb	10		0.05	20	0.02
Pd	20		0.25	1	0.06
Pr	5000				0.01
Pt	40		0.5	20	0.08
Rb	2		0.05		0.02
Re	500			20	0.06
Rh	4			20	0.02
Ru	70			4	0.05
S				50	500
Sb	30	0.1	0.2	60	0.02
Sc	20			0.2	0.08
Se	70	0.02	0.2	60	0.5
Si	60		0.4	3	10
Sm	2000				0.04
Sn	100		0.2	40	0.03
Sr	2		0.02	0.05	0.02
Ta	1000			20	0.02
Tb	600				0.01
Te	20	0.02	0.1	50	0.04
Th					0.02
Ti	50		1	0.5	0.06
Tl	9		0.1	40	0.02
Tm	10				0.01
U	10000			10	0.01
V	40		0.2	2	0.03
W	1000			20	0.06
Y	50			0.2	0.02
Yb	5				0.03
Zn	0.8		0.01	1	0.08
Zr	300			0.8	0.03

All detection limits are given in micrograms per litre and were determined using elemental standards in dilute aqueous solution. Atomic absorption (Model 5100) and ICP emission (Plasma II) detection limits are based on a 95% confidence level (2 standard deviations) using instrumental parameters optimized for the individual element. ICP emission detection limits obtained during multielement analyses will typically be within a factor of 2 of the values shown. Cold vapour mercury AA detection limits were determined using an MHS-20 mercury/hydride system with an amalgamation accessory. Furnace AA (Zeeman/5100) detection limits were determined using STPF conditions and are based on 100–μL sample volumes. ICP-MS (ELAN) detection limits were determined using operating parameters optimized for full mass range coverage and a 98% confidence level (3 standard deviations). ICP-MS detection limits using operating conditions optimized for individual elements are frequently better than the values shown. ICP-MS detection limits for fluorine and chlorine were determined using the ELAN's negative ion detection capabilities.

this feature offers a considerable improvement in precision over arc/spark emission methods. Relative precisions are in the range of 0.5% to 2%. The high temperatures of plasmas lead to a high degree of atomic excitation and thus high sensitivity for measurements, which may often be made at levels below 0.1 ppm and with a linear response up to 1 000 ppm or more. Figure 8.12 shows a typical calibration curve. The high temperatures also ensure that virtually all compounds are broken down to their constituent elements. Thus matrix effects are largely eliminated. Furthermore, background interferences from molecular species, which are so often a problem in other methods of atomic spectrometry, are removed. As the plasma is not in contact with the electrodes or jet no contamination is introduced from this source. A further consequence of the high temperature is the large number of emission lines which are excited. If the matrix is complex, line overlap may lead to spectral interferences. This situation is to be contrasted with atomic absorption spectrophotometry where such interferences are very rare. Other disadvantages are the costly instrumentation and high running costs, especially as large volumes of argon are required. With some sample types the difficulties of effecting complete dissolution are also limiting factors.

APPLICATIONS OF PLASMA EMISSION SPECTROMETRY

The potential for the employment of plasma emission spectrometry is enormous and it is finding use in almost every field where trace element analysis is carried out. Some seventy elements, including most metals and some non-metals, such as phosphorus and carbon, may be determined individually or in parallel. As many as thirty or more elements may be determined on the same sample. Table 8.4 is illustrative of elements which may be analysed and compares detection limits for plasma emission with those for ICP–MS and atomic absorption. Rocks, soils, waters and biological tissue are typical of samples to which the method may be applied. In geochemistry, and in quality control of potable waters and pollution studies in general, the multi-element capability and wide (10^5) dynamic range of the method are of great value. Plasma emission spectrometry is well established as a routine method of analysis in these areas.

8.4 Inductively Coupled Plasma–Mass Spectrometry (ICP–Mass)

SUMMARY

Principles
    Qualitative analysis by separation of atomic ions in a mass spectrometer.
    Quantitative analysis from magnitude of ion current.

Instrumentation

Ions produced in an ICP-torch interfaced to a quadrupole mass spectrometer. Sample introduction by nebulizer, laser vaporization or electrical heating.

Applications

In principle to all elements at low (ppb) concentrations. Valuable for first look analysis, at solid samples. Isotopic ratio measurements can distinguish between sources of elements in tracer studies and environmental samples.

Disadvantages

Complex, expensive instrumentation. Precision sometimes poor. Interference from polyatomic ions.

PRINCIPLES

The ICP torch provides a rich source of free atoms and ions from the elements comprising the sample. In ICP–MS part of the sample stream from a point close to the centre of the fireball is directed to a mass spectrometer. The resulting mass spectrum can be used to identify elements from the mass numbers of the ion peaks and the peak size for quantitative analysis. Moreover the whole spectrum can be displayed at the same time providing qualitative analysis for a wide range of elements from one display only. A typical spectrum is shown in figure 8.13.

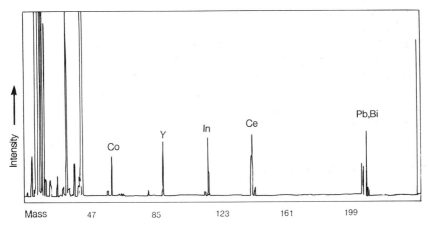

Figure 8.13. ICP–MS spectrum of Be, Al, Co, Y, In, Ce, Pb, Bi and U (100 mg dm^{-3}) (courtesy of Dr K. Jarvis, Plasma Mass Spectrometry Laboratory, Royal Holloway College)

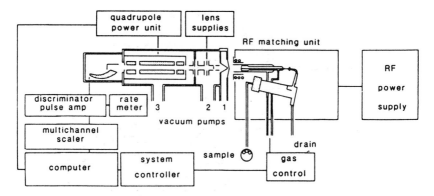

Figure 8.14.   Schematic diagram of a typical ICP–MS system. In some systems the
multichannel scaler may be omitted

INSTRUMENTATION

The layout of an ICP–Mass Spectrometer is shown schematically in figure 8.14 and comprises three essential parts; the ICP torch, the interface and the mass spectrometer. The ICP torch differs little from that discussed earlier and the mass spectrometer is very similar to those used for organic mass spectrometry and discussed in chapter 9. Typically a quadrupole instrument would be used. The construction of the interface is shown in figure 8.15 and is based on the use of a pair of watercooled cones which divert a portion of the sample stream into the ion optics of the mass spectrometer whence the mass spectrum is produced by standard mass spectrometer operation.

APPLICATIONS

ICP–MS may be applied to the determination of elements across the whole of the periodic table from lithium to the actinides. With certain exceptions, limits of detection are of the same order or better than those for graphite–furnace–AA or ICP–AES. Table 8.4 makes a comprehensive comparison.

A particular advantage of ICP–MS derives from its ability to display a complete mass spectrum at one time. Combined with sample introduction by laser ablation it constitutes a very powerful tool for first look analysis, e.g. in geological prospecting or ecological surveys. ICP–MS is applicable to the whole range of areas where minor or trace elements are to be determined.

The ability to provide data on the isotope composition of analyte elements, is a further special feature of the technique. This has already been exploited in the identification of sources of environmental lead

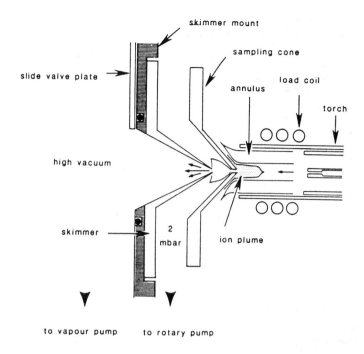

Figure 8.15.   Cross-section of typical ICP ion source showing plasma torch and ion
extraction interface. Extraction and skimmer sizes are slightly exaggerated for clarity

contamination and poisoning. The radiogeneric source of lead means that
different deposits may have different isotopic composition. In the UK this
feature has been used to distinguish between lead from paint or water pipes
on the one hand and petrol anti-knock agents on the other (figure 8.16). In
geochemistry isotopic ratios can be used to indicate the date of crystallization
of rocks. The ratio is fixed at the time of crystallization, but can change
subsequently as a result of radioactive decay. With a measurement of isotopic
ratios, and a knowledge of the radioactive half-life, the original date of
crystallization can be estimated. Ratios such as $^{87}Rb : ^{87}Sr$, $^{147}Sm : ^{144}Nd$,
$^{207}Pb : ^{204}Pb$, and $^{206}Pb : ^{204}Pb$ are used.

Tracer studies in which chemically similar species are studied on the basis
of containing a radioisotope are discusssed in chapter 10. It is fairly obvious
that, with detection techniques readily available for the measurement of non-
radioactive isotopes, the principle can be extended to non-radioactive
systems. Where *in vivo* studies are concerned there are clear safety reasons for
so doing. Although some progress is being made in this direction, it is being
limited by the high cost and poor availability of isotopically enriched tracers.

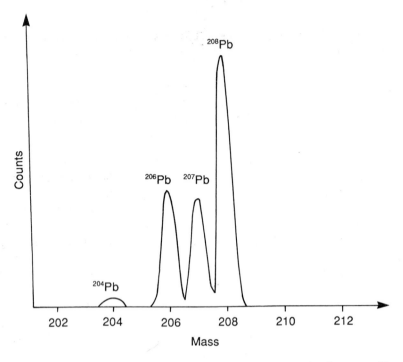

Figure 8.16. ICP–MS spectrum showing the naturally occurring isotopes of lead

Current disadvantages include the high cost of purchase and running of instruments, whilst polyatomic ions, e.g. Ar_2O^+, ArN^+, ArO^+ can interfere by obscuring some elemental peaks.

8.5 Flame Emission Spectrometry

SUMMARY

Principles

Emission of electromagnetic radiation in the visible and ultraviolet regions of the spectrum by atoms after electronic excitation in flames.

Instrumentation

Flame photometer or spectrophotometer incorporating nebulizer and burner, filters, prism or grating monochromator, photocell or photo-multiplier detection system.

Applications

Quantitative determination of metals in solution, especially alkali and alkaline earths in clinical samples. Relative precision 1 to 4 %.

Disadvantages

Intensity of emission is very sensitive to changes in flame temperature. Spectral interferences and self-absorption are common problems.

Flame emission spectrometry closely resembles arc/spark emission spectrometry in principle and instrumentation. The difference lies in the substitution of a chemical flame for the high-temperature electric discharges. The general characteristics of flames and the analytically relevant processes occurring within them are discussed later. It is sufficient to note here the lower temperature of flames, typically 2 000 to 3 000 K, and the consequently lower energy available to induce electronic excitation. It follows that the flame-induced emission spectrum of an element will be much less complex than the corresponding spectrum resulting from an electric discharge or plasma. Flame emission spectrometry is a particularly useful technique for the determination of volatile elements with low excitation energies such as the alkali and alkaline earth metals. This is in part due to a high sensitivity resulting from the relatively low degree of ionization induced by cool flames and a longer residence time for the analyte atoms in the flame as compared to an arc or spark.

INSTRUMENTATION

A typical instrumental arrangement for flame emission measurements is shown in figure 8.17(a). The sample, in the form of a solution, is drawn into a *nebulizer* where it is converted into a fine mist or aerosol. From there it passes into the flame along with air or oxygen and a fuel gas. Following thermal excitation, the radiation emitted as excited atoms relax is viewed by a photocell or photomultiplier. The current generated in the detector circuit may be read directly or, more conveniently, converted to a meter or digital readout in analyte concentration. It should be noted that the more stable emission resulting from flame excitation facilitates the almost instantaneous measurement of line intensities and this should be contrasted with the need to measure time-integrated intensities where arc/spark excitation is employed. Low-temperature flames produce spectra that are sufficiently simple to allow the use of a narrow band-pass filter to isolate the required emission line for quantitative measurement, such an instrument being termed a *flame photometer*. If better resolution is needed to isolate lines in more complex spectra,

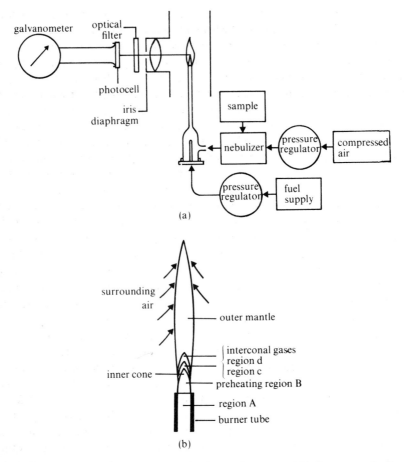

Figure 8.17.   (a) Schematic diagram of a flame photometer. (b) Structure of a flame

or to minimize interference from background emission, a *flame atomic emission spectrometer* incorporating prism or grating dispersion is necessary and the technique is known as flame atomic emission spectrometry.

FLAME CHARACTERISTICS

Flames used in analytical measurements are similar to those produced by Bunsen burners with the added provision of a means of introducing the sample directly into the combustion zone. Support (oxidant) and fuel gases are fed to a *nebulizer* along with the sample solution. The mixed gases and sample aerosol then pass through the jets of the burner where ignition occurs.

For flame emission measurements, burners of the Meker type with a circular orifice covered by a grille are used whereas in atomic absorption spectrometry, a slit burner is preferred. In both cases, the flame consists of two principal *zones* or *cones*, figure 8.17(b). The *inner cone* or *primary reaction zone* is confined to the region just above the burner orifice; it is here that combustion, atomization and thermal excitation occur. The combustion products consist largely of CO, H_2, CO_2, N_2, H_2O and free radicals such as $OH^{\cdot}$. Above the primary zone is the *outer cone* or *secondary* reaction zone. In this region, cooling occurs as a result of mixing with the surrounding atmosphere. This may lead in turn to the entrainment of impurities, such as sodium compounds, which will increase background emission from the flame. A second problem concerns the build-up of water in the outer cone resulting in considerable emission by $OH^{\cdot}$ radicals between 280 and 350 nm. One method of minimizing this problem has been to use *divided flames*. A cone of gas, which may be the existing support gas or a noble gas, is injected between the inner and outer cones, thus effectively 'lifting off' the outer reaction zone, reducing the flame background and stabilizing the important inner cone. The maximum temperature of a non-divided flame is reached at the junction of the inner and outer cones although the dividing line is not always clear and an inter-conal zone may often be identified. The optical axis of the instrument is usually aligned on this zone but the optimum position for maximum sensitivity for each element varies and should be chosen by experimental observation.

The fuel gas will burn towards the burner orifice with a velocity of between 1 and 50 m s^{-1}. To prevent blow-back, it is necessary for the rate of fuel supply to exceed this *burning velocity*. Details of burning velocities and flame temperatures for some common fuel/support gas combinations are given in table 8.5. It will be seen that high temperatures can be attained with, for example, hydrogen/oxygen flames but at the expense of a high burning velocity which leads to a low residence time for the analyte. A good compromise is reached when acetylene is the fuel gas, as temperatures of 2 000 to 3 000 K may be achieved at moderate burning velocities.

Table 8.5 Characteristics of some typical gas mixtures

FUEL GAS	SUPPORT GAS	BURNING VELOCITY/m s^{-1}	FLAME TEMPERATURE/K
propane	Air	0.8	1 900
hydrogen	Air	4.4	2 000
acetylene	Air	2.7	2 450
hydrogen	Nitrous Oxide	3.0	2 850
acetylene	Nitrous Oxide	5.0	2 950
hydrogen	Oxygen	37	2 800
acetylene	Oxygen	25	3 100

FLAME PROCESSES

Flame atomization and excitation can be divided into a number of stages. Firstly, the heat of the flame evaporates solvent from the droplets of sample aerosol leaving a cloud of small particles of the solid compounds originally present in the solution. These are then vaporized and molecular associations broken down releasing free atoms (*atomization*) some of which undergo electronic excitation or ionization. The population of free atoms in the flame will be an equilibrium value dependent on the rates of nebulization and atomization from the original sample solution and the rates at which the atoms are removed with exhaust gases, by chemical reactions with the flame gases or by ionization. The various factors are summarized as follows:

1. *Rate of fuel flow.* This affects the rate of nebulization and the residence time of the atoms within the flame.
2. *The viscosity of the solvent.* This affects the rate of nebulization.
3. *The chemical nature of the solvent.* This may lead to the formation of stable solvated species and a modification of flame conditions depending on the degree of inflammability.
4. *Other chemical species in the solution or flame.* These may form stable non-volatile compounds with the analyte.
5. *Flame temperature.* This will control the rate of solvent evaporation, the break-up of molecular associations containing the analyte and the extent of ionization of analyte atoms. This last factor is related to flame temperature by the Maxwell–Boltzmann equation (p. 272).

Careful control and optimization of the above factors is necessary in all techniques involving flames, i.e. *flame photometry* and *flame atomic emission spectrometry, atomic absorption spectrometry* (p. 317) and *atomic fluorescence spectrometry* (p. 330). The first two require a maximization of the number of excited atoms in the flame and the number which relax by the emission of electromagnetic radiation. Conversely, atomic absorption and atomic fluorescence spectrometry depend upon the number of ground-state atoms in the flame which are capable of being excited by incident radiation from an external source. An additional factor affecting the emission process is the possibility of relaxation of excited atoms by *non-radiative transitions.* These are caused by collision with other particles within the flame thereby dissipating their excess energy and resulting in a decrease in emission intensity.

EMISSION SPECTRA

The comparatively low thermal energy of flames results in the production of simple atomic and molecular spectra. Atomic lines arising from transitions

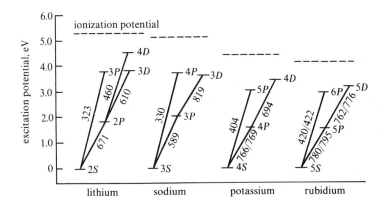

Figure 8.18.   Energy level diagrams for the easily excited atomic lines of lithium, sodium, potassium and rubidium. Wavelengths are given in nanometres for the spectral lines produced by transitions between the different levels. The ionization potential is indicated by the dashed line above the respective diagrams

to the ground state from the first two or three excited levels will predominate, the most intense one originating from the lowest excited level. Some typical transitions and the corresponding emission wavelengths are shown in figure 8.18. The degree of ionization is very small in low-temperature flames ($< 2\,500$ K) except for some of the alkali metals, although it may approach 100% at flame temperatures exceeding 3 000 K (table 8.7). The number of spectral lines observed from ionized species is therefore highly dependent upon the analyte and the operating conditions.

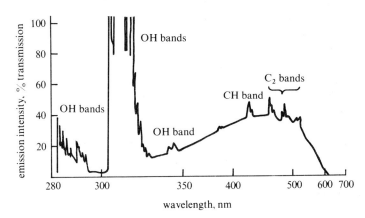

Figure 8.19.   Flame background spectrum of an acetylene–oxygen flame containing an organic aerosol

Band spectra due to such molecular species as CaOH, BaOH and LaO and from flame combustion products including $OH^{\cdot}$, CO, O_2, $CH^{\cdot}$, H_2O and $C_2^{\cdot}$ may also be observed. Background emission by the flame (figure 8.19) includes contributions from molecular species and continuum radiation from incandescent particles and depends upon the combination of fuel and support gases used. The sample solvent and matrix will further augment background radiation.

QUANTITATIVE MEASUREMENTS AND INTERFERENCES

The intensity of a spectral line is related to the solution concentration of the analyte in a similar complex manner to that described for arc/spark emission (p. 291) although the degree of ionization α will generally be much less (p. 318). Thus, intensity and concentration are directly proportional. However, the intensity of a spectral line is very sensitive to changes in flame temperature because such changes can have a pronounced effect on the small proportion of atoms occupying excited levels compared to those in the ground state (p. 272). Quantitative measurements are made by reference to a previously prepared calibration curve or by the method of standard addition. In either case, the conditions for measurement must be carefully optimized with reference to the choice of emission line, flame temperature, concentration range of samples and linearity of response. Relative precision is of the order of 1 to 4%. Flame emission measurements are susceptible to interferences from numerous sources which may enhance or depress line intensities.

Spectral interferences may arise from the close proximity of other emission lines or bands to the analyte line or by overlap with it. They can often be eliminated or minimized by increasing the resolution of the instrumentation, e.g. changing from a filter photometer to a grating spectrophotometer. Alternatively, another analyte line can be selected for measurements. Correction for background emission is also important and is made by monitoring the emission from a blank solution at the wavelength of the analyte line or by averaging measurements made close to the line and on either side of it.

Self-absorption is a phenomenon whereby emitted radiation is re-absorbed as it passes outwards from the central region of the flame (cf. arc/spark spectrometry). It occurs because of interaction with ground state atoms of the analyte in the cooler outer fringes of the flame and results in attenuation of the intensity of emission. It is particularly noticeable for lines originating from the lowest excited level and increases with the concentration of the analyte solution, figure 8.20.

The presence of species in the flame other than those of the analyte may alter the emitted intensities of analyte lines through chemical interactions. Thus, easily ionized elements in hot flames will suppress the ionization of the

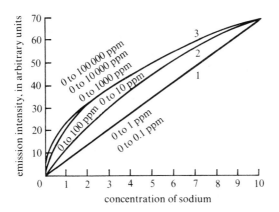

Figure 8.20.   Emission intensity of sodium in the acetylene–oxygen flame at 589 nm
showing the effect of self-absorption on calibration curves

analyte atoms thereby increasing the intensity of atom lines. The effect can be
used to advantage in eliminating variations due to sample composition, the
matrix effect, and improving sensitivity. This is achieved by adding large
amounts of an easily ionizable element to samples and standards. Lithium
salts are examples of such radiation buffers (cf. arc/spark emission spectrometry). Inorganic anions generally lower the emitted intensity of metallic
analyte lines by compound formation in the sample aerosol which reduces
the population of atoms in the flame. Sulphate, nitrate, phosphate and
aluminate are notable examples whose effects have been well-studied in the
determination of the alkaline earth metals. The addition of *releasing* or
chelating agents, e.g. EDTA, which protect the metal from the interfering
ion, is the recognized way of eliminating the effect.

Organic solvents enhance emitted intensities mainly because of a higher
resultant flame temperature (water has a cooling effect), a more rapid rate of
feed into the flame because of the generally lower viscosity, and the formation
of smaller droplets in the aerosol because of reduced surface tension. The
resultant enhancement of spectral line intensity may be 3- to over 100-fold.
Conversely, the presence of salts, acids and other dissolved species will
depress the intensity of emission from the analyte and underlines the need for
careful matching of samples and standards.

APPLICATIONS OF FLAME PHOTOMETRY AND FLAME ATOMIC EMISSION
SPECTROMETRY

Flame emission spectrometry is used extensively for the determination of
trace metals in solution and in particular the alkali and alkaline earth metals.

The most notable applications are the determinations of Na, K, Ca and Mg in body fluids and other biological samples for clinical diagnosis. Simple filter instruments generally provide adequate resolution for this type of analysis. The same elements, together with B, Fe, Cu and Mn, are important constituents of soils and fertilizers and the technique is therefore also useful for the analysis of agricultural materials. Although many other trace metals can be determined in a variety of matrices, there has been a preference for the use of atomic absorption spectrometry because variations in flame temperature are much less critical and spectral interference is negligible. Detection limits for flame emission techniques are comparable to those for atomic absorption, i.e. from <0.01 to 10 ppm (table 8.6). Flame emission spectrometry complements atomic absorption spectrometry because it operates most effectively for elements which are easily ionized, whilst atomic absorption methods demand a minimum of ionization (table 8.7).

Table 8.6   Some detection limits for atomic absorption, fluorescence and flame emission methods

ELEMENT	SPECTRAL LINE/nm	FLAME	METHOD	DETECTION LIMIT/ppm
Ag	328	C_2H_2–air	Abs	0.01
As	194	H_2–N_2 (diff.)	Flu	0.25
Al	396	C_2H_2–N_2O	Em	0.01
Ba	553	C_2H_2–N_2O	Em	0.001
Ca	423	C_2H_2–N_2O	Em	0.001
Cd	229	C_2H_2–air	Abs	0·002
Cd	229	H_2–air	Flu	0.001
Cr	359	C_2H_2–air	Abs	0.01
Cu	325	C_2H_2–air	Abs	0·01
Fe	248	C_2H_2–air	Abs	0.01
K	766	C_2H_2–O_2	Em	0.001
Li	671	C_2H_2–N_2O	Em	0.000 1
Mg	285	C_2H_2–air	Abs	0·001
Mn	279	C_2H_2–air	Abs	0.01
Na	589	H_2–O_2	Em	0.001
Pb	283	C_2H_2–air	Abs	0.1
Se	196	propane–air	Flu	0.15
Sr	461	C_2H_2–N_2O	Em	0.001
Te	214	propane–air	Flu	0.05
Zn	214	C_2H_2–air	Abs	0.01
Zn	214	H_2–O_2	Flu	0.000 1

8.6 Atomic Absorption Spectrometry

SUMMARY

Principles

Absorption of electromagnetic radiation in the visible and ultraviolet regions of the spectrum by atoms resulting in changes in electronic structure. Observed by passing radiation characteristic of a particular element through an atomic vapour of the sample. Sample vaporized by aspiration of solution into a flame or evaporation from electrically heated surface.

Instrumentation

Sources emitting radiation characteristic of element of interest (hollow-cathode lamp). Flame or electrically heated furnace or carbon rod. Monochromator, photomultiplier, recorder.

Applications

The most widely used technique for the quantitative determination of metals at trace levels (0.1 to 100 ppm) in a wide range of materials. Relative precision 0.5 to 2%.

Disadvantages

Sample must be in solution or at least volatile. Individual source lamps required for each element.

When electromagnetic radiation characteristic of electronic transitions in the outer orbitals of atoms of a particular element is passed through an atomic vapour of that element, the radiation at certain frequencies is attenuated. The absorbed radiation excites electrons from the ground state to various higher energy levels (excited states) and the degree of absorption is a quantitative measure of the concentration of ground-state atoms in the vapour. The energy changes involved correspond to radiation in the UV and visible regions of the spectrum. As only atoms in the ground state will respond in this way, the conditions used for volatilizing and decomposing the sample to produce an atomic vapour must induce the minimum of ionization. This can be achieved by flame excitation where temperatures seldom exceed 3 000 K. Reference to the Maxwell–Boltzmann equation p. 272 and table 8.7 shows that for most elements, practically 100% of atoms will be in the ground state even in a moderately hot flame such as air–acetylene (2 400 K). The only exceptions are the easily ionized alkali and alkaline earth metals where the energies of the

Table 8.7   Degree of ionization of elements in flames

ELEMENT	AIR–PROPANE 2 200 K	OXYGEN– HYDROGEN 2 450 K	OXYGEN– ACETYLENE 2 800 K	NITROUS OXIDE– ACETYLENE 3 230 K
Li	<0.01	0.9	16	68
Na	0.3	5.0	26	82
K	2.5	32	82	98
Rb	13.5	44	90	99
Cs	28.3	70	96	100
Be				0
Mg				3
Ca	<0.01	1.0	7.3	43
Sr	<0.1	2.7	17	71
Ba	1.0	8.6	43	92
Al				17

first excited states lie relatively close to those of the ground states. Even in these cases, over 90% of such atoms are likely to remain in the ground state if cooler flames, e.g. air–propane, are used (p. 319). The situation should be contrasted with that encountered in flame photometry which depends on the emission of radiation by the comparatively few excited atoms present in the flame. However, because of fundamental differences between absorption and emission processes it does not follow that atomic absorption is necessarily a more sensitive technique than flame emission.

ABSORPTION OF CHARACTERISTIC RADIATION

The extent to which radiation of a particular frequency is absorbed by an atomic vapour is related to the length of the path traversed and to the concentration of absorbing atoms in the vapour. This is analogous to the Beer–Lambert law relating to samples in solution (p. 355). Thus, for a collimated, monochromatic beam of radiation of incident intensity I_0 passing through an atomic vapour of thickness l,

$$I_v = I_0 \, e^{-k_v l} \tag{8.3}$$

where I_v is the intensity of the transmitted radiation at frequency v and k_v is the corresponding absorption coefficient. The value of k_v is determined by the concentration of atoms which can absorb at frequency v and is given by the expression

$$\int k_v \, dv = \frac{\pi e^2}{mc} N_v f \tag{8.4}$$

where m and e represent the mass and charge of the electron, N_v is the number of atoms per cm^3 capable of absorbing radiation of frequency v (i.e. ground state atoms) and f is the *oscillator strength*, defined as the number of electrons per atom capable of being excited by the incident radiation. Hence, for transitions from the ground state, the *integrated absorption* is proportional to N_v, which approximates to the concentration of the element in the sample.

Measurement of integrated absorption requires a knowledge of the absorption line profile. At 2 000 to 3 000 K, the overall line width is about 10^{-2} nm which is extremely narrow when compared to absorption bands observed for samples in solution. This is to be expected since changes in molecular electronic energy are accompanied by rotational and vibrational changes, and in solution collisions with solvent molecules cause the individual bands to coalesce to form band-envelopes (p. 362). The overall width of an atomic absorption line is determined by:

1. The natural width (about 10^{-5} nm).
2. Doppler broadening, caused by the thermally induced movement of atoms relative to the spectrometer. (This is analogous to the apparent change in pitch of a train whistle as it approaches and passes an observer.)
3. Collisional or pressure broadening and resonance broadening. These are caused by collisions between unlike and like atoms respectively in the sample vapour. Only the former is significant in flames.
4. Stark and Zeeman broadening caused by electric and magnetic fields respectively set up within the sample vapour and which perturb atomic energy levels.

In flames, only Doppler and, to a lesser extent, collisional broadening contribute significantly to the overall linewidth.

To make accurate measurements of the integrated absorption associated with such narrow lines requires that the linewidth of the radiation source be appreciably smaller than that of the absorption line. In practice, this could be achieved with a continuum source only if expensive instrumentation of extremely high resolving power were used, and it is doubtful whether conventional photomultiplier detectors would be sufficiently sensitive at the resulting low radiation intensities. An alternative arrangement is to measure k_v at the centre of the absorption line, where it reaches a maximum value, using a *sharp line* radiation source (*vide infra*) characteristic of the element of interest. Assuming that Doppler broadening is the only significant line broadening effect, then the maximum value of k_v is given by

$$k_{max} = \frac{2\lambda^2}{\Delta\lambda} \left(\frac{\ln 2}{\pi}\right)^{\frac{1}{2}} \frac{\pi e^2}{mc^2} N_v f \qquad (8.5)$$

where $\Delta\lambda$ is the Doppler linewidth at wavelength λ. Thus k_{max} is directly

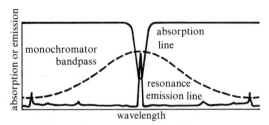

Figure 8.21. Profiles of an absorption line, an emission line from a sharp line
source and the bandpass of a monochromator

proportional to N_v, the sample concentration. The profiles of an absorption
line, an emission line from a sharp line source and the bandpass of a monochromator are shown in figure 8.21. It will be seen that radiation from the
sharp line source is absorbed at the centre of the absorption line and that the
amount absorbed represents a substantial proportion of the total radiated
intensity. By contrast, radiation passed by a monochromator (bandwidth
1 to 20 nm) is absorbed over the entire width of the absorption line, which
invalidates equation (8.5), and the fraction absorbed is extremely small.

INSTRUMENTATION

A diagram of a single beam atomic absorption spectrometer is shown in figure
8.22. It consists of a sharp-line radiation source characteristic of the element
of interest, usually a *hollow-cathode* lamp, a solution nebulizer and burner, or
an electrically heated furnace, and a monochromator, photomultiplier and
recording system. A deuterium continuum radiation source used for background correction (see below) is also shown. The purpose of the monochromator is to isolate a particular emission line from a number of characteristic
lines emitted by the hollow-cathode lamp. The components are aligned so as
to allow radiation from the lamp to pass directly through the flame and then
via the monochromator to the detector. An absorption measurement is made
by comparing the intensities of radiation reaching the detector with and
without the sample solution being aspirated into the flame (I_v and I_0 in
equation (8.3)). In practice, the reading for the latter is set to zero absorbance
whilst a blank solution is aspirated, the reading for the sample then being
given directly in absorbance units. The procedure is strictly analogous to that
used for absorbance measurements in uv and visible molecular spectrometry
(p. 352). To avoid interference from emission by excited atoms in the flame
and from random background emission by the flame, the output of the lamp
is modulated, usually at 50 Hz, and the detection system tuned to the same
frequency. Alternatively, a mechanical 'chopper' which physically interrupts
the radiation beam, can be used to simulate modulation of the lamp output.

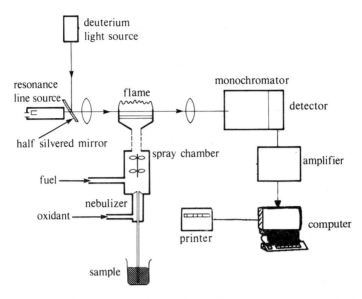

Figure 8.22.   Practical system for flame atomic absorption spectrometry including a
deuterium background corrector

Broad band absorption by volatile molecular species can often give rise to a
high and variable background. Various instrumental strategies have been
used for making the appropriate corrections. One popular method has been
the use of a deuterium continuum source as shown in figure 8.22. A half-
silvered mirror enables light from both the sharp line source and the
continuum source to be directed through the sample vapour at the same time.
This mirror is coated with small reflecting circles, thus leaving a light path for
the hollow cathode radiation, and with the continuum source modulated 180°
out of phase with the sharp line source the two signals may be electronically
distinguished. The continuum radiation will be diminished to a negligible extent by sharp line absorption but to a much greater extent by the
broad band absorption. Thus its intensity can be used to compute a background correction for the analytical measurement. An alternative way of
producing broad band radiation is the *Smith–Hieftje* technique. This uses the
hollow cathode lamp as a source, by running it at high currents when broad
band radiation is emitted. The lamp is pulsed alternately at high current, and
at lower currents for sharpline emission.

A further approach to correction for broad band interference utilizes the
Zeeman effect. Under the effect of a strong magnetic field atomic orbitals can
be split into sets with energies higher or lower than the original. The orbitals
responsible for the broad band absorption remain largely unaffected.

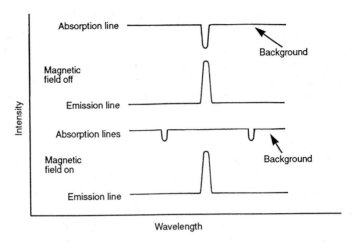

Figure 8.23. Zeeman effect splitting of atomic absorption lines. (Redrawn from *Concepts Instrumentation and Techniques in Atomic Absorption Spectrophotometry* (R. D. Beaty and J. D. Kerber). Perkin-Elmer, Norwalk, Connecticut, USA)

Measurement of absorption at the wavelength of a characteristic atomic absorption line in the absence of a magnetic field will include both background and atomic absorption contributions. Under the influence of the magnetic field only the background absorption will be recorded, allowing correction for it to be made. High levels of background correction are possible by the Zeeman effect approach. The Zeeman effect is illustrated schematically in figure 8.23.

More sophisticated instruments employ the double beam principle (p. 275), which may be operated in two different ways. One method is to split the incident beam into two, directing one half through the sample and the other through a reference burner, the signals from two separate but matched detectors then being compared. This method is cumbersome and expensive as it requires duplication of part of the optical system and the detector. An alternative and preferable method is to direct the sample and reference signals in rapid sequence on to a single detector by means of a rotating mirror, thus eliminating the need for the second matched detector. These different instrument configurations are sometimes known as 'double-beam-in-space' and 'double-beam-in-time' respectively. The double beam principle is applied to atomic absorption measurements in a limited way and reference burners are not normally employed. Hence corrections for changes in lamp intensity and detector sensitivity are facilitated but those for flame background are not.

Sharp-line Sources
Ideally, the emission line used should have a half-width less than that of the

corresponding absorption line otherwise equation (8.4) will be invalidated. The most suitable and widely used source which fulfils this requirement is the hollow-cathode lamp, although recently, interest has been shown in microwave-excited electrodeless discharge tubes. Both sources produce emission lines whose half-widths are considerably less than absorption lines observed in flames because Doppler broadening in the former is less and there is negligible collisional broadening.

Hollow-cathode Lamps

A diagram of a hollow-cathode lamp is shown in figure 8.24. It consists of a sealed glass envelope with a quartz end-window, and containing a hollowed-out cylindrical cathode of some 2 mm internal diameter together with a tungsten wire anode. The cathode is fabricated from the element to be determined. By reducing the pressure inside the envelope to about 1.3 kN m^{-2} and passing a current of 5 to 50 mA at an applied potential of about 300 V, a low-pressure glow-discharge confined to the inside of the cathode and characteristic of the cathode material is produced. The action of the gas used to fill the envelope is to bombard the cathode thereby vaporizing atoms from the surface, a process known as 'sputtering'. The resulting emission spectrum contains lines from both the cathode material and the filling gas. The choice of gas is restricted to those whose emission lines do not coincide with lines from the cathode itself and which will not ionize the sputtered atoms. In practice, neon or argon are preferred.

The emission spectrum of the cathode material includes a number of intense, sharp lines arising from transitions between excited states and the ground state, so-called *resonance radiation*. Generally, only a few resonance lines per element are suitable for quantitative work and there will be variation in the ranges of concentration over which absorbance measurements can be made. The maximum value of the absorption coefficient k_{max} is directly proportional to the oscillator strength f and indirectly proportional to the

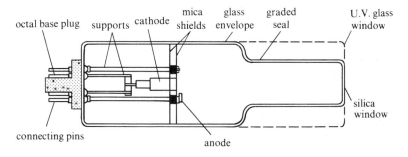

Figure 8.24. Diagram of hollow-cathode lamp

Doppler linewidth, $\Delta\lambda$ (equation (8.4)), hence the most sensitive lines are those which combine a narrow linewidth with a large oscillator strength. A wide range of linear response is attained if the width of the emission line is less than one-fifth that of the absorption line. Guides to the appropriate choice of emission line are given in most standard texts on atomic absorption spectrometry.

Hollow-cathode lamps are currently available for over sixty elements. Several multi-element lamps have been constructed and are useful for routine determinations, but they have proved to be of doubtful performance up to now. More successful with regard to multi-element analysis have been computer controlled automated systems, which enable a 'programme' of sequential measurements to be made with instrumental parameters being adjusted to the optimum for each element to be measured.

Electrodeless Discharge Tubes

Radiation is derived from a sealed quartz tube containing a few milligrams of an element or a volatile compound and neon or argon at low pressure. The discharge is produced by a microwave source via a waveguide cavity or using rf induction. The emission spectrum of the element concerned contains only the most prominent resonance lines and with intensities up to one hundred times those derived from a hollow-cathode lamp. However, the reliability of such sources has been questioned and the only ones which are currently considered successful are those for arsenic, antimony, bismuth, selenium and tellurium using rf excitation. Fortunately, these are the elements for which hollow-cathode lamps are the least successful.

SAMPLE VAPORIZATION

The production of a homogeneous atomic vapour from a sample is achieved by aspirating a solution into a flame or evaporating small volumes in an electrically heated tube furnace or from the surface of a carbon rod. In all cases, the thermal energy supplied must (a) evaporate the solvent and (b) dissociate the remaining solids into their constituent atoms without causing appreciable ionization.

Flame Vaporization

The sample solution is drawn first into a nebulizer by the flow of support gas where it forms a mist or aerosol. Fuel gas is introduced and the mixture passed to a spray chamber where large droplets condense and run to waste. Alternatively a small volume of solution (10–100 μl) may be injected directly into the orifice of the nebulizer using a micropipette.

The resulting homogeneous mixture of sample droplets and gases passes to the burner for combustion. With the former method an equilibrium concentration of free atoms will be established in the flame, whilst in the latter the concentration rises to a maximum and then falls to zero. The signal from the photomultiplier will vary in the same manner. The direct injection method enables small samples to be analysed, compared with the 2–3 cm^3 required by continuous aspiration. The burner consists of a metal block containing a row of circular holes or one or more slots about 10 cm long. It is aligned along the optical axis of the instrument and just below the beam from the lamp so as to provide a flame of long absorption path and hence maximum sensitivity. The design of the burner coupled with the use of a separate nebulizer and spray chamber ensure the formation of a non-turbulent or laminar flame which results in stable operation, good precision and easy choice of optimum vaporizing conditions. The most generally useful flame is air–acetylene, with a moderate burning velocity and a temperature of about 2 500 K, although the cooler air–propane and the hotter nitrous oxide–acetylene flames are also used. The last is particularly useful for samples containing refractory materials. Where maximum sensitivity is sought, the use of a separated flame may be advantageous because of reduced flame emission by species such as OH˙ and CO˙ resulting in a lower level of noise in the detector circuit. This is particularly effective for the determination of elements whose resonance lines occur below 200 nm, e.g. arsenic and selenium. A more detailed discussion of flame characteristics is given on p. 312.

Flameless Vaporization

If the production of an atomic vapour can be achieved without using a flame, there are a number of potential advantages:

1. The elimination of anomalous results arising from interactions between the sample and components of the flame.
2. Increased sensitivity arising from a longer residence time within the beam of radiation from the lamp. Residence times in flames are low because of strong vertical thermal currents (p. 311).
3. Increased sensitivity because of a higher proportion of the analyte being converted to free atoms. (The conversion may be as low as 0.1 % for flame atomization.)
4. The ability to handle very small samples such as clinical specimens. A nebulizer, spray chamber, burner arrangement consumes several cm^3 of sample per minute, most of which runs to waste.

Two forms of flameless atomizer are in use, i.e. the graphite tube or L'Vov furnace and the carbon rod or filament. Of these the first has proved to be the most generally effective and popular. It is widely used in a variety of modifications. In both cases, the temperature is raised rapidly to about 2 500 K by the passage of a heavy current for a period of 1 to 2 minutes. Tube furnaces, which are usually 5 or 10 cm × 3 mm, may be flushed through with argon before vaporizing the sample so as to prevent the formation of refractory oxides and oxidation of the graphite. The axis of the furnace is aligned along the optical path of the radiation from the lamp and as the vaporized sample is contained within this region, maximum sensitivity can be attained. The sample (1 to 50 mm³) is deposited on the bottom inner surface of the tube near the centre and the temperature raised to about 2 500 K from cold within one or two minutes. The heating cycle can be controlled so as to allow solvents to evaporate or organic residues to be ashed before raising the temperature rapidly to that required to produce an atomic vapour. Thus the absorbance signal from a flameless source will constitute a peak whose height or area is related to the analyte concentration (figure 8.25). Refinements to the simple tube design have been dictated by precision losses deriving from uneven heating across the tube. Whilst the graphite is heated rapidly by the flow of the electric current, the centre of the tube is heated by radiant energy only, and its temperature may lag considerably behind that of the graphite. Hence elements volatilized from the internal surface of the tube may condense in the cooler centre. To overcome this problem the use of a L'Vov shelf or platform has been introduced. By depositing the sample on this, rather than the tube walls, it is heated essentially by radiant energy alone and the condensation problem is removed. Figure 8.26 shows the arrangement of a L'Vov shelf. Although figure 8.25 shows the tube being heated by a longitudinal passage of electric current, transverse heating is also used. A more uniform temperature profile is produced by the latter, with the consequent

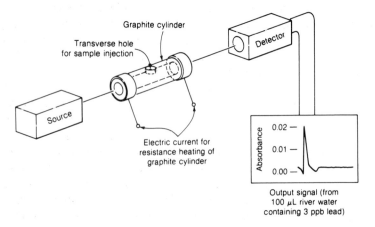

Figure 8.25.   Graphite furnace for atomic absorption analysis and typical output signal

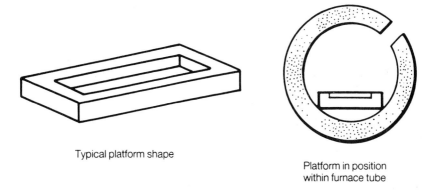

Typical platform shape

Platform in position
within furnace tube

Figure 8.26. Platform shape and orientation within a furnace tube

improvement in precision. The carbon rod is a simpler device having a small shallow recess for the sample machined in the top surface. It is positioned with the sample recess just below the optical axis, otherwise its operation is identical to that of a graphite furnace. The absolute sensitivity of flameless atomizers is very high, as little as 10^{-14} g for volatile elements such as zinc, magnesium and cadmium, but relative sensitivities are only marginally better than those obtained by flame vaporization because the sample size is much smaller (table 8.4).

Vaporization by Reduction and Hydride Generation
In a limited number of cases an element or one of its compounds is sufficiently volatile for it to be separated from the matrix and introduced into the light path without the use of a nebulizer or electrothermal atomizer. Mercury is a unique metal in that it has a significant vapour pressure at room temperature. Furthermore, the vapour is monatomic and unreactive. Thus it may be entrained in a gas stream and measured by the atomic absorption of this *cold vapour*. A typical analytical method would first involve the use of an oxidizing mixture such as sulphuric acid and potassium permanganate to destroy organic matter. The mercuric sulphate thus produced is then reduced with stannous chloride to produce mercury vapour, which is swept through a narrow quartz cell about 15 cm long with a diameter of about 0.75 cm. In some methods the vapour is swept straight through the cell giving a transient absorption peak similar to that produced by electrothermal vaporization. Alternatively, the vapour may be circulated in a closed loop leading to a stable signal when all the mercury has been volatilized into circulation. Other methods involve collection as a gold or copper amalgam prior to measurement.

Elements such as antimony, arsenic, bismuth, germanium, tin, selenium, tellurium and lead, although not volatile in the elemental form, have volatile hydrides. At different times a variety of strong reducing agents has been used to produce these hydrides, but sodium borohydride is generally accepted as the most suitable reagent. Wet oxidation to destroy organic material is a common prerequisite to the production of the hydrides. The formation of some hydrides is comparatively slow and the best sensitivity can be obtained only when a collection vessel is used before measurement. Once formed, the hydrides are entrained in argon and swept into the light path where they are decomposed to produce an elemental vapour. The decomposition is achieved either by direct injection into a 'cool' hydrogen flame or in a long path (15 cm) silica cell with external heating. Both air/acetylene flames and electrothermal heating have been used.

QUANTITATIVE MEASUREMENTS AND INTERFERENCES

Quantitative measurements may be made using a previously prepared calibration curve or by the method of standard addition. In either case, operating conditions must first be optimized with regard to the expected concentration range of samples and linearity of response. This involves appropriate choice of resonance line (usually made by reference to tables), adjustment of lamp current, flame temperature and sample aspiration rate, burner alignment and monochromator slit width. Standard solutions are best prepared by appropriate dilution of 1 000 ppm stock solutions and should be matched as closely as possible in gross composition to those of the samples. The relative precision of atomic absorption measurements is good, and in most cases 0.5–2 % is attainable without difficulty where flame atomization is used. Precisions for flameless methods are, however, often much poorer as a result of some of the interferences discussed below. Calibration curves invariably show curvature towards the concentration axis when absorbances exceed one. This non-linearity is caused by unabsorbed radiation reaching the detector or when the half-width of the emission line from the lamp approaches or exceeds that of the absorption line (p. 319). Unabsorbed radiation may reach the detector from numerous sources, including emission lines of the cathode material close to the chosen resonance line or from the filling gas, scattered radiation in the monochromator and radiation by-passing the flame or sample vapour.

Interferences in atomic absorption measurements can arise from spectral, chemical and physical sources. *Spectral interference* resulting from the overlap of absorption lines is rare because of the simplicity of the absorption spectrum and the sharpness of the lines. However, broad band absorption by molecular species can lead to significant background interference. Correction for this may be made by matrix matching of samples and standards, or by use of a standard addition method (p. 32). Instrumental correction for background absorption using a double beam instrument or a continuum source has already been discussed (p. 322). An alternative is to assess the background absorption on a non-resonance line two or three band-passes away from the analytical line and to correct the sample absorption accordingly. This method assumes the molecular absorption to be constant over several band passes. The elimination of spectral interference from the emission of radiation by the heated sample and matrix has been discussed on page 320.

Chemical effects include stable compound formation and ionization, both of which decrease the population of free atoms in the sample vapour and thereby lower the measured absorbance. Examples of compound formation include reactions between alkaline earth metals and oxyanions such as aluminates, silicates and phosphates, as well as the formation of stable oxides of aluminium, vanadium, boron etc. In the former case, releasing agents, typically strontium or lanthanum salts which themselves form stable oxy-salts, are added in excess. Stable oxides can sometimes be dissociated by using a hotter flame. The addition of easily combusted complexing agents such as EDTA to the sample solution and the use of fuel rich flames with a low oxygen content are ways of minimizing oxide formation. The use of inert atmospheres for flameless vaporization has previously been mentioned (p. 325). Ionization of sample atoms can be suppressed by adding an ionization buffer, i.e. an easily ionizable metal such as lithium or lanthanum, in excess. This is particularly necessary if the hot (3 200 K) nitrous oxide-acetylene flame is used. The degree of ionization for several elements in this flame is included in table 8.7. It should be noted that lanthanum salts serve the dual function of releasing agents and ionization buffers.

Two important types of *physical interference* are observed. One of these derives from the formation of solid solutions of one element within another, e.g. chromium in iron. The effect of this is to modify the volatilization profile of the analyte, and to make it dependent upon the matrix composition. Use of a releasing agent will usually obviate the problem for flame volatilization, but these *interelement effects* have proved much more intractable for flameless methods. Careful matrix matching of samples and standards can be used to reduce the interference. It should be remembered that the analytical measurement for a flameless technique is of the height of a transient peak (p. 326) and that the peak shape will be changed by alterations in the volatilization profile. Measurement of peak areas rather than peak heights can improve reproducibility. However, even with the application of all these remedies, interelement effects remain major contributors to the poorer precisions observed in flameless atomic absorption measurements. Where aspiration of a sample into a flame is employed, variable aspiration rates due to changes in the surface tension and viscosity with solution composition can occur. Careful matching of samples and standards with regard to acidity, total salt concentration and other major constituents is needed to overcome this source of error.

Atomic absorption spectrometry is one of the most widely used techniques for the determination of metals at trace levels in solution. Its popularity compared with that of flame emission is due to its relative freedom from interferences by inter-element effects and its relative insensitivity to variations in flame temperature. Except for the routine determination of alkali and alkaline earth metals, for which flame photometry is often preferred, over sixty elements can be determined in almost any matrix by atomic absorption. Examples include heavy metals in body fluids, polluted waters, foodstuffs, soft drinks and beer, the analysis of metallurgical and geochemical samples and the determination of many metals in soils, crude oils, petroleum products and plastics. Detection limits generally lie in the range 0.01 to 1 ppm ($mg\,dm^{-3}$) (table 8.4) but these can be improved by chemical preconcentration procedures involving solvent extraction or ion exchange.

ATOMIC FLUORESCENCE SPECTROMETRY

Radiation absorbed by atoms under conditions used in atomic absorption spectrometry may be re-emitted as *fluorescence*. The fluorescent radiation is characteristic of the atoms which have absorbed the primary radiation and is emitted in all directions. It may be monitored in any direction other than in a direct line with radiation from the hollow-cathode lamp which ensures that the detector will not respond to the primary absorption process nor to unabsorbed radiation from the lamp. The intensity of fluorescent emission is directly proportional to the concentration of the absorbing atoms but it is diminished by collisions between excited atoms and other species within the flame, a process known as *quenching*. Nitrogen and hydrocarbons enhance quenching, and flames incorporating either should be avoided or their effect modified by dilution with argon.

The instrumentation required for atomic fluorescence measurements is simpler than that used for absorption. As the detector is placed so as to avoid receiving radiation directly from the lamp, it is not strictly necessary to use a sharp-line source or a monochromator. Furthermore, fluorescence intensities are directly related to the intensity of the primary radiation so that detection limits can be improved by employing a high-intensity discharge lamp.

Atomic fluorescence spectrometry has a number of potential advantages when compared to atomic absorption. The most important is the relative ease with which several elements can be determined simultaneously. This arises from the non-directional nature of fluorescence emission, which enables separate hollow-cathode lamps or a continuum source providing suitable primary radiation to be grouped around a circular burner with one or more detectors.

Basically, atomic fluorescence is a simpler and more versatile technique than atomic absorption but suffers from its susceptibility to quenching effects and to background noise arising from scattering of radiation by particles in the flame. The latter is particularly serious for refractory materials and in high-temperature flames. Detection limits for some elements are lower than by atomic absorption or flame emission measurements, e.g. elements with resonance lines around 200 nm or below (As, Se, Te). Instruments based upon the use of a chemical flame as the atom reservoir have not proved to be successful. The introduction of the ICP torch renewed interest in atomic fluorescence and new instruments based on the ICP torch as a source of free atoms were constructed. However, these seem to have been only slightly more satisfactory than earlier instruments. Some detection limits are included in table 8.6.

8.7 X-ray Emission Spectrometry

SUMMARY

Principles

Characteristic X-rays emitted from transitions involving K and L electrons; wavelength and emission intensity measurements for qualitative and quantitative elemental analysis respectively.

Instrumentation

Excitation of sample by bombardment with electrons, radioactive particles or 'white X-rays'. Dispersive crystal analysers dispersing radiation at angles dependent upon energy (wavelength), detection of radiation with gas ionization or scintillation counters. Non-dispersive semiconductor detectors used in conjunction with multichannel pulse height analysers. Electron beam excitation together with scanning electron microscopes.

Applications

Non-destructive elemental analysis of solid or liquid samples for major and minor constituents. Used in routine analysis of metallurgical and mineral samples. Most suited to the determination of heavy elements in light matrices (e.g. Br or Pb in petroleum). Well suited for on-stream, routine analysis. Electron beam excitation methods valuable in surface studies in combination with electron microscopy. Detection limits generally in the range 10 to 100 ppm. Relative precision 5 to 10%.

Disadvantages

Matrix absorption, secondary fluorescence and scattering phenomena limit sensitivity and precision in many cases, especially with dense matrices. The sensitivity falls off with atomic number; elements with $Z < 15$ are particularly difficult to analyse. Analysis is characteristic of surface layers (5 to 500 μm depth) only for a solid specimen, instruments are often large, complicated and costly.

X-RAY PROCESSES

Electronic transitions within the inner shells of an atom involve energy changes consistent with the absorption or emission of electromagnetic radiation of high energy (short wavelength). Analytical techniques based on an exploitation of these changes have a number of potential advantages:

1. The possible electronic rearrangements are limited in number, and the energy levels are widely spaced. This leads to the production of simple uncomplicated spectra, which are characteristic of the elements present within the specimen.
2. The electrons involved are in the core of the atom and should be much less affected by external influences such as chemical bonding which greatly modify the properties of the outer electrons. For light elements however the K and L electrons may also be the valency electrons and significant effects may be observed.
3. The radiations involved with these low lying energy changes are energetic and highly penetrating. In these circumstances the analysis of solid samples without prior treatment becomes a real possibility and time consuming stages such as dissolution may well be eliminated.

The Emission of Primary X-rays

Primary X-rays may be produced by the bombardment of a target with a stream of high energy particles such as 20 to 50 keV electrons or nuclear particles from a radioactive source such as ^{241}Am. The impact of the bombarding particles on the target is non-selective and produces a wide range of energy transitions and consequently a continuum of X-ray emissions. Because of the wide range of wavelengths which are involved the emissions are sometimes known as 'white X-rays'. Superimposed on the continuum will be emission peaks characteristic of transitions involving the K and L shells of the elements within the target (figure 8.27). The wavelengths of the peaks are

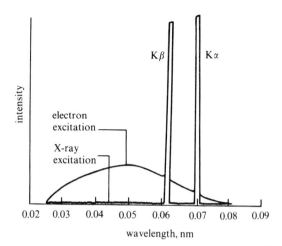

Figure 8.27.   Spectra of X-rays emitted from a molybdenum target as a result of electron (25 keV) and X-ray (35 keV) bombardment

related to the atomic numbers of the emitting atoms by the Moseley equation (8.6) and thus may be used for the qualitative analysis of a target material,

$$\lambda^{-\frac{1}{2}} = a(Z - b) \qquad (8.6)$$

where a is a constant and b has a specific value for all lines in one series, i.e. $b(K_\alpha$ lines$) = 1.0$, and $b(L_\alpha$ lines$) = 7.4$.

Furthermore, under controlled bombardment conditions, peak intensity measurements may be used for a quantitative determination of the appropriate element. Measurements of the characteristics and intensity of primary X-rays produced by electron bombardment constitute the basis of *electron probe microanalysis*. Figure 8.28 illustrates the complex nature of the reactions initiated by the impact of an electron beam on a target. As a consequence of this complexity it has proved extraordinarily difficult to make fully quantitative measurements, and it is only recently with the widespread application of dedicated computers that this has become possible.

In electron probe microanalysis the surface characteristics of the specimen are also of importance. Figure 8.29 illustrates how an uneven surface can lead to variable attenuation of the emitted X-rays and the importance of the angle between detector and specimen surface being as near 90° as possible. Such a high *take off angle* will minimize the surface effects.

The Absorption and Fluorescent Emission of X-rays
When primary X-rays are directed on to a secondary target, i.e. the sample, a proportion of the incident rays will be absorbed. The absorption process

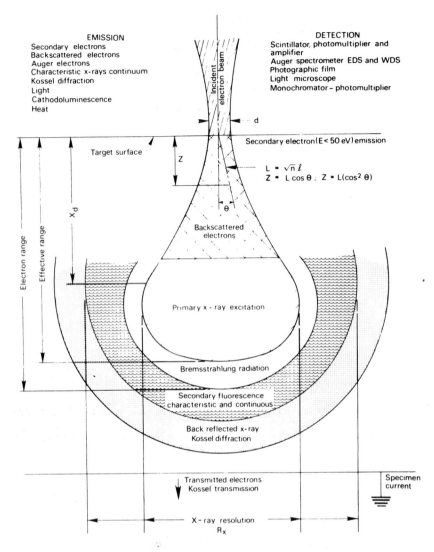

EMISSION
Secondary electrons
Backscattered electrons
Auger electrons
Characteristic x-rays continuum
Kossel diffraction
Light
Cathodoluminescence
Heat

DETECTION
Scintillator, photomultiplier and amplifier
Auger spectrometer EDS and WDS
Photographic film
Light microscope
Monochromator – photomultiplier

Incident electron beam

Secondary electron (E < 50 eV) emission

Target surface

$L = \sqrt{n}\,\ell$
$Z = L\cos\theta$; $Z = L(\cos^2\theta)$

Backscattered electrons

Electron range

Effective range

X_d

Primary x-ray excitation

Bremsstrahlung radiation

Secondary fluorescence characteristic and continuous

Back reflected x-ray Kossel diffraction

Transmitted electrons Kossel transmission

Specimen current

X-ray resolution R_x

Figure 8.28. Diagram showing the effects of the interaction of an electron beam with a solid specimen

involves the ejection of inner (K or L) electrons from the atoms of the sample. Subsequently the excited atoms relax to the ground state, and in doing so many will lose their excess energy in the form of secondary X-ray photons as electrons from the higher orbitals drop into the 'hole' in the K or L shell. Typical transitions are summarized in figures 8.30 and 8.31. The re-emission

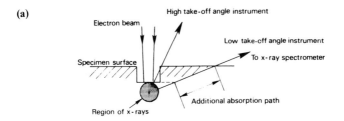

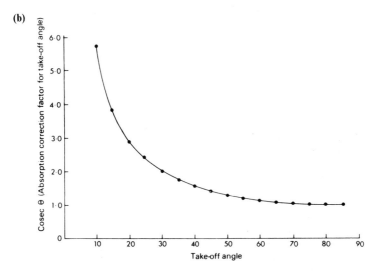

Figure 8.29.   (a) Diagram showing the additional absorption path that X-rays must take when operation is at a low take-off angle, as compared to operation at a high take-off angle. (b) Relationship between absorption correction factor and take-off angle

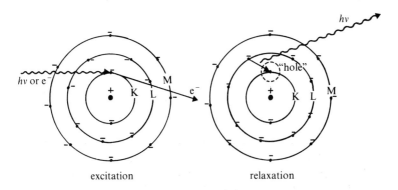

Figure 8.30.   Schematic representation of an X-ray fluorescence process

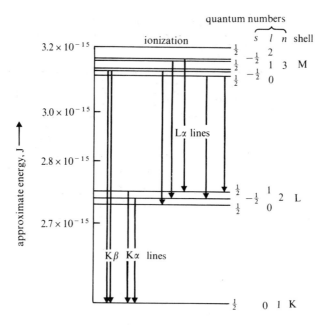

Figure 8.31.   X-ray transitions in the molybdenum atom. Lines are designated by the letter referring to the lower quantum group and α where $\Delta n = -1$ and β where
$$\Delta n = -2$$

of X-rays in this way is known as *X-ray fluorescence* and the associated analytical method as *X-ray fluorescence spectrometry*. The relation between the two principal techniques of X-ray emission spectrometry is summarized in figure 8.32.

Not all excited atoms will relax by the re-emission of X-rays and the proportion that do is known as the *fluorescence yield factor* φ. The amount of the primary radiation absorbed I_a and the intensity of the fluorescent emission I_F are thus related by

$$I_F = \varphi I_a \qquad\qquad (8.7)$$

φ may be as high as 0.5 for heavy elements but falls off rapidly to about 0.01 for light elements ($Z < 15$). The competitive relaxation process known as the *Auger effect* involves radiationless transitions and the ejection of valency electrons. X-ray fluorescence spectrometry has developed as a technique of outstanding importance, whilst absorption methods are much less valuable. However, it will be seen that because of the absorption effect any emitted radiation is attenuated by the sample matrix and absorption processes thus have an important bearing on the interpretation and employment of primary

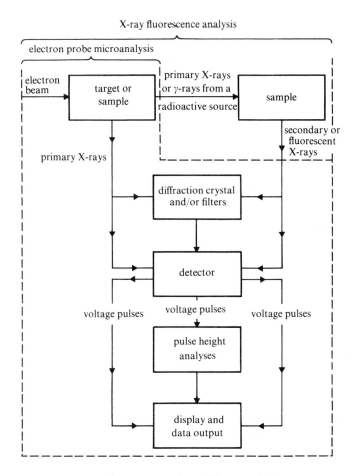

Figure 8.32. Schematic representation of X-ray emission spectrometry

emission, and fluorescence measurements. Furthermore, radiant energy thus absorbed by the matrix will be re-emitted in a process known as *secondary fluorescence*. The secondary radiation may of course undergo further absorption and re-emission by the matrix in *tertiary fluorescence*. As a consequence of these complex matrix effects the analytical signal will be diminished and the background increased. For good quality quantitative analysis it is clearly very important to assess and make correction for them. In doing this microcomputers are extensively employed.

The absorption of X-rays by a target is described by an exponential expression akin to the Beer–Lambert law (p. 355)

$$I = I_0 \exp(-\mu \rho b) \tag{8.8}$$

where I_o is the intensity of an incident beam of monochromatic X-rays and I is its attenuated intensity after penetrating a distance b into a target of density ρ. μ is the *mass absorption coefficient* and depends upon the atomic number of the element and the energy (wavelength) of the radiation only. The value of μ increases rapidly with the atomic number of the element e.g.

$$\mu_C = 5.5 \text{ m}^2 \text{ kg}^{-1}, \qquad \mu_{Pb} = 241 \text{ m}^2 \text{ kg}^{-1}$$

When μ for an element is plotted as a function of the incident energy or the incident radiation, figure 8.33, it shows a smooth decrease as the energy increases (wavelength decreases). Sharp discontinuities or *absorption edges* appear when the ionization energy for a K or L electron is reached. At this point a large increase in μ occurs as the X-rays are absorbed in producing ionization. The energy of X-ray photons emitted in L to K transitions will be less than that required for complete ionization, figure 8.31, and the K_α line for an element will appear at an energy below that of the absorption edge corresponding to K-ionization. Consequently the target element will have a low absorption for its own fluorescent radiation.

The matrix however may bring about considerable attenuation, and for a multi-element target the mass absorption coefficients are additive.

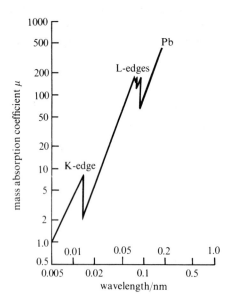

Figure 8.33.   Log–log plot of mass absorption coefficient against wavelength for lead showing K and L absorption edges

$$\mu(\text{overall}) = \mu_1 W_1 + \mu_2 W_2 + \mu_3 W_3 + \cdots$$

$$= \sum_i \mu_i W_i \qquad (8.9)$$

where μ_i and W_i are respectively the mass absorption coefficient and the mass fraction of element i.

The relation between the intensities of the incident radiation I_o and the fluorescent radiation I_F is complex, i.e.

$$I_F = \frac{(I_o \varphi \mu_x W_x)}{(\mu + \mu')} \qquad (8.10)$$

where φ is the fluorescence yield factor, equation (8.7), μ_x the mass absorption coefficient of element x for the incident radiation. W_x is the mass fraction of element x, and μ and μ' are the mass absorption coefficients of the matrix for incident and fluorescent radiations respectively (derived from equation (8.9)). It will be seen from equation (8.10) that even for a simple two component mixture the signal from one component will be modified by the other, and the extent of the modification will depend upon the mass fractions of the two elements. Working curves need to be prepared (figure 8.34) to enable corrections to be applied. Where the matrix is more complex this simple solution is not possible and great care needs to be taken in preparing standards for calibration purposes. For a heavy matrix, measurements may sometimes be improved by dissolution of the sample in a light solvent such as water or in

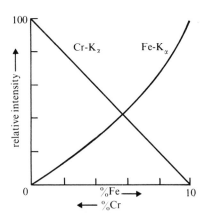

Figure 8.34.   Analytical curves for the determination of iron and chromium in a mixture of the two

a solid such as borax by a fusion procedure. Samples may also be mounted as thin layers on filter paper discs. In this way, matrix absorption problems may be minimized but the dilution of the sample can result in a reduction of the overall sensitivity of the measurement. The maximum depths within the sample at which X-ray emission can be observed will vary with the density of the matrix and the extent to which the emitted rays are attenuated. Similarly the depths to which the incident X-rays or electrons penetrate will be limited and varied. In general X-ray fluorescence data will be characteristic of the surface layers of atoms only (5 to 500 μm) and it is important to know whether or not these are truly representative of the specimen as a whole. For electron bombardment penetration may be limited to 1 to 2 μm.

In addition to absorption problems, measurements will be affected by secondary fluorescence and scattered radiations which will enter the detector and increase the general background. Detection limits under optimum conditions (a heavy element in a light matrix) may be as low as 10 ppm. Quantitative analysis is however difficult below the 20 to 100 ppm region if a reasonable precision (5 % or better) is to be obtained.

INSTRUMENTATION

The instruments used in X-ray emission spectrometry reflect the principles set out in chapter 7. Radiation characteristic of the specimen is produced by electron or radiation bombardment. Monochromatic radiation is then presented to the detector by a diffraction device or by use of a series of narrow bandpass filters. Alternatively pulse height analysis (p. 463) can be applied to

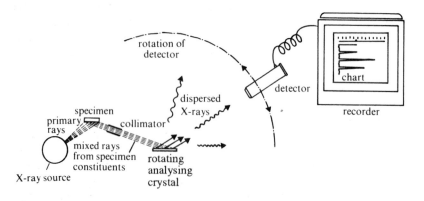

Figure 8.35.   Schematic layout of a dispersive X-ray fluorescence spectrometer

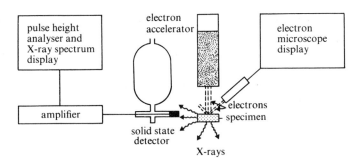

Figure 8.36.   Schematic layout of an electron microscope—electron probe analyser

a series of pulses which have been generated with a size proportional to the radiation energy. Typical X-ray spectrometry arrangements are shown in figures 8.35 and 8.36.

Excitation

Primary X-rays are produced by the bombardment of a suitable target with a stream of accelerated electrons. A typical X-ray generator uses an evacuated tube into which the target (e.g. tungsten) projects as a cooled anode together with a tungsten filament cathode. At a potential of 20 to 50 keV electrons are emitted from the cathode and bombard the target resulting in electronic excitations. The generated X-rays pass through a window made of a light material such as beryllium or an organic polymer.

Electron probe microanalysis functions by direct examination of the primary X-rays produced when the specimen is used as a target for an electron beam. Focused electron beams allow a 'spot analysis' of a 1 μm^3 section of the specimen. One current development employs the electron beam within a scanning electron microscope to provide both a visual picture of the surface of the sample and an elemental analysis of the section being viewed. Spectra obtained from primary X-rays always have the characteristic emission peaks superimposed on a continuum of background radiation, figure 8.27. This feature limits the precision, sensitivity and resolution of electron probe microanalysis.

For the majority of X-ray fluorescence methods primary X-rays with a wide spectrum of energies are directed on to the specimen. The secondary radiation emitted is viewed through a collimator at an angle of 90° to the incident beam. X-ray emission may also be stimulated by bombardment with nuclear radiation from a suitable radioactive source (e.g. ^{55}Fe, ^{241}Am, ^{238}Pu). The range of energies available is more limited in this case and curtails the flexibility of the method. However, a number of specific element analysers have been constructed in which radioactive excitation is used. One variable energy device currently available uses [241]Am and a series of six interchangeable targets (Cu, Rb, Mo, Ag, Ba, Tb). In this way a number of excitation lines can be produced. The advantages of these radioactive sources lie in their simplicity and small size as well as their independence of electrical power. Thus they are suited to portable instruments. Spectra obtained from fluorescent emission show much better resolution than primary X-ray spectra as they lack the continuum background, figure 8.27. Consequently sensitivity and precision are rather better for fluorescence methods than they are for electron excitation techniques.

The Detection and Analysis of Emitted X-rays

The characteristic X-radiations emitted by a sample may be analysed in a number of different ways. In a procedure akin to filter photometry (p. 275) the absorption edge phenomenon is exploited to construct narrow bandpass filters which can be employed to present the radiation of a single characteristic line to the detector, figure 8.37. Dispersive analysis of X-rays is achieved using a crystal to effect diffraction. The angle, θ, of diffraction is related to the wavelength, λ, of the radiation and the spacing, d, of the crystal planes by the Bragg equation in which n has integral values.

$$n\lambda = 2d \sin \theta \qquad (8.11)$$

To provide for a comprehensive range of wavelengths a variety of crystals with different spacings are required (table 8.8).

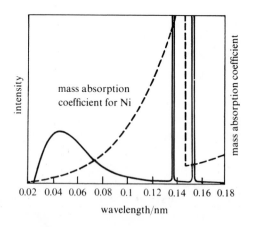

Figure 8.37.   Primary X-ray spectrum for Cu with the absorption edge of a nickel filter superimposed showing how it may be used to isolate the long wavelength emission peak

Table 8.8   Some typical crystals used in dispersive X-ray analysis

CRYSTAL	$2d$/nm	WAVELENGTH RANGE/nm
Lithium fluoride (LiF)	0.403	0.0351–0.384
α-Quartz (SiO$_2$)	0.852	0.0742–0.812
Pentaerythritol (PET) C(CH$_2$OH)$_4$	0.874	0.0762–0.834
Rubidium hydrogen phthalate (RAP) o—C$_6$H$_4$(CO$_2$H)(CO$_2$Rb)	2.612	0.228–2.492

The analysing crystal is rotated to direct radiations of varying wavelengths on to the detector in turn, figure 8.35. It follows that the detector must also move and it will be seen from the Bragg equation (8.11) that it must do so through twice the angle of rotation of the crystal. As a result, X-ray spectra are often plotted with the intensity as a function of 2θ, figure 8.38. Any detector sensitive to ionizing electromagnetic radiations will serve for monochromatic radiation presented by a filter system or dispersing crystal. Such detectors and their characteristics are discussed in detail in chapter 10. Geiger–Müller, scintillation and proportional counters have all been employed, with the latter pair proving the most useful. In these two devices the voltage pulses generated have sizes proportional to the energies of the incident

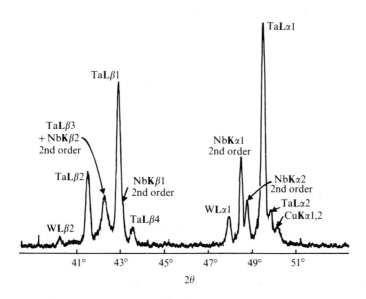

Figure 8.38.   X-ray spectra of niobium and tantalum

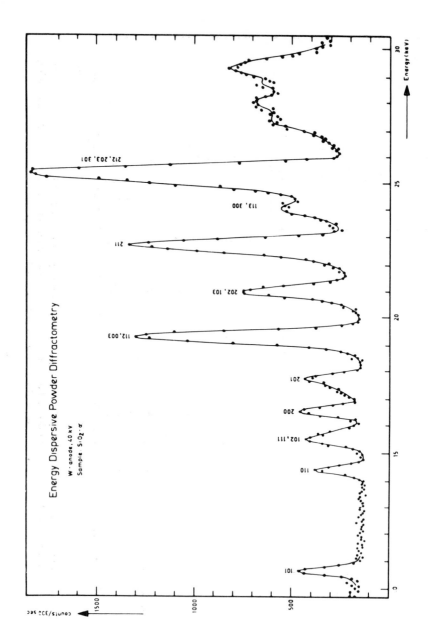

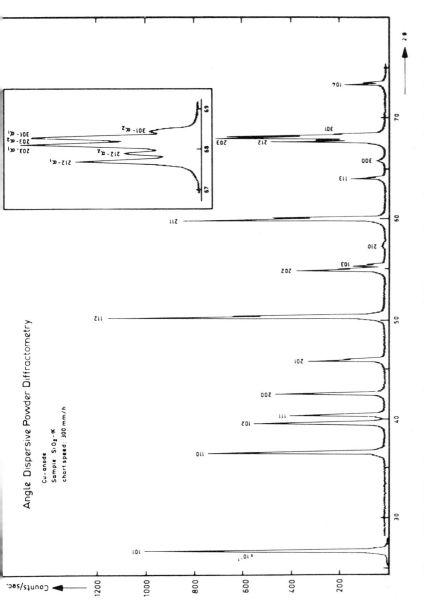

Figure 8.39. Diffraction spectra of α-quartz, recorded by energy dispersive and angle dispersive detectors, contrasting the different resolutions. The energy dispersive spectrum was recorded in 5 minutes while the angle dispersive record required 54 minutes

photons, whence a measure of pulse height analysis may be applied to reduce the background and improve the resolution of the spectrum.

The alternative approach to detection and analysis incorporates a solid state detector and a multichannel pulse height analysis system. The crystals used are of silicon (of the highly pure *intrinsic* type), or the lithium drift principle (p. 461) is utilized. All emitted radiations are presented to the detector simultaneously and a spectrum is generated from an electronic analysis of the mixture of voltage pulses produced. Chapter 10 contains a more detailed account of pulse height analysis and solid state detectors. Production of an X-ray spectrum in this way is sometimes known as *energy dispersive analysis of X-rays* (EDAX) and where an electron microscope is employed as SEM-EDAX.

Filter based instruments lack flexibility and are generally employed for the routine analysis of specific elements, whilst instruments utilizing a dispersing crystal have the greatest resolution. There is in principle no limit to this resolution as the detector may be placed at a large distance from the crystal. Limitations on the size of the instrument and the low radiation intensities at long distances mean, however, that the best resolution is about 8×10^{-18} J (50 eV). By contrast the solid state detector and its associated analyser has no mechanical movement but can provide a resolution of about 2.4×10^{-16} J (150 eV). Improvement in semiconductor technology is steadily increasing the resolving power but figure 8.39 shows the advantage still enjoyed by dispersive analysis. An earlier disadvantage of solid state detectors was their relatively low efficiency, but improvement in crystal design and the fabrication of larger crystals ($50-100 \text{ cm}^3$) have reduced this problem.

The random nature of the ionizing events recorded by the detector must also be borne in mind. To achieve measurements with standard deviations of 1% it is necessary to record at least 10^4 counts. For signals of low intensity this may take several hours to accumulate. This point is discussed further in chapter 10.

APPLICATIONS OF X-RAY EMISSION SPECTROMETRY

Electron probe and X-ray fluorescence methods of analysis are used for rather different but complementary purposes. The ability to provide an elemental 'spot analysis' is the important characteristic of electron probe methods, which thus find use in analytical problems where the composition of the specimen changes over short distances. The examination of the distribution of heavy metals within the cellular structure of biological specimens, the distribution of metal crystallites on the surface of heterogeneous catalysts, or the differences in composition in the region of surface irregularities and faults in alloys, are all important examples of this application. Figure 8.40

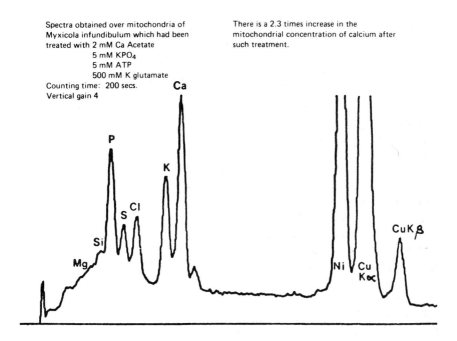

Spectra obtained over mitochondria of
Myxicola infundibulum which had been
treated with 2 mM Ca Acetate
          5 mM KPO$_4$
          5 mM ATP
          500 mM K glutamate
Counting time: 200 secs.
Vertical gain 4

There is a 2.3 times increase in the
mitochondrial concentration of calcium after
such treatment.

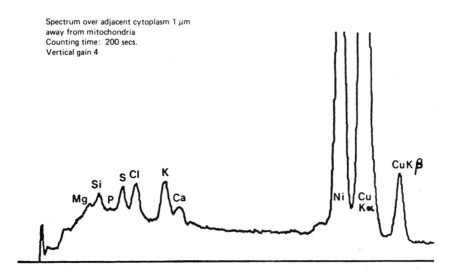

Spectrum over adjacent cytoplasm 1 μm
away from mitochondria
Counting time: 200 secs.
Vertical gain 4

Figure 8.40.   Electron probe—SEM X-ray spectra demonstrating the ability of the
technique to distinguish between targets in close juxtaposition

Table 8.9   Precision of silicate analyses

ELEMENT	APPROX. COMP./ wt. %	STD. DEV./Wt. %		
		X-RAY	WET	ARC/SPARK
O	48.0	0.4	0.3	—
Na	2.6	0.02	0.15	0.11
Mg	0.5	0.01	0.15	0.12
Al	9.3	0.04	0.19	0.53
Si	30.0	0.10	0.14	1.1
K	3.7	0.03	0.21	0.15
Ca	2.5	0.01	0.07	0.14
Ti	0.22	0.003	—	—
Fe	3.1	0.02	0.21	0.14

illustrates the analysis of parts of a biological cell just 1 μm apart. Combination of electron probe analysis with electron microscopy enables visual examination to be used to identify the areas of interest prior to the analytical measurement.

X-ray fluorescence measurements have a wide application in routine industrial analysis particularly in mineral processing and metallurgy. The simplicity of operation, and the elimination of time consuming wet chemical steps can have a dramatic effect on the time taken to complete an analysis. Niobium–tantalum ores for instance present a special problem as a result of the chemical similarity of the two elements. The analysis of such samples by wet chemical procedures has typically taken 10 days for completion. X-ray fluorescence, however, provides a result in less than an hour because it readily distinguishes between elements with widely differing atomic numbers, figure 8.38. A further important characteristic of this method of analysis is its non-destructive nature, which together with the previously mentioned factors produces an attractive basis for automated 'on stream' analysis in a chemical plant. Lead or bromine containing petroleum additives are monitored in this way as are the major constituents of cement, e.g. CaO, SiO_2, Al_2O_3 and Fe_2O_3.

Table 8.9 shows an analysis of a silicate rock and compares the precision of X-ray fluorescence analysis with wet chemical methods and arc/spark emission spectrometry.

Problems

1. List the advantages and disadvantages of dc-arc spectrometry and inductively coupled plasma spectrometry, when used to determine trace metals in solid samples. Indicate the main reasons for preferring ICP spectrometry, in most cases.

2. Flame atomic absorption spectrometry has achieved very wide use as a routine method for the determination of trace metals in solution. However, for alkali metals flame photometry has remained popular. Why is this?

3. A sample of stainless steel (0.320 g) was weighed out and dissolved in nitric acid. The resulting solution was made up to $1\,dm^3$ with water. Five standards and the sample solution were analysed for nickel consecutively on a flame atomic absorption spectrophotometer with the following results:

Solution	Ni concentration/ppm	Absorbance
1	2	0.126
2	4	0.250
3	6	0.374
4	8	0.500
5	10	0.626
Sample	—	0.220

Calculate the amount of nickel in the steel.

4. The solution prepared in question 3 was used to investigate the effect of the matrix on the measurement of nickel. Five aliquots $(10\,cm^3)$ of the steel solution were measured out and standard amounts of nickel added to each. The volume of each was then adjusted to $20.0\,cm^3$. When these solutions were measured the following results were obtained.

Solution	Added Ni concentration in solution/ppm	Absorbance
1	2	0.260
2	4	0.383
3	6	0.510
4	8	0.635
5	10	0.762

Calculate the concentration of nickel in the stainless steel according to these data. Suggest an explanation for any difference from the value obtained in question 2.

5. Contrast the type of information obtainable in the determination of trace metals by X-ray fluorescence and electron probe analysis.

Further Reading
BOUMANS, P. W. J. M., *ICP Emission Spectrometry*, Wiley, 1987.
DATE, A. R. and GRAY, A. L., *Applications of Inductively Coupled Plasma Mass Spectrometry*, Blackie, 1989.
EBDON. L., *An Introduction to Atomic Absorption Spectroscopy*, Heyden and Son Ltd., London, 1982.
LAWES, G., *Scanning Electron Microscopy and X-Ray Microanalysis*, Wiley, Chichester, 1987.
METCALFE, E., *Atomic Absorption and Emission Spectroscopy*, Wiley, Chichester, 1987.

SLAVIN, M., *Emission Spectrochemical Analysis*, Wiley-Interscience, New York, 1971.
WHISTON, C., *X-Ray Methods*, Wiley, Chichester, 1987.
WILLARD, H. H., MERRITT, L. L. Jr., DEAN, J. A. and SETTLE, F. A. Jr., *Instrumental Methods of Analysis*, 7th Ed., Wadsworth Publishing Co., 1988.

Chapter 9

Molecular Spectrometry

The nature of molecular energy is complex, comprising contributions from translational, rotational and vibrational motions, from electrons occupying molecular orbitals, and from nuclear spins. The range of these various energies covers a large part of the electromagnetic spectrum so that the techniques of molecular spectrometry are both varied and comprehensive in the analytical information they provide. The separation between quantized levels for each type of molecular energy also varies, the relative magnitudes being in the order $\Delta E_{\text{electronic}} > \Delta E_{\text{vibrational}} > \Delta E_{\text{rotational}} > \Delta E_{\text{nuclear spin}}$. Translational levels are too closely spaced to be considered quantized, while the very small differences between nuclear spin levels arise only when the molecules are subjected to a strong magnetic field.

Both absorption and emission may be observed in each region of the spectrum, but in practice only absorption spectra are studied extensively. Three techniques are important for analytical purposes: *visible and ultraviolet spectrometry* (electronic), *infrared spectrometry* (vibrational) and *nuclear magnetic resonance spectrometry* (nuclear spin). The characteristic spectra associated with each of these techniques differ appreciably in their complexity and intensity. Changes in electronic energy are accompanied by simultaneous transitions between vibrational and rotational levels and result in broad-band spectra. Vibrational spectra have somewhat broadened bands because of simultaneous changes in rotational energy, whilst nuclear magnetic resonance spectra are characterized by narrow bands.

A fourth technique used for the characterization of molecules is *mass spectrometry*. It is included in this chapter because the structural information it provides is similar to that obtained from the other techniques although the principle is entirely different. It is a destructive method in which the *fragmentation pattern* of sample molecules is used to determine empirical formulae and molecular weights, and to identify structural features.

351

INTENSITY OF ABSORPTION BANDS

The intensity of a spectral transition is related to the magnitude of the change in dipole moment and the relative populations of the energy levels between which the transition occurs (p. 272). In addition, spectroscopic selection rules may 'forbid' certain transitions, although in some cases, 'forbidden' transitions may give rise to weak bands because of perturbation effects which relax the rules. The largest changes in dipole moment are associated with electronic transitions, and visible or ultraviolet spectrometry is therefore generally the most sensitive of the three techniques for quantitative analysis. Nuclear magnetic resonance spectrometry is the least sensitive in this respect because populations of nuclear spin energy levels are very similar.

DISSIPATION OF ABSORBED ENERGY

Most absorbed energy is converted into translational energy which appears as heat, the process occurring within a fraction of a second and enabling spectra to be recorded rapidly or re-recorded as necessary. However, after the absorption of radiation in the ultraviolet region of the spectrum, energy may be re-emitted, usually at a longer wavelength, in the form of *fluorescence*. This occurs in a number of complex organic molecules such as quinine, anthracene and fluorescein and in some inorganic compounds where the lifetime of the excited state is prolonged by resonance stabilization. The measurement of fluorescence radiation, which provides a sensitive means of quantitative analysis, is known as *fluorimetry* (p. 373).

INSTRUMENTATION

The basic instrumentation used for spectrometric measurements has already been described in chapter 7 (p. 274). The natures of sources, mono-chromators, detectors, and sample cells required for molecular absorption techniques are summarized in table 9.1. The principal difference between instrumentation for atomic emission and molecular absorption spectrometry is in the need for a separate source of radiation for the latter. In the infrared, visible and ultraviolet regions, 'white' sources are used, i.e. the energy or frequency range of the source covers most or all of the relevant portion of the spectrum. In contrast, nuclear magnetic resonance spectrometers employ a narrow waveband radio-frequency transmitter, a tuned detector and no monochromator.

Recording spectrophotometers in the infrared, visible and ultraviolet regions necessarily employ the double-beam principle of operation (figures 9.1(a) and (b) and p. 274), but for quantitative measurements at fixed

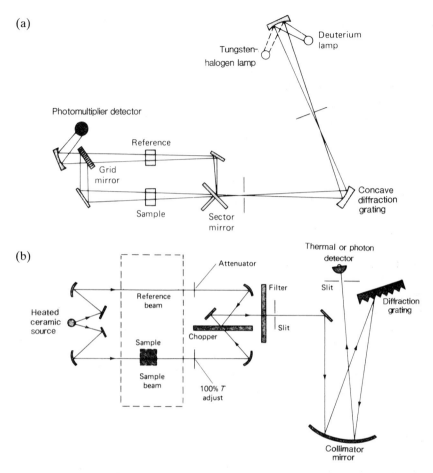

Figure 9.1. (a) Double-beam recording UV/visible spectrophotometer. (b) Double-beam recording IR spectrophotometer (courtesy of Beckman Instruments, Inc., Fullerton, California)

wavelength single-beam instruments are often preferred. Diode array uv/ visible spectrophotometers provide a means of acquiring a full-range spectrum in under 1 second. The optics are reversed in the sense that the monochromator disperses the radiation onto an array of photodiodes *after* it has passed through the sample (figure 9.2 (a)). The array, which replaces the usual photomultiplier detector, consists of up to about 300 photodiodes each of which receives a narrow band (2 to 10 nm wide) of radiation from the monochromator through the uv or the uv and visible regions. This facilitates the continuous monitoring of a wide spectral range and, using computer control, the recording, storage and display of a spectrum on a VDU screen in

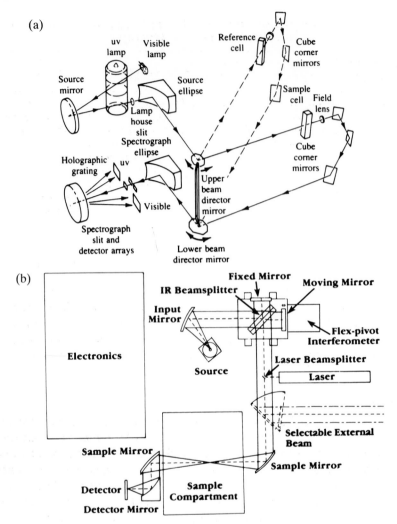

Figure 9.2. (a) UV/visible diode array spectrometer (Hewlett–Packard HP8450A) using the reversed-optics configuration (courtesy of Hewlett–Packard). (b) Fourier-transform IR spectrometer (Nicolet brochure)

about 0.1 seconds. This type of instrument has found particular use as an hplc detector (p. 126). Cheap, simple instruments for quantitative work in the visible region use coloured filters with a relatively wide bandpass. These are variously known as *filter-photometers, absorptiometers or colorimeters*. Infrared instruments are now predominantly of the Fourier transform type employing an interferometer (figure 9.2(b) and p. 278), the cheaper and simpler dispersive spectrophotometers having declined in popularity. Most

Table 9.1 Components of instruments used for ultraviolet, visible, infrared and magnetic resonance spectrometry

SPECTROMETRIC TECHNIQUE	ENERGY SOURCE	CELL MATERIAL	MONOCHROMATOR	DETECTOR
ultraviolet	hydrogen or deuterium lamp	quartz	diffraction grating or quartz prism	phototube, photomultiplier or diode array
visible	tungsten-halogen lamp	glass	diffraction grating or glass prism (simple instruments may use coloured filters only) diffraction grating	
infrared	heated ceramic	alkali halide or water-resistant material		thermal or photon
magnetic resonance	rf transmitter	glass	—	rf receiver

ultraviolet, visible and infrared instruments now benefit from micro-processor or microcomputer control, many displaying recorded spectra on a VDU screen. The software available with these instruments for spectral manipulation, storage and library searches is becoming increasingly sophisticated (p. 529 *et seq.*).

QUANTITATIVE ANALYSIS

Quantitative methods based on the absorption of electromagnetic radiation involve measurement of the reduction in intensity of the radiation on passage through an absorbing medium, i.e. the sample. The degree of absorption is determined by comparing the intensity of the transmitted beam when no absorbing species is present, i.e. a blank, with that transmitted by the sample. For monochromatic, collimated radiation passing through a homogeneous liquid sample, the reduction in intensity of the incident radiation can be related to the concentration of absorbing species and to the thickness of the absorbing medium, both relations being embodied in the *Beer–Lambert law*.

Derivation of the Beer–Lambert Law
Lambert's law, which concerns the thickness of an absorbing medium, states that successive equal thicknesses absorb equal fractions of monochromatic incident radiation. This leads to an exponential decrease in the intensity of the radiation as it passes through the layer. In mathematical terms,

$$I = I_0 e^{-kl} \tag{9.1}$$

where I_0 and I are the incident and transmitted intensities respectively, l is the thickness of the absorbing medium, and k is a constant determined by the wavelength of the radiation and the nature of the sample. Rearrangement of the above equation and converting to log to the base 10,

$$2.303 \log_{10} \frac{I_0}{I} = kl \qquad (9.2)$$

or

$$\log_{10} \frac{I_0}{I} = k'l \qquad (9.3)$$

Beer's law deals in a similar way with the concentration C of an absorbing species and leads to the relation

$$\log_{10} \frac{I_0}{I} = k''C \qquad (9.4)$$

Combining the two gives the *Beer–Lambert* law which may be expressed in the form

$$\log_{10}(I_0/I) = A = \varepsilon Cl \qquad (9.5)$$

where $\log_{10}(I_0/I)$ is defined as the *absorbance* A and ε is a constant known as the *molar absorptivity*.[1] The value of ε (the absorbance of a 1 M solution in a 1 cm cell) depends upon the nature of the absorbing species and on the wavelength of the incident radiation. Absorbance is thus seen to be directly proportional both to the concentration of the absorbing species and to the thickness of the absorbing medium. It is related to *transmittance* T defined as I/I_0 (the fraction of radiation transmitted) by the equation

$$A = \log_{10} \frac{1}{T} \qquad (9.6)$$

Instruments used for quantitative measurements may incorporate a transmittance, a percentage transmittance or an absorbance scale. If the molecular weight of the substance is unknown, $E_{1\%}^{1\,cm}$, representing the absorbance of a 1% solution in a 1 cm cell, may be used in place of ε. Absorbance is an additive property so that at a particular wavelength, the total absorbance of a solution containing a number of absorbing components is given by

$$A_{total} = \varepsilon_1 C_1 l + \varepsilon_2 C_2 l + \varepsilon_3 C_3 l + \cdots \qquad (9.7)$$

assuming there is no interaction between the individual solutes. Measured absorbances are increased by scattering of the radiation at solution–cell and cell–air interfaces and by large molecules or particulate matter in the solution, but these effects can be cancelled out by measuring the absorbance of all solutions relative to a blank and using closely matched cells.

---

[1] Strictly the *molar absorption coefficient*.

Use of the Beer–Lambert Law

Although applicable to measurements in all regions of the electromagnetic spectrum, only in visible, ultraviolet and infrared spectrometry are quantitative measurements based on the Beer–Lambert law used extensively. The usual procedure is to prepare a calibration graph, or Beer's law plot, by plotting absorbance against concentration for a series of standards. This should give a straight line passing through the origin and a slope equal to the product εl. Measurements are generally made at a maximum in the absorbance curve to maximize sensitivity and to minimize errors in setting the instrument at the chosen wavelength. This also minimizes apparent deviations from Beer's law for incident radiation of wide bandwidth (*vide infra*). The concentrations of unknowns can then be read directly from the graph or calculated using a factor, i.e. the absorbance reading is divided by the slope εl. The composition of mixtures of two or more absorbing materials can be established by measuring standards and samples at two or more wavelengths, preferably corresponding to the absorbance maximum of each component. The concentrations of the components can be calculated from a set of simultaneous equations once the respective molar absorptivities at each wavelength are known, e.g. for two components

$$A_{\text{mixture}} = \varepsilon_1^{\lambda 1} C l + \varepsilon_2^{\lambda 1} C l \text{ at } \lambda_1$$

$$\text{and } A_{\text{mixture}} = \varepsilon_1^{\lambda 2} C l + \varepsilon_2^{\lambda 2} C l \text{ at } \lambda_2$$

The values of $\varepsilon_1^{\lambda 1}$, $\varepsilon_2^{\lambda 1}$ etc, are calculated from calibration graphs for the separate components.

Deviations from the Beer–Lambert Law

There are no known exceptions to the Lambert law for homogeneous samples. Beer's law is a limiting case applicable only to dilute solutions and monochromatic radiation. Deviations may be observed in practice, but these are all 'apparent' in the sense that the limiting conditions have been 'contravened' either chemically or instrumentally. Such deviations show as a curvature of the calibration graph, but this does not necessarily preclude the use of a particular method unless the deviation is non-reproducible.

Apparent deviations may be summarized as follows:

1. At concentrations greater than about 0.01 M, refractive index changes and the perturbing effect of solute molecules or ions on the charge distribution of their neighbours both affect the value of ε. Positive or negative deviations may result.
2. Solutes involved in chemical equilibria, i.e. dissociation, association or complex formation, or in interaction with solvent molecules, may show marked spectral changes with concentration, e.g.

(a) The position of the OH stretching band in the infrared spectrum of benzyl alcohol run in carbon tetrachloride is concentration sensitive because of polymerization:

$$4C_6H_5CH_2OH \underset{\text{dilute}}{\overset{\text{concentrated}}{\rightleftharpoons}} (C_6H_5CH_2OH)_4$$

monomer, 3 700 cm^{-1} tetramer, 3 400 cm^{-1}

Absorbance measurements at 3 700 cm^{-1} show negative deviations at high concentrations, while those made at 3 400 cm^{-1} show positive deviations.

(b) Dichromate and chromate ions in aqueous solutions interconvert to a degree which is pH dependent:

$$Cr_2O_7^{2-} + H_2O \rightleftharpoons 2CrO_4^- + 2H^+$$

348 nm   372 nm

Unless standards are prepared in buffered media, positive or negative deviations may result from measurements made at 348 nm or 372 nm respectively Alternatively, measurements can be made at the *isosbestic point*, i.e. where the absorbance curves of each form intersect, and where absorbance is not a function of equilibrium concentrations but only of the overall concentration. Solutions of weak acids and bases should also be measured at their isosbestic points for the same reason.

3. Negative deviations occur if the radiation used is polychromatic, as in the case of filter-photometers, and if measurements are made where the value of ε changes appreciably across the width of the band, figure 9.3. However, measurements made at the maximum of a broad band show little or no deviation.

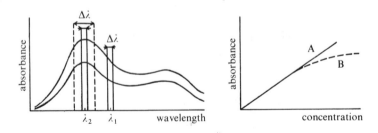

Figure 9.3. Effect of bandpass and choice of wavelength on a Beer's law plot. Curve A represents a calibration curve using a narrow bandpass monochromator at λ_2. Curve B represents a calibration curve using a wide bandpass filter at λ_2 or a narrow bandpass monochromator at λ_1

4. 'Stray' light passing through the optical system is generally not absorbed by the sample to the same extent as that of the selected wavelength. Its presence leads to negative deviations.

PRECISION OF ABSORPTION MEASUREMENTS

The precision of absorption measurements depends upon the degree of sophistication of the instrumentation and on the chemical species involved, both of which can affect the apparent validity of the Beer–Lambert law. Where a single absorbing species exhibiting a broad flat maximum is to be determined, adequate results can often be achieved with a simple filter-photometer. In the visible region, this technique is known as *colorimetry*. The inherent disadvantage of colorimetric procedures using simple filter instruments with a broad bandpass lies in the invalidation of the Beer–Lambert law and the lack of compatibility between results from different instruments. Another limitation is the inability to determine a true absorption curve. For these reasons, it is often desirable to use a spectrophotometer which uses a prism or grating monochromator, the technique being known as *spectrophotometry*.

Random errors associated with the measurement of absorption impose a limitation on the degree of precision attainable. At low absorbances, noise in the measuring circuit becomes the governing factor while at high absorbances, the small amount of radiation reaching the detector necessitates high sensitivity (amplifier gain) settings. If the error in measuring the radiation intensity is constant over the operating range of a detector, it can be shown that the relative error in the absorbance reading passes through a minimum value. The absorbance corresponding to the minimum error can be calculated by differentiating the Beer–Lambert equation twice and setting the second differential to zero. Thus,

$$A = \log_{10} \frac{1}{T} = \varepsilon C l \tag{9.8}$$

Converting to a natural logarithm and differentiating

$$-\frac{0.434}{T} dT = \varepsilon l \cdot dC$$

whence by substitution and rearrangement

$$\frac{dC}{C} = \frac{0.434}{T \log_{10} T} dT \tag{9.9}$$

where dC/C represents the relative error in concentration and dT is the constant error or uncertainty in measuring T (instrumental error).

Converting to a natural logarithm and differentiating a second time gives

$$\frac{d}{dT}\left(\frac{0.434}{T\log_{10}T}\,dT\right) = -\frac{1}{(T\ln T)^2}\,(1+\ln T) \tag{9.10}$$

Equating to zero to find the minimum in the curve

$$\frac{1}{T^2(\ln T)^2} - \frac{1}{T^2\ln T} = 0 \tag{9.11}$$

whence

$$-2.303\log_{10}T = 1 \tag{9.12}$$

and

$$A = 1/2.303 = 0.434 \tag{9.13}$$

Hence, for an absorbance of 0.434, there is a minimum in the relative error of the measurement. Figure 9.4 (curve A) shows how the relative error varies with absorbance ($dT = 1\%$) for a simple instrument incorporating a photo-voltaic detector (p. 280). Measurements outside the range 0.2 to 0.8 are subject to a rapidly increasing relative error. The minimum in the error curve for instruments with photomultiplier detectors is 0.86 because the instrumental reading error dT is proportional to $T^{\frac{1}{2}}$. The curve rises more gradually at high absorbances thereby extending the useful working range to at least 2.

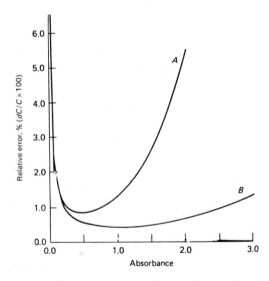

Figure 9.4.   Variation of relative error with absorbance. Constant instrumental error 1%. Curve A: photovoltaic detector; Curve B: photomultiplier detector

Spectrophotometric measurements normally have a relative precision of 2 to 5%.

Each of the major techniques of molecular spectrometry, including mass spectrometry, will now be examined in more detail. Exercises in the interpretation of spectral data in relation to the identification and structural analysis of organic compounds are given at the end of the chapter.

9.1 Visible and Ultraviolet Spectrometry

SUMMARY

Principles

Absorption of electromagnetic radiation in the visible and ultraviolet regions of the spectrum resulting in changes in the electronic structure of ions and molecules.

Instrumentation

Filter-photometer or spectrophotometer incorporating prism or grating monochromator, phototube photomultiplier or diode array, glass, quartz or plastic cells.

Applications

Probably the most widely used technique for quantitative trace analysis. Used as an adjunct to other spectrometric techniques in the identification and structural analysis of organic materials. Relative precision 0.5 to 5%.

Disadvantages

Samples should be in solution. Mixtures can be difficult to analyse without prior separation of the constituents.

Absorption of radiation in the visible and ultraviolet regions of the electromagnetic spectrum results in electronic transitions between molecular orbitals. The energy changes are relatively large, corresponding to about 10^5 J mol^{-1}. This corresponds to a wavelength range of 200 to 800 nm or a wavenumber range of 12 000 to 50 000 cm^{-1}. All molecules can undergo electronic transitions, but in some cases absorption occurs below 200 nm where atmospheric absorption necessitates the use of expensive vacuum instrumentation.

Because of the large energy changes involved in electronic transitions there are always simultaneous changes in rotational and vibrational energies. The effect of this upon the spectrum of a gaseous diatomic molecule is seen by considering a potential energy diagram, figure 9.5, which relates the potential energy of a vibrating molecule to the internuclear distance. The two curves represent the ground and first excited electronic states, and the horizontal lines the quantized vibrational levels associated with each. For simplicity, rotational levels are not shown. The horizontal displacement of the curves indicates a difference in equilibrium bond length between the electronic states. According to the Franck–Condon principle, an electronic transition occurs very much more rapidly than vibrational or rotational motion and may be represented in the diagram as a vertical line between two vibrational levels in the ground and first excited electronic states respectively. At room temperature all molecules will be in the ground electronic state and probably in the lowest vibrational level. Transitions to any vibrational level in the first excited electronic state are allowed by the spectroscopic selection rules so that the spectrum will consist of a series of closely spaced converging lines, the intensities varying with the respective transition probabilities. Each line may

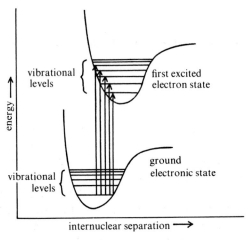

Figure 9.5. Electronic and vibrational transitions in a gaseous diatomic molecule

be further resolved into lines on either side of it due to transitions between the even more closely spaced rotational levels which are associated with each vibrational level. Such *vibrational* and *rotational fine structure* lines are not usually observed for samples run in solution because of physical interactions between solute and solvent molecules which cause collisional broadening of the lines. The resulting overlapping bands coalesce to give one or more broad *band-envelopes*. These are characterized by the position of each maximum λ_{max} and the corresponding intensity or molar absorptivity ε. For polyatomic molecules and metal complexes, the complete spectrum may contain several bands arising from a number of electronic transitions and their associated rotational and vibrational fine structures, figure 9.6. The origins of these bands are discussed in the following sections.

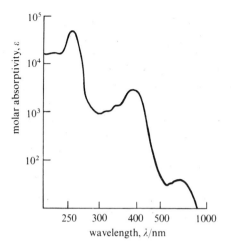

Figure 9.6.   uv/visible absorption spectrum of bis(8-hydroxyquinoline)Cu(II)

POLYATOMIC ORGANIC MOLECULES

According to molecular orbital theory, the interaction of atomic orbitals leads to the formation of bonding and antibonding molecular orbitals. Depending on the nature of the overlapping atomic orbitals, molecular orbitals may be of the σ type, the electron density being concentrated along the internuclear axis, or of the π type where the electron density is concentrated on either side of the internuclear axis. Electron density probability contours for electrons occupying σ and π (bonding) and σ^* and π^* (antibonding) orbitals are shown in figure 9.7(a). The relative energies of these orbitals and that of a non-bonding orbital n, which may be occupied by electrons not participating in bonding, are given in figure 9.7(b).

In most organic compounds the bonding and non-bonding orbitals are filled and the antibonding orbitals are vacant. From the diagram it will be seen that the lowest energy and therefore the longest wavelength transitions are from non-bonding orbitals to antibonding π orbitals, i.e. $n \rightarrow \pi^*$. These give rise to bands in the near uv and visible regions. Other allowed transitions in order of increasing energy (shorter wavelength) are $n \rightarrow \sigma^*$ and $\pi \rightarrow \pi^*$, which have comparable energies, and $\sigma \rightarrow \sigma^*$. The latter occur in the far uv or vacuum region below 200 nm and are of little use analytically. Consequently saturated hydrocarbons which are transparent in the near uv and visible regions make useful solvents. Intense bands (large ε) are produced by $\sigma \rightarrow \sigma^*$ and $\pi \rightarrow \pi^*$ transitions, whereas those arising from $n \rightarrow \sigma^*$ and $n \rightarrow \pi^*$

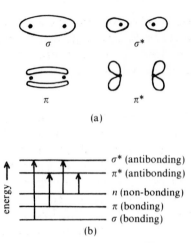

Figure 9.7. Shapes and relative energies of molecular orbitals. (a) Bonding and antibonding orbitals. (b) Relative energies of orbitals and possible transitions between them

transitions are characteristically weak because of unfavourable selection rules.

Unsaturated groups, known as *chromophores*, are responsible for $n \to \pi^*$, and $\pi \to \pi^*$ absorption mainly in the near uv and visible regions and are of most value for diagnostic purposes and for quantitative analysis. The λ_{max} and ε values for some typical chromophores are given in table 9.2. The positions and intensities of the absorption bands are sensitive

Table 9.2 *Absorption characteristics of some typical chromophores

CHROMOPHORE	EXAMPLE	TRANSITION	λ_{max}/nm	$\varepsilon/mol^{-1}\ m^2$
$\diagdown C = C \diagup$	ethylene	$\pi \to \pi^*$	165	1 500
$\diagdown C = O$	acetone	$\pi \to \pi^*$	188	90
		$n \to \pi^*$	279	1.5
—N=N—	azomethane	$n \to \pi^*$	347	0.45
—N=O	nitrosobutane	$\pi \to \pi^*$	300	10
		$n \to \pi^*$	665	2
⬡	benzene	$\pi \to \pi^*$	200	800
			255	21.5

* See footnote to Table 9.3.

to substituents close to the chromophore, to conjugation with other chromophores, and to solvent effects. Saturated groups containing heteroatoms which modify the absorption due to a chromophore are called *auxochromes* and include $-OH$, $-Cl$, $-OR$ and $-NR_2$.

Conjugation Effects

Absorption bands due to conjugated chromophores are shifted to longer wavelengths (*bathochromic* or *red shift*) and intensified relative to an isolated chromophore. The shift can be explained in terms of interaction or delocalization of the π and π^* orbitals of each chromophore to produce new orbitals in which the highest π orbital and the lowest π^* orbital are closer in energy. Figure 9.8 shows the conjugation of two ethylene chromophores to form 1,3-butadiene. The $\pi \rightarrow \pi^*$ transition in ethylene occurs at 165 nm with an ε value of 1 500 whereas in 1,3-butadiene the values are 217 nm and 2 100 respectively.

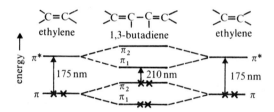

Figure 9.8.   Effect of conjugation on absorption of ethylene

If two unlike chromophores are conjugated and one group has non-bonding electrons, the $n \rightarrow \pi^*$ transition is also shifted bathochromically because the energy of the antibonding orbital is lowered. Thus the weak $n \rightarrow \pi^*$ band in a saturated carbonyl compound is shifted from below 300 nm to above 300 nm with an increase in ε. Conjugation of additional chromophoric groups moves λ_{max} progressively towards the visible region and increases ε. For example, tetradecahexaene (six double bonds) absorbs at the blue end of the visible region and appears yellow whilst with further conjugation, as in the carotenes (ten or more double bonds), the compound may appear orange, red, purple, or even black. Benzene can be regarded as a special case of a conjugated triene. It has a relatively weak $\pi \rightarrow \pi^*$ band near 255 nm which shifts bathochromically and intensifies on chromophoric substitution. Of more importance is that substitution produces a new and intense band between 200 and 300 nm which arises from a $\pi \rightarrow \pi^*$ transition in the extended conjugated system. Absorption data for some conjugated chromophores including chromophorically substituted benzenes are given in table 9.3.

Table 9.3* Absorption characteristics of some conjugated chromophores and chromophorically substituted benzenes

CHROMOPHORE	EXAMPLE	TRANSITION	λ_{max}/nm	$\varepsilon/mol^{-1} m^2$
C=C—C=C	butadiene	$\pi \to \pi^*$	217	2 100
(C=C)$_6$	tetradecahexaene	$\pi \to \pi^*$	360	6 300
(C=C)$_{11}$	γ-carotene	$\pi \to \pi^*$	460	6 500
C=C—C=O	crotonaldehyde	$\pi \to \pi^*$	217	1 600
		$n \to \pi^*$	321	2
C=O (benzene)	acetophenone	$\pi \to \pi^*$	240	1 300
			278	110
		$n \to \pi^*$	320	5
NO₂ (benzene)	nitrobenzene	$\pi \to \pi^*$	252	1 000
			280	100
		$n \to \pi^*$	333	13
naphthalene ring	naphthalene	$\pi \to \pi^*$	221	10 000
			286	930
			312	28

* The values for ε are in $mol^{-1} m^2$ to conform to SI. Most other published data are still given in $mol^{-1} l\ cm^{-1}$ and are therefore an order of magnitude greater.

Effect of Auxochromes

In general, auxochromic substitution of chromophores causes bathochromic shifts and increases in intensity for $\pi \to \pi^*$ transitions, and *hypsochromic* or *blue* shifts (to shorter wavelengths) for $n \to \pi^*$ transitions. The shifts are explainable in terms of *mesomeric* (*resonance*) effects caused by interaction of lone pair electrons associated with such auxochromes as $-OH$, $-Cl$, $-NH_2$ with the π system of the chromophore. This leads to increases in the energies of π and π^* orbitals, the π being raised by more than the π^*, but leaves the energy of the non-bonding orbital unchanged. Empirical rules have been devised by Woodward, Fieser and Scott[1] to enable the additive effects of auxochromic substitution on the absorption of aromatic and other conjugated systems to be predicted. The rules for diene absorption are reproduced in table 9.4.

Table 9.4   Rules for diene absorption (9.1)

	WAVELENGTH, nm
base value for heteroannular diene	214
base value for homoannular diene	253
increments added for:	
double bond extending conjugation	30
alkyl substituent or ring residue	5
exocyclic double bond	5
polar groups OAc	0
OAlk	6
SAlk	30
Cl, Br	5
$N(Alk)_2$	60
solvent correction	0
calculated λ_{max} Total	

Solvent Effects

Absorption bands arising from $n \to \pi^*$ transitions suffer hypsochromic shifts on increasing the solvent polarity, whilst those of $\pi \to \pi^*$ transitions are shifted bathochromically. Explanations lie in the fact that the energy of the non-bonding orbital is lowered by hydrogen bonding in the more polar solvent thus increasing the energy of the $n \to \pi^*$ transition, but the energy of the π^* orbital is decreased relative to the π orbital. The positions and intensities of $\pi \to \pi^*$ bands in such compounds as phenols and amines exhibit a marked sensitivity to pH changes because of changes in the interaction of non-bonding electrons with the π-system.

METAL COMPLEXES

Complexes of metals with organic and inorganic ligands which absorb in the visible region of the spectrum are of great importance in quantitative analysis. Transitions giving rise to coloured complexes are of three types:

1. *d–d* transitions within a transition metal ion. These are usually of low intensity and of little use for determination at trace levels.

2. Excitations within an organic ligand. These are typical $n \to \pi^*$ and $\pi \to \pi^*$ transitions which are affected by the presence of a metal.

3. Charge-transfer transitions, involving the transfer of an electron between two orbitals one of which is associated predominantly with the ligand and the other with the metal.

The last two types give rise to many strongly coloured complexes suitable for trace analysis.

Bands due to *d-d* transitions are responsible for the colours of transition metal ions in aqueous solutions. Absorption of radiation results in the movement of electrons between filled and half-filled or empty metal *d*-orbitals which differ in energy because of the electrostatic field created by coordination of the ligands. Various colours are produced depending on the metal and the nature of the coordinating ligand. The absorption band shifts towards the uv with increasing strength of ligand field, the order for some of the more common ligands being $I^- < Br^- < Cl^- < S^{2-} < OH^- < H_2O < NCS^- < EDTA^{4-} < NH_3 < $ ethylenediamine $< o$-phenanthroline $< NO_2^- \ll CN^-$. This is known as a *spectrochemical series*. In most cases, the bands are of low intensity because the transitions are spectroscopically forbidden by rules of symmetry. The fact that they occur at all is probably due to vibrational distortions which relax the rules. This does not apply to tetrahedral and square planar complexes which have no centre of symmetry and which generally have quite intense absorption bands.

A large number of metal complexes involve organic ligands in which the absorption bands of the ligand are modified to a varying degree by coordination to the metal. The effect on the spectrum of the ligand depends on whether the metal–ligand bonds are predominantly covalent or ionic. In complexes where bonding to the metal is essentially ionic small shifts in bands due to $n \rightarrow \pi^*$ and $\pi \rightarrow \pi^*$ transitions are observed with little change in intensity, the spectrum of the metal complex being similar to that of the protonated ligand. Examples of this type include metal complexes with hydroxynaphthylazo dyes such as that formed between magnesium and eriochrome black T, figure 9.9(a). The absorption maximum of the ligand dianion at 650 nm is shifted to 530 nm by interaction with Mg^{2+}. Similarly, a series of coloured complexes are formed between Cu, Co, Ni, Fe(III), Ag, Bi, Hg(I), Pd and Pt and dithiooxaminde or rubeanic acid, figure 9.9(b). The stronger the complex formed, the more the absorption band is shifted towards the uv.

Where the metal–ligand bond is strongly covalent and possibly includes back bonding from the metal into vacant ligand orbitals, the spectrum of the ligand may be significantly changed. Ionic complexes formed between Ca, Sr and Ba and the dyes metalphthalein, calcein and thymolphthalexone are strongly coloured with similar absorption spectra to the dissociated form of the ligands whereas complexes with metals forming more covalent bonds are shifted into the near uv because of modification of the conjugated chromophores. Dithizone (diphenylthiocarbazone), figure 9.9(c), forms highly coloured complexes with metals which can back-donate into ligand orbitals. The range of absorption maxima extends from 620 nm for the reagent

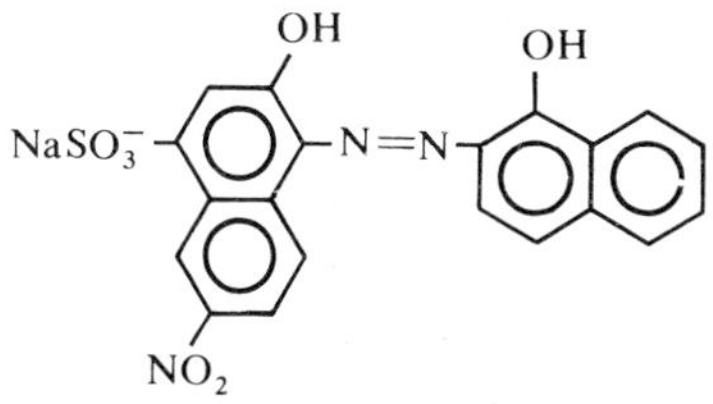

OH   OH

NaSO$_3^-$ ⟨benzene⟩–N=N–⟨naphthalene⟩

NO$_2$

(a) eriochrome black T

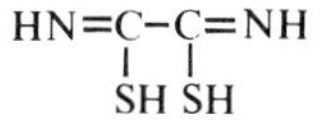

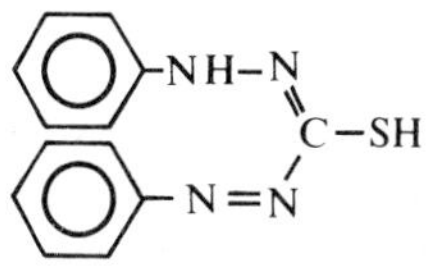

HN=C–C=NH
   | |
   SH SH

⟨benzene⟩–NH–N
              ‖
              C–SH
⟨benzene⟩–N=N

(b) dithiooxamide (rubeanic acid)   (c) dithizone (diphenylthiocarbazone)

Figure 9.9.

mono-anion to 490 nm for the Hg(II) complex and 460 nm for the Ag(I) complex both of which are essentially covalent.

The origins of charge-transfer bands are quite different to those for the complexes already discussed. Associated mainly with covalent, and especially transition metal complexes, they are due to transitions between σ or π ligand orbitals and empty or anti-bonding metal orbitals. Transitions from metal σ orbitals to vacant ligand orbitals are also sometimes involved. These transitions are allowed by symmetry rules and consequently give much more intense bands than most d–d transitions. The intense red colour of the Fe(III)-thiocyanate complex, the orange colour of the titanium(IV)-peroxy complex and the purple permanganate ion are all a result of the transfer of electrons from metal to ligand orbitals. Such transitions are often associated with metals in high oxidation states and have been described as photo-chemical oxidation-reduction reactions. Organic ligands forming charge-transfer complexes include sulphosalicylic acid, o-phenanthroline and other aromatic amines which have been used as reagents in the determination of Fe, Cu(I), Zn and Cd.

QUALITATIVE ANALYSIS—THE IDENTIFICATION OF STRUCTURAL FEATURES

Visible and uv spectrometry are of secondary importance to other spectral methods for the identification and structural analysis of unknown compounds. This is a direct consequence of the broad bands and rather simple spectra which make differentiation between structurally related compounds difficult.

As an adjunct to infrared, magnetic resonance and mass spectrometry, however, it can play a useful role. It can be particularly helpful in confirming the presence of acidic or basic groups in a molecule from the changes in band position and intensity associated with changes in pH (p. 367).

QUANTITATIVE ANALYSIS—ABSORPTIOMETRY

The use of visible and uv spectrometry for quantitative analysis by comparing the absorbance of standards and samples at a selected wavelength is perhaps the most widespread of all analytical techniques. It is also one of the most sensitive. The analysis of mixtures of two or more components is facilitated by the additivity of absorbances. This has been discussed earlier (p. 355). Other applications include measurement of the absorption of complexes as a function of solution conditions or time to establish their composition, and to determine thermodynamic and kinetic stability for analytical purposes or for more fundamental studies.

CHOICE OF COLORIMETRIC AND SPECTROPHOTOMETRIC PROCEDURES

Not all chromogenic compounds are suitable for quantitative measurements, and the choice of system and procedure depends largely on the chemistry of the species to be determined. Points to be considered in the selection of a procedure include:

1. Stability of the absorbance with respect to time (30 minutes should be the minimum) and to minor variations in pH, ionic strength and temperature.
2. Degree of selectivity of a complexing agent including the effect of other species likely to be present and the effect of an excess of the reagent. Calculations based on conditional constants (p. 40) may help to establish optimum conditions.
3. Conformity to the Beer–Lambert law and the value of the molar absorptivity.

One of the major problems lies in the extent of interference from other constituents of a sample. This can often be obviated by a prior separation using chromatography or solvent extraction (p. 55) or by the use of masking agents (p. 41), pH control or changes in oxidation state. Standards should always be matched to the gross composition of the sample as closely as possible, and calibration curves frequently checked. The precision of absorption measurements has already been discussed (p. 359).

A valuable means of comparing two or more methods with regard to

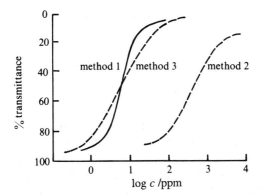

Figure 9.10. Ringbom calibration curves

precision, conformity to the Beer–Lambert law and the range of concentrations usefully measured is to plot calibration data in the form of a Ringbom[2] curve, i.e. transmittance versus log concentration. The straight line portions of the S-shaped curves, figure 9.10, indicate optimum concentration ranges and the relative precision, assuming a 1% constant instrumental error, is $230/S$ where S is the slope of the straight line portion of the curve, sometimes called the *sensitivity*. Hence, in the figure, method 1 is seen to be more sensitive and to give better precision than method 3, but has a narrower optimum range. Method 2 is useful at higher concentration levels only.

Photometric Titrations

The change in absorbance during a titration in which one of the reactants or products absorbs in the visible or uv region can be used to locate the end point. Plots of absorbance as a function of volume of titrant added give two straight line portions of differing slope intersecting at the end point. Curvature near the end point is caused by incomplete reaction, whilst curvature elsewhere is caused by apparent deviations from the Beer–Lambert law or by dilution errors. Corrections for changes in volume will eliminate the last problem, but using a concentrated titrant is simpler. Photometric titrations have similar advantages to amperometric and conductometric titrations (pp. 253, 261) when compared to visual indicator (p. 186) and potentiometric titrations (p. 239) or to direct measurements:

1. Data near the end point are unimportant provided straight-line portions can be established and extrapolated. Very dilute solutions can be titrated.
2. Precision and accuracy are better than for a direct absorbance measure-

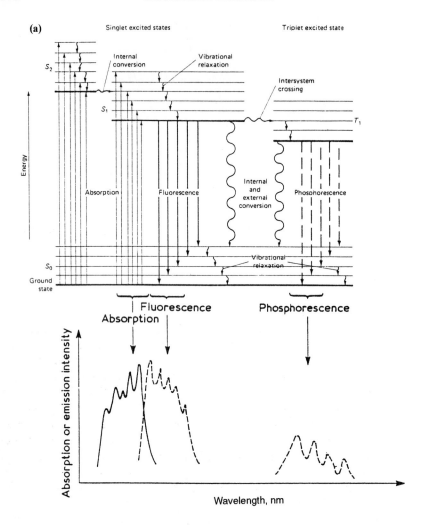

Figure 9.11(a).   Relaxation mechanisms from excited electronic states

ment as the titration curve averages the data; relative precision is around 0.5%.

3. The substance to be determined need not itself absorb whilst a degree of background absorbance or turbidity can be tolerated.

The method is particularly suited to complexometric titrations of mixtures or where there is no suitable visual indicator.

Figure 9.11(b). Effect of structural rigidity on quantum yield, Φ_F

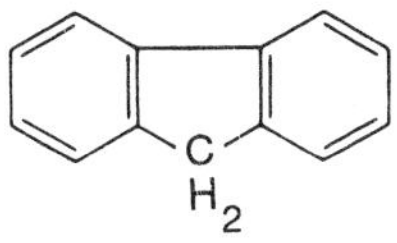

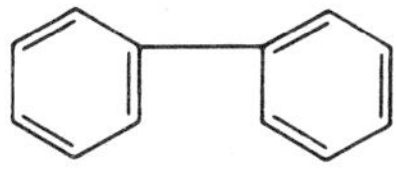

FLUORIMETRY

The absorption of electromagnetic radiation by molecular species in solution in the UV/visible region is followed by *relaxation* from excited electronic states to the ground state mostly by a combination of *radiationless* processes. *Vibrational relaxation*, where the excess energy is rapidly dissipated as heat in 10^{-11} to 10^{-15} s by collisions between solute and solvent molecules, may be followed by the slower (10^{-9} to over 1 s) mechanisms of *internal conversion* and *intersystem crossing*. Of analytical significance is the possibility of *photoluminescence*, a general term for relaxation in which radiation of a longer wavelength (lower energy) than that originally absorbed is re-emitted after about 10^{-8} s or longer and following vibrational relaxation. *Fluorescence* and *phosphorescence* are two distinct forms of photoluminescence, both giving rise to analytical techniques of very high sensitivity. A summary of relaxation mechanisms is shown diagrammatically in figure 9.11(a).

A *fluorescence emission spectrum* arises from transitions between the lowest vibrational level of the first excited electronic state and different vibrational levels of the ground state. Excited species return to the ground state very rapidly after excitation, i.e. within about 10^{-8} s, so that emission of radiation ceases as soon as the exciting radiation is removed. In solution, due to collisional broadening, one or more fairly broad overlapping bands are observed. The overall appearance is an approximate mirror image of the absorption or *excitation* spectrum of the species but displaced to longer wavelengths due to the initial rapid loss of energy by vibrational relaxation which precedes the fluorescent emission. The shortest wavelength (highest energy) emission, which arises from the transition to the lowest vibrational level in the ground electronic state, is called *resonance fluorescence*. Solute-solvent interactions usually result in the position of the resonance band being displaced to a slightly longer wavelength than the corresponding excitation band, although in theory they should coincide being due to transitions between the same vibrational levels. *Phosphorescence* can be observed only after a molecular system has undergone radiationless relaxation by inter-system crossing. It is of longer wavelength (lower energy) than fluorescence and persists for seconds or even minutes after excitation. Analytical methods based on phosphorescence will not be discussed further.

The number of molecular species capable of relaxing by fluorescence is limited mainly to rigid organic and mostly aromatic structures for which radiationless relaxation mechanisms are often comparatively slow. The intensity of fluorescent emission is dependent on a *quantum efficiency factor* or *yield*, Φ_F, which can vary between zero (no fluorescence) and unity (*all* excited molecules relax by fluorescence). Quantum efficiency is determined by a number of structural factors, including the presence and positions of hetero-atoms in the molecule, and the solvent used. It generally increases with degree of rigidity and extended conjugation especially in fused aromatic systems, cf. biphenyl and fluorene (figure 9.11(b)). Structures containing heteroatoms or are ionized *may* undergo rapid relaxation by intersystem crossing thereby reducing or inhibiting fluorescence, e.g. pyridine, nitrobenzene and iodo-benzene exhibit no fluorescence whereas aniline, phenol and chlorobenzene all fluoresce.

Some examples of fluorescing species and their analytical uses are included in table 9.5.

There is a linear relation between the concentration, C, of a fluorescent analyte and the intensity of emission, I_F, given by the equation

$$I_F = 2.303 \, \Phi_F \, I_o \, \varepsilon C l \tag{9.14}$$

where Φ_F and ε are the quantum yield and molar absorptivity respectively of the fluorescing species, I_o is the intensity of the incident (exciting) radiation and l is the path length of the sample cell. If I_o and l are constant and the absorbance ($\varepsilon C l$) is small, i.e. the analyte concentration is low, then

$$I_F = k \, C \tag{9.15}$$

Calibration curves are linear over several orders of magnitude but eventually show curvature and even reversal due to *quenching* effects. These are caused by partial or total absorption of the emitted radiation by unexcited analyte molecules, dissolved oxygen and other species, particularly if they are para-magnetic. Unlike absorptiometry, sensitivity can be improved by increasing the intensity of the exciting radiation, I_o.

Fluorimeters have similar components to UV/visible spectrometers but differ in the geometry of the radiation beams and in the need to be able to select or scan both excitation and emission wavelengths. In most designs, the fluorescence, which is emitted from the sample equally in all directions, is

Table 9.5 Applications of quantitative visible and ultraviolet spectrometry

ELEMENT OR COMPOUND DETERMINED	REAGENT	EXAMPLE OF APPLICATION
Absorptiometry		
Fe	*o*-phenanthroline	natural waters, petroleum products
Cu	neocuproine	minerals, alloys
Mn	oxidation to MnO_4^-	steels
Cr	diphenylcarbazide	alloys, minerals
Hg, Pb	dithizone	food products, fish
P, PO_4^{3-}	reduction to molybdenum blue	fertilizer residues, soils
aspirin	—	analgesic preparations
cholesterol	Liebermann–Burchard reaction	body fluids
sulphonamides	diazo derivatives	drug preparations
DDT	nitrated derivative	soils, fish

ELEMENT OR COMPOUND DETERMINED	FLUORESCENT EMISSION MAXIMUM (nm)	EXAMPLE OF APPLICATION
Fluorimetry		
vitamin A	500 nm	foodstuffs, vitamin tablets
LSD	>430 nm	body fluids
amphetamine	282–300 nm	⎱ drug preparations
codeine, morphine	345 nm	⎰ and body fluids
PAH	320–550 nm	environmental samples
amines and amino acids as derivatized by ninhydrin, *o*-phthaldehyde (OPA), dansyl chloride or fluorescamine	440–550 nm	pre- or post-column derivatization for hplc separations
Al as alizarin garnet red complex	580 nm	water samples and soils
Cd as 2-(o-hydroxyphenyl) —benzoxazole complex	—	water samples and soils
B as benzoin complex	450 nm	water samples and soils

measured at 90° to the direction of the excitation beam. This prevents incident, scattered and reflected radiation from reaching the detector. *Filter fluorimeters* are single-beam instruments capable of high sensitivity for quantitative analysis but lacking in the versatility needed for investigative qualitative scans. *Spectrofluorimeters* employ two monochromators which allows independent scanning of excitation or emission spectra over the whole of the UV/visible region (200 to 800 nm) in addition to making quantitative measurements at fixed wavelengths. Both filter fluorimeters and spectrofluorimeters can be fitted with flow-through microcells for use as detectors for high performance liquid chromatography or capillary electrophoresis but purpose-designed detectors are preferable (p. 126).

APPLICATIONS OF UV/VISIBLE SPECTROMETRY AND FLUORIMETRY

Quantitative analysis by UV visible absorption spectrometry is practised by almost every analytical laboratory at one time or another. Most inorganic, organic and biochemical substances can be determined either directly or after the formation of an absorbing derivative or complex. A selection of typical applications is given in table 9.5. The technique is among the most sensitive and is predominantly used for the determination of minor, trace and ultra-trace level constituents. Its applications to the determination of metals in trace amounts were particularly widespread because of the many intensely coloured complexes known and the degree of selectivity introduced by proper choice of organic reagent and masking reactions or by solvent extraction. In recent years however, atomic absorption and ICP spectrometry (p. 296, 317) have largely superseded the use of UV/visible spectrometry due to their simplicity, freedom from interferences and excellent precision.

Fluorimetry is used much less extensively than absorptiometry because of the limited number of naturally fluorescing species, although its range of applications can be extended by forming fluorescent derivatives or complexes of non-fluorescent analytes. Quantitative applications predominate because, like absorption spectra, fluorescence spectra consist of only a few broad bands. It is inherently more selective and up to three orders of magnitude more sensitive than absorptiometry, detection limits extending down to ppb levels. This has made it particularly valuable for the determination of trace contaminants in foodstuffs and pharmaceuticals and for the determination of fluorescent substances in clinical and forensic samples. The range of possible analytes includes non-transition metals which form fluorescent neutral chelate complexes (charged complexes and those of transition metals do not fluoresce due to rapid relaxation by internal conversion and intersystem crossing), polynuclear aromatic hydrocarbons (PAH), lysergic acid diethyl-amide (LSD), penicillin, chlorophyll and other plant pigments, numerous alkaloids, steroids, some vitamins, amino acids, proteins and enzymes. Some examples are given in table 9.5.

9.2 Infrared Spectrometry

SUMMARY

Principles

Absorption of electromagnetic radiation in the infrared region of the spectrum resulting in changes in the vibrational energy of molecules.

Instrumentation

Fourier transform spectrometer or double-beam spectrophotometer incorporating prism or grating monochromator, thermal or photon detector, alkali halide cells.

Applications

Very widespread use, largely for the identification and structural analysis of organic materials; useful for quantitative analysis but less widely used than uv and visible spectrometry.

Disadvantages

Difficult to analyse mixtures. Special cells required for aqueous samples.

Absorption of radiation in the infrared region of the electromagnetic spectrum results in changes in the vibration energy of molecules. Energy changes are typically 6×10^3 to 42×10^3 J mol^{-1}, which corresponds to 250 to 4 000 cm^{-1} (2.5 to 40μm), although some occur between 4 000 cm^{-1} and the beginning of the visible region at about 12 500 cm^{-1} (0.8 μm or 800 nm) in a region known as the near infrared (NIR). A molecule can absorb energy only if there is a net change in the *dipole moment* during a particular vibration, a condition fulfilled by virtually all polyatomic molecules. The absorption spectra of such molecules are often very complex and the underlying reason for this complexity is best understood by first considering the spectrum of a heteronuclear diatomic molecule in the gaseous state.

DIATOMIC MOLECULES

The classical analogy of a vibrating diatomic molecule is that of two weights connected by a spring. The potential energy of such a vibrating system is related to the displacement of the weights relative to each other along the

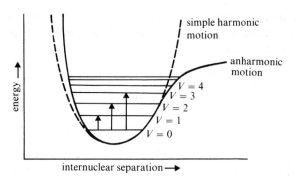

Figure 9.12. Vibrational energy levels of a diatomic molecule

axis of the spring. If the motion is *simple harmonic* the relation is given by $E = \frac{1}{2}fx^2$ where E denotes potential energy, x the displacement and f the force constant or stiffness of the spring. The equation describes a parabola, figure 9.12, and shows that identical changes in potential energy occur on stretching or compressing the spring. A diatomic molecule behaves rather differently in that the forces of internuclear repulsion arising when the bond is compressed build up rapidly whilst, on stretching, the bond weakens and may finally disrupt. The resulting vibrational motion is *anharmonic* in nature and the mathematical relation between potential energy and displacement is necessarily more complex. The corresponding curve is modified in the manner shown in figure 9.12.

The concept of discrete or quantized energy levels can be superimposed on this diagram by representing them as a series of horizontal lines, the spacing of which becomes closer with increasing energy due to the anharmonic nature of the vibration. In quantum mechanical terms, these levels are labelled $V = 0,1,2,3 \ldots$, where V is the vibrational quantum number, and the general mathematical expression for the potential energy of the system expressed in wavenumbers is given by

$$\varepsilon_{vib} = (V + \tfrac{1}{2})\,\sigma_e - (V + \tfrac{1}{2})^2 \sigma_e\, \chi_e \ \mathrm{cm}^{-1} \tag{9.16}$$

σ_e denoting the equilibrium vibration frequency and χ_e the anharmonicity constant. This expression may be used to predict the appearance of the absorption spectrum of a diatomic molecule. At room temperature most molecules are in the ground vibrational state ($V = 0$) so that only transitions from this level need be considered. The spectroscopic selection rule allows any change in the value of V, so for transitions originating in the ground state,

$$\Delta\varepsilon_{0 \to n} = n\sigma_e\,[1 - (n+1)\chi_e]\,\mathrm{cm}^{-1} \tag{9.17}$$

where n may be 1,2,3. . . .

Substitution of appropriate values of n into the equation leads to a set of expressions giving the wavenumber of each transition, e.g. the first three and their designations are

$$\sigma_{spec} = \Delta\varepsilon_{0\to1} = \sigma_e(1 - 2\chi_e) \quad \textit{fundamental}$$

$$\sigma_{spec} = \Delta\varepsilon_{0\to2} = 2\sigma_e(1 - 3\chi_e) \quad \textit{first overtone}$$

$$\sigma_{spec} = \Delta\varepsilon_{0\to3} = 3\sigma_e(1 - 4\chi_e) \quad \textit{second overtone}$$

where σ_{spec} is the wavenumber of the spectroscopic absorption.

The transition probability falls off rapidly with increasing n, and the spectrum therefore consists of two or three lines of diminishing intensity, each one being separated from the next by rather less than the value of σ_e. Although basically a simple spectrum simultaneous rotational transitions produce a pattern of fine structure lines on either side of the vibrational transition. Rotational fine structure is not usually resolved for samples run in the liquid state or in solution. This is due to collisional broadening and the resulting spectrum would have the appearance shown in figure 9.13. For a given

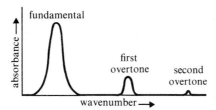

Figure 9.13.   Infrared spectrum of a diatomic molecule in solution

molecule, and using the classical analogy considered earlier, the positions of the fundamental and hence the overtones, are determined by the atomic masses m_1 and m_2, and the stiffness of the bond. These are related by the equation

$$\sigma_e = \frac{1}{2\pi c}\left(\frac{f}{\mu}\right)^{\frac{1}{2}} \tag{9.18}$$

where f is the force constant of the bond and μ the reduced mass, given by $(m_1 m_2)/(m_1 + m_2)$.

It is important to note that the stiffer the bond or the lighter the atoms, the higher is the wavenumber of the fundamental and vice versa. Qualitative conclusions can therefore be reached from the relative positions of absorption bands in the infrared region (*vide infra*).

POLYATOMIC MOLECULES

For polyatomic molecules, whose spectra are usually recorded as liquids, solids or in solution, the complexity of the spectrum increases rapidly with the number of atoms N in the molecule. The number of fundamentals, or *normal modes* of vibration is given by $3N - 6$ and $3N - 5$ for bent and linear molecules respectively. A *normal mode* of vibration is defined as the movement of all the atoms of the molecule in phase. In the case of polyatomics, the number of fundamentals includes both *stretching* and *bending* vibrations, the latter involving changes of bond angle. There are associated overtones with each fundamental and, in addition, other absorption bands may arise due to interactions between certain of the fundamentals and overtones. These are known as *combination* and *Fermi resonance* bands. The overall appearance of the spectrum is that of a large number of resolved and partially resolved bands of varying intensity, figures 9.17 to 9.24. The most intense bands are fundamentals in which there is a large change of dipole moment during the vibration. Conversely, a band may be weak or absent if the dipole change is small or zero. Many bands can be assigned to the vibration of particular chemical groups within a molecule, the wavenumber being largely independent of the rest of the structure. Such *characteristic vibration frequencies* are of paramount importance in obtaining structural information and in the identification of unknown compounds. For the most part it is in the analysis of organic and organometallic compounds that infrared spectrometry has had its greatest impact.

CHARACTERISTIC VIBRATION FREQUENCIES

The occurrence of characteristic vibration or group frequencies can be explained in terms of relative masses and of force constants using the classical analogy of weights and springs as depicted in figure 9.14.

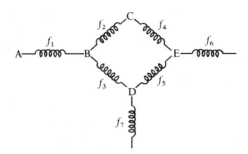

Figure 9.14.　Classical analogy of polyatomic molecule

(a) *Mass Effect*

If the masses $A, B, C, D \ldots$ are all similar and the force constants $f_1, f_2, f_3 \ldots$ are of the same magnitude, the vibrations of individual atoms are strongly coupled with the result that no band can be assigned solely to any particular group of atoms. If, however, the mass of atom A is considerably smaller than those of the other atoms, one mode of vibration will involve the stretching of the bond between A and the rest of the molecule. The system can be considered as approximating to a diatomic molecule A—X, and the wavenumber of the vibration is then given by

$$\sigma_{A-X} = \frac{1}{2\pi c} \left(\frac{f_{A-X}}{\mu_{A-X}}\right)^{\frac{1}{2}}$$

where f_{A-X} is the force constant of the A—X bond and

$$\mu_{A-X} = \left(\frac{AX}{A+X}\right) \simeq A$$

Hence, to a good approximation

$$\sigma_{A-X} = \frac{1}{2\pi c} \left(\frac{f_{A-X}}{A}\right)^{\frac{1}{2}}$$

and the wavenumber of the vibration is therefore independent of the masses of the other atoms in the molecule. This type of stretching vibration is exemplified by hydrogen atoms linked to carbon, oxygen, nitrogen or sulphur, and results in absorption bands around 3 000 cm^{-1}. The wavenumber ranges of a

Table 9.6 Wavenumbers of characteristic vibrations
(a) X—H Vibrations

X—H	STRETCH/cm^{-1}	BEND/cm^{-1}
C—H	2 700–3 300	1 300–1 500
		600– 900
O—H	3 000–3 700	1 200–1 500
N—H	3 000–3 700	1 500–1 700
S—H	2 550–2 600	700– 900

(b) Saturated and unsaturated groups

GROUP	STRETCH/cm^{-1}
C—C, C—N, C—O	400–1 300
C=C, C=N, C=O	1 550–1 900
C≡C, C≡N	2 100–2 400

number of such groups are given in table 9.6(a). Also shown in the table are the wavenumbers of the bending modes, and it should be noted that the energies involved, and hence the wavenumbers, are significantly lower than those of the corresponding stretching vibrations.

(b) *Force Constant Effect*

Considering the same model as that shown in figure 9.14, if the force constant of the A—B bond is significantly higher than those for the rest of the molecule, one stretching vibration will occur at a higher wavenumber as σ_e is directly proportional to f_{A-B}. This situation arises in molecules with unsaturated groups such as C=C, C≡C and C=O because the magnitude of the force constant increases significantly with bond order. Examples of the wavenumber ranges for some unsaturated and saturated groups are compared in table 9.6(b). Differences between the force constants of single bonds are not sufficiently great to enable characteristic group frequencies to be identified. Prominent bands can be observed, however, for groups such as C—O where the dipole change during the vibration is large. For groups with multiple bonds the change in dipole and hence the intensity of the band is determined by the position of the group in the molecule. Thus little or no absorption due to the multiple bond is observed for symmetrically substituted alkenes or alkynes, whereas a terminal unsaturated group gives rise to a strong absorption.

FACTORS AFFECTING GROUP FREQUENCIES

Hydrogen Bonding

Compounds containing proton donor groups such as O—H and N—H can be involved in intra- or intermolecular hydrogen bonding in the presence of proton acceptors, e.g. O, N, halogens, C=C. The stiffness of the X—H bond is thereby lessened resulting in a lowering of the stretching frequency, and the band broadens and often intensifies. Conversely, the frequency of the bending mode is raised but the effect is much less pronounced. *Intermolecular* hydrogen bonding is suppressed at elevated temperatures but it is favoured by a high solute concentration. *Intramolecular* hydrogen bonding is also reduced at elevated temperatures but it is unaffected by changes in solute concentration. These effects are particularly significant in the spectra of alcohols, phenols, carboxylic acids and amines.

Adjacent groups

The wealth of structural information that can be obtained from infrared spectra arises largely from differences in band position, shape and intensity that are observed when comparing similar but not identical compounds.

Predictable shifts occur on altering the substituents or structure adjacent to a group giving rise to a particular absorption. These can be explained in terms of an *inductive effect*, a *mesomeric* or *resonance effect* and *steric factors* such as ring strain. The effects are typified by the carbonyl group, whose stretching frequency occurs within the range 1 650 to 1 850 cm^{-1}. Shifts are usually related to the position of the carbonyl band in pure acetone (1 715 cm^{-1}). The proximity of electron withdrawing substituents causes an inductive withdrawal of electron density from around the oxygen atom thereby shortening the bond, figure 9.15. This increases the force constant and hence the wavenumber of the vibration, table 9.7. If the adjacent group contains a weakly electronegative atom with lone-pair electrons or electrons occupying π-orbitals, resonance interaction with electrons in the carbonyl group leads to a reduction in bond order, figure 9.15. The correspondingly smaller force

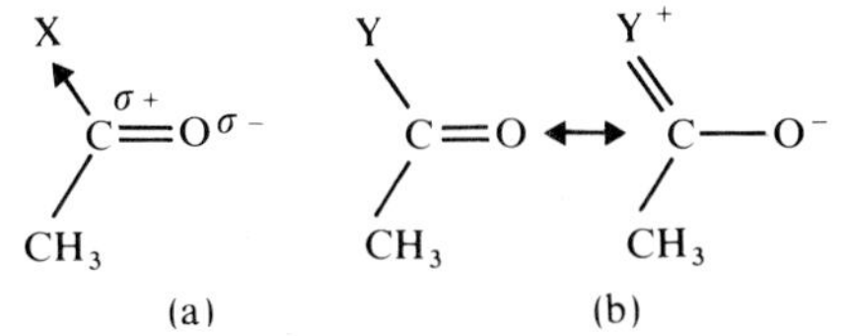

Figure 9.15. Effect of substituents of the carbonyl bond. (a) Inductive effect, X = H, OH, OR, halogen. (b) Resonance effect, Y = NH_2, NHR, NR_2, phenyl

constant results in a lower value for the wavenumber of the vibration, table 9.7. Steric hindrance may inhibit the resonance effect by reducing the coplanarity of a conjugated system whilst carbonyl groups which form part of a strained ring system absorb at high wavenumbers because of the constraint imposed upon the carbon atom, table 9.7.

Table 9.7 Effect of substituents on the wave number of the carbonyl stretching vibration

COMPOUND	C=O STRETCH/cm^{-1}
saturated ketones	1 710–1 720
carboxylic acids, monomer	1 750–1 770
carboxylic acids, dimer	1 710–1 720
saturated esters	1 725–1 745
acid chlorides	1 790–1 815
aliphatic amides	1 630–1 700
aromatic ketones	1 680–1 700
aromatic and α, β-unsaturated esters	1 715–1 730
γ-lactones	1 760–1 780

Coupled Vibrations

Groups in which two or more atoms of an element are bonded to a common atom are mechanically coupled resulting in a band which is generally resolved into two components. These correspond to *symmetric* and *asymmetric* modes of vibration. They also execute a variety of bending vibrations each of which may be observed at a characteristic wavenumber. Examples include CH_3, CH_2, NH_2, NO_2 and COO^-. The stretching and bending modes of a methylene group are illustrated in figure 9.16. It should be noted that asymmetric vibrations occur at a higher wavenumber than symmetric and that the absorption is more intense. Another form of vibrational coupling occurs between the C—H groups in both aromatic and olefinic compounds resulting in bands which are characteristic of the substitution pattern. In the case of *out-of-plane bending*, they are of a high intensity and are especially useful for diagnostic purposes (*vide infra*).

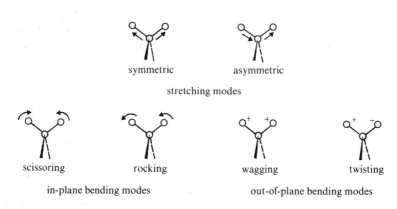

Figure 9.16.   Stretching and bending modes of a methylene group

QUALITATIVE ANALYSIS—THE IDENTIFICATION OF STRUCTURAL FEATURES

The value of infrared spectrometry as a means of identification of unknown compounds and to investigate structural features is immense. Spectra are used in an empirical manner by comparison of samples with known materials and by reference to charts of group frequencies. A simplified correlation chart is shown in table 9.8. The interpretation of infrared spectra is best considered by discussing the prominent features of a representative series of compounds.

The spectrum of a saturated hydrocarbon, *n*-decane is shown in figure 9.17. Four prominent bands arise from C—H stretching and bending vibrations: asymmetric and symmetric C—H stretching ($\sim 2\,900\,cm^{-1}$), CH_2 bending and asymmetric CH_3 bending ($\sim 1\,460\,cm^{-1}$), symmetric CH_3

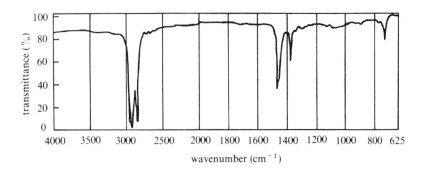

Figure 9.17.   Infrared spectrum of *n*-decane

bending ($\sim 1\ 380\ \text{cm}^{-1}$) and rocking of all the CH_2 groups in the chain ($\sim 720\ \text{cm}^{-1}$). The position of the latter depends upon chain length and is not observed if there are fewer than four methylene groups together. The C—H stretching bands may be further resolved into components due to CH_2 and CH_3 groups. It should be noted that no bands are observed above $3\ 000\ \text{cm}^{-1}$ indicating the absence of an aromatic, olefinic or acetylenic structure.

Contrast this spectrum with those of 1,2-dimethylbenzene, figure 9.18, and 1-octene, figure 9.19. In both spectra, additional C—H stretching vibrations appear above $3\ 000\ \text{cm}^{-1}$, and characteristic C—H out-of-plane bending vibrations are seen between $650\ \text{cm}^{-1}$ and $1\ 000\ \text{cm}^{-1}$. The bands at $910\ \text{cm}^{-1}$ and $990\ \text{cm}^{-1}$ in the spectrum of 1-octene signify the presence of a vinyl group. This is confirmed by a sharp band at $1\ 640\ \text{cm}^{-1}$ due to stretching of the alkene double bond. The spectrum of *o*-xylene is more complex and includes skeletal vibrations of the ring and C—H in-plane bending between $1\ 000$ and $1\ 600\ \text{cm}^{-1}$, together with weak overtone bands in the region between $1\ 600$

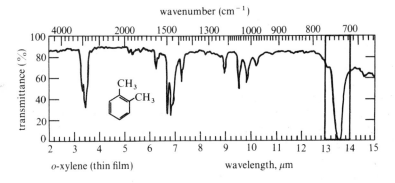

Figure 9.18.   Infrared spectrum of 1,2-dimethylbenzene

Table 9.8 Simplified infrared spectral correlation charts

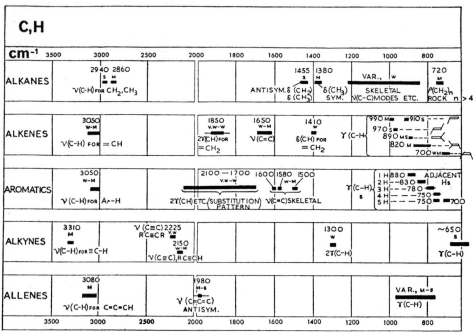

Table 9.8—*Continued*

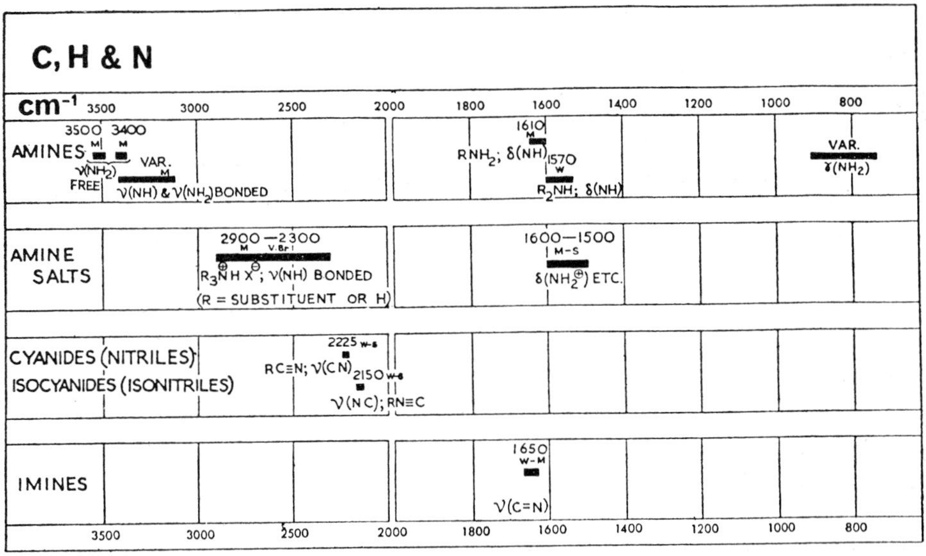

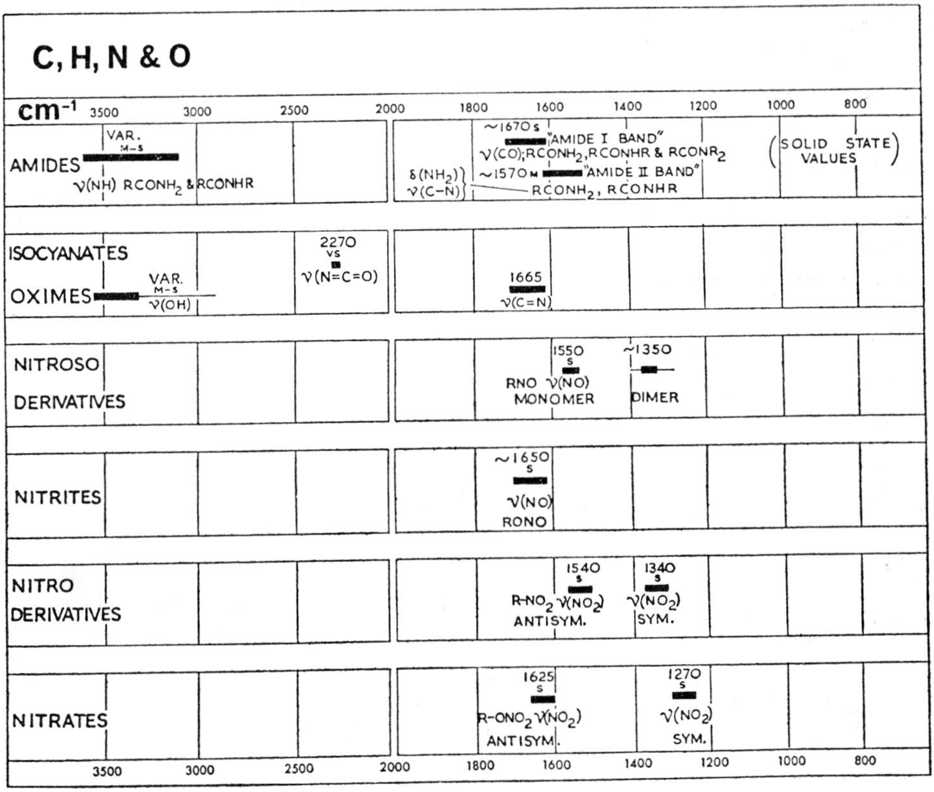

C, H & N

cm⁻¹									
3500	3000	2500	2000	1800	1600	1400	1200	1000	800

AMINES
3500 3400 M — ν(NH₂) FREE — VAR. M — ν(NH) & ν(NH₂) BONDED
1610 M — RNH₂; δ(NH) — 1570 w — R₂NH; δ(NH)
VAR. — γ(NH₂)

AMINE SALTS
2900 — 2300 M v.Br — R₃NH X⊖; ν(NH) BONDED (R = SUBSTITUENT OR H)
1600 — 1500 M-S — δ(NH₂⊕) ETC.

CYANIDES (NITRILES) ISOCYANIDES (ISONITRILES)
2225 w-s — RC≡N; ν(CN) — 2150 w-s — ν(NC); RN≡C

IMINES
1650 w-м — ν(C=N)

3500	3000	2500	2000	1800	1600	1400	1200	1000	800

C, H, N & O

cm⁻¹									
3500	3000	2500	2000	1800	1600	1400	1200	1000	800

AMIDES
VAR. M-S — ν(NH) RCONH₂ & RCONHR
δ(NH₂) ν(C–N) — ~1670 s "AMIDE I BAND" ν(CO); RCONH₂, RCONHR & RCONR₂
~1570 M "AMIDE II BAND" RCONH₂, RCONHR
(SOLID STATE VALUES)

ISOCYANATES
2270 vs — ν(N=C=O)

OXIMES
VAR. M-S — ν(OH)
1665 — ν(C=N)

NITROSO DERIVATIVES
1550 s — RNO ν(NO) MONOMER
~1350 — DIMER

NITRITES
~1650 s — ν(NO) RONO

NITRO DERIVATIVES
1540 s — R-NO₂ ν(NO₂) ANTISYM.
1340 s — ν(NO₂) SYM.

NITRATES
1625 s — R-ONO₂ ν(NO₂) ANTISYM.
1270 s — ν(NO₂) SYM.

3500	3000	2500	2000	1800	1600	1400	1200	1000	800

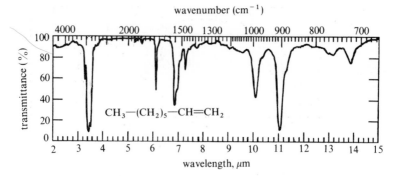

Figure 9.19.   Infrared spectrum of 1-octene

and $2\,000\ cm^{-1}$. The presence of 1,2 disubstitution is apparent from the single strong band at $740\ cm^{-1}$ due to C—H out-of-plane bending.

Compare, also, the spectrum of *n*-decane with those of *n*-propyl alcohol, figure 9.20 and *n*-undecanal, figure 9.21. The spectrum of the alcohol is characterized by vibrations due to the O—H and C—O groups. The O—H stretching band at $\sim 3\,350\ cm^{-1}$ is broadened by hydrogen bonding while the strong band at $\sim 1\,070\ cm^{-1}$ is due to a complex C—O stretching vibration. In the aldehyde a strong band assigned to stretching of the carbonyl group appears at $\sim 1\,730\ cm^{-1}$ and a sharp peak due to stretching of the aldehydic C—H ($\sim 2\,720\ cm^{-1}$) serves to differentiate it from other types of carbonyl compound.

Notes to table 9.8

The charts are arranged in order of increasing element content. Correlations given in one chart (e.g. those for CH_2 and CH_3) are not repeated in subsequent charts. The frequency limits within which the band of a particular grouping is usually found are indicated by the black strips and extensions of the range to include unusual examples are shown as thin lines, e.g. relative intensities are given in a very approximate fashion (see below). Both the position and the intensity of some absorptions are dependent on state, solvent, etc. and the actual frequency quoted is that most commonly observed.

Abbreviations

v = stretch (v is also used for quoting an absorption frequency, e.g. v_{max} 1 700 cm^{-1}); δ = in-plane bend; γ = out-of-plane bend; ρ = rock; ω = wag; τ = twist; sym = symmetrical; antisym = antisymmetrically (asymmetric); var. = variable (in frequency); br = broad; w = weak (in absorption); m = medium; s = strong; vw = very weak; vs = very strong; w–m = weak to medium (i.e. typical intensity range for that particular band)

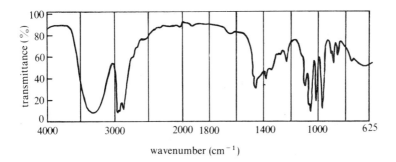

Figure 9.20.   Infrared spectrum of *n*-propyl alcohol

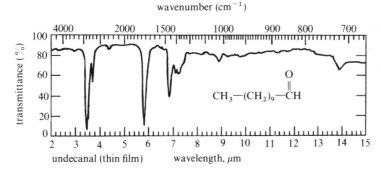

Figure 9.21.   Infrared spectrum of *n*-undecanal

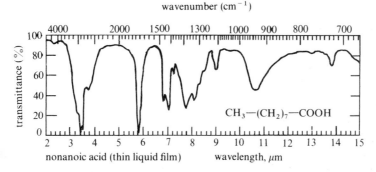

Figure 9.22.   Infrared spectrum of *n*-nonanoic acid

The spectrum of a carboxylic acid can be compared with that of an ester in figures 9.22 and 9.23. Such acids exist primarily as dimers, and the strongly hydrogen bonded structures give rise to a very broad O—H stretching band between 2 500–3 500 cm^{-1} and another at around 940 cm^{-1} due to O—H

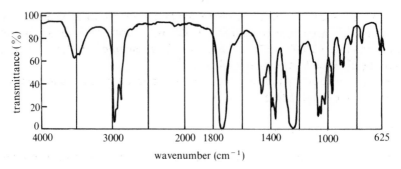

Figure 9.23.   Infrared spectrum of *n*-propyl acetate

out-of-plane bending. An intense carbonyl stretching band occurs near 1 710 cm^{-1}, similar to that of an aliphatic ketone because of hydrogen bonding, and the C—O stretch appears as a broad absorption between 1 200 and 1 300 cm^{-1}. The ester can be distinguished by the absence of O—H bands around 3 000 and 940 cm^{-1} and the shift of the carbonyl band to ~1 740 cm^{-1} caused by replacement of the acidic proton with an alkyl group. (N.B. carbonyl absorption in a monomer acid occurs at ~1 760 cm^{-1}.) Furthermore the C—O stretching band between 1 200 and 1 300 cm^{-1} is considerably intensified. It should be noted that in the spectra of acids the C—H stretching band is frequently observed as a sharp peak above the broad O—H band.

The final spectrum, figure 9.24, is that of *p*-nitroaniline. Although complex it is immediately recognizable as being aromatic (C—H above 3 000 cm^{-1}, skeletal and overtone bands between 1 400 and 2 000 cm^{-1} and a C—H out-of-plane bending vibrations at 835 cm^{-1} indicating *p*-substitution). Asymmetric and symmetric N=O stretching bands are very prominent at

p-nitroaniline $O_2NC_6H_4NH_2$
m.w. 138.13, m.p. 148.5–149.5°

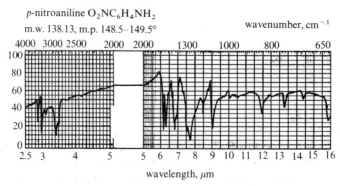

Figure 9.24.   Infrared spectrum of *p*-nitroaniline

1 475 and 1 310 cm^{-1}, whilst evidence for an amine group is given by the doublet N—H stretching band near 3 400 cm^{-1} and the in-plane N—H bending band near 1 600 cm^{-1}. A strong C—N stretching vibration overlaps the symmetric N=O band near 1 300 cm^{-1}.

Some further examples of infrared spectra are given in the section on the combined use of spectral data (p. 438).

QUANTITATIVE ANALYSIS

Absorption in the infrared region of the spectrum can be expected to obey the Beer–Lambert law (p. 355) for bands which are well resolved and in the absence of chemical effects such as hydrogen bonding. It is important that the slit width of the instrument is kept constant for a series of measurements as band intensities are highly sensitive to changes. This is because slit widths used in infrared spectrometry are of the same order of magnitude as bandwidths. A commonly used method for establishing the blank or background absorbance is to record the spectrum in the vicinity of the band of interest and to construct a *baseline*, figure 9.25, for all standards and samples. The baseline absorbance in measured at the point where a perpendicular line from the peak maximum intersects the baseline and this absorbance is subtracted from the peak absorbance in all cases.

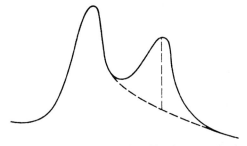

Figure 9.25.   Construction of a baseline for quantitative analysis

Most cells used in infrared spectrometry have sodium chloride windows and the path length is likely to vary with use because of corrosion. For quantitative work, therefore, the same cell should be used for samples and standards. In general, quantitative analysis in the infrared region of the spectrum is not practised as widely as in the ultraviolet and visible regions, partly because of the additional care necessary to obtain reliable results and partly because the technique is generally considered to be less sensitive and less precise; a precision of 3 to 8% can be expected. However, the computer-aided analysis of solids by reflectance measurements, particularly in the near infrared region (NIR), is of growing importance. Although the OH, CH and

NH overtones that occur in this region are weak (small molar absorptivities), radiation sources are more powerful, detectors more sensitive and the radiation penetrates more deeply into the sample surface than in the mid-infrared region giving a more reliable measure of bulk composition. Quantitative calculations involve chemometrics (p. 33) in the form of computer controlled pattern recognition. Sets of equations generated from absorption measurements on many standards at multiple wavelengths throughout the NIR region are used in comparisons with corresponding data from the samples. Unfortunately, this approach is susceptible to the unexpected presence of a major impurity.

SAMPLING PROCEDURES

Solid samples can be handled in a number of ways. A *mull*, or paste is made by grinding a few milligrams of sample in an agate mortar with a drop of a liquid paraffin (e.g. Nujol) or hexachlorobutadiene. The paste is formed into a thin film by compressing it between sodium chloride or potassium bromide plates which are mounted in a suitable sample holder. A complete spectrum of the sample can be obtained from separate mulls in Nujol and hexachlorobutadiene. The C—H bands in the former obscure the sample bands in that region but these can be observed in the spectrum of the chlorinated butadiene mull. An alternative procedure is to grind a few milligrams of sample with about 200 mg of dried potassium bromide and to compress the mixture into a thin, homogeneous disc using a hydraulic press. Liquid samples can be examined *neat* as a thin film between sodium chloride or potassium bromide plates or as a solution in a cell of path-length 0.1 to 10 mm. Three solvents in widespread use and which are largely transparent in the infrared region, are carbon tetrachloride, chloroform and carbon disulphide. Spectra of the surface of solids and liquid samples can also be obtained either by attenuated total reflectance (ATR) or by diffuse reflectance. The former involves placing the sample in contact with the sides of an optically flat thin prismatic crystal of thallium bromide/iodide (KRS-5) through which radiation from the source is passed. By directing the beam into one end of the crystal at the correct angle of incidence it can be made to undergo multiple internal reflections before emerging from the other end and passing through the spectrometer to the detector. At each reflection, the beam penetrates the surface of the sample to a depth of a few micrometres. The intensity of the beam is attenuated according to the absorption characteristics of the sample so enabling an absorption spectrum to be recorded. Diffuse reflectance spectra are obtained using a specially designed curved or plain mirror system to collect the radiation diffusely reflected from the surface of a sample over a wide solid angle and focusing it onto the detector. It is particularly suitable for powdered solid samples, which are simply placed in a small cup, and for liquids or gases adsorbed onto the surface of a non-absorbing powdered substrate such as alkali halide. As little as a few micrograms of material can give an acceptable spectrum.

Infrared microscopy combines an optical microscope with an FT-ir spectrometer enabling pico- to femtogram (10^{-12} to 10^{-15} g) quantities of substances to be characterized or very small areas of larger samples to be analysed. Beam-condensing optics focus the radiation onto an area of the sample identified using the optical microscope and either reflectance or transmittance spectra can be recorded. The highly-sensitive MCT detector (p. 281) is normally used as its size can be matched to that of the radiation beam to maximize its response.

Gaseous samples require long path length cells to produce absorption bands of reasonable intensity; up to several metres of optical path are obtainable from cells incorporating mirrors which produce multiple reflections. For gc-ir light pipes provide the best sensitivity (p. 111).

APPLICATIONS OF INFRARED SPECTROMETRY

The unique appearance of an infrared spectrum has resulted in the extensive use of infrared spectrometry to characterize such materials as natural products, polymers, detergents, lubricants, fats and resins. It is of particular value to the petroleum and polymer industries, to drug manufacturers and to producers of organic chemicals. Quantitative applications include the quality control of additives in fuel and lubricant blends and to assess the extent of chemical changes in various products due to ageing and use. The food and grain industries use quantitative reflectance measurements in the NIR for the determination of water, fats, proteins and sugars. Non-dispersive infrared analysers are used to monitor gas streams in industrial processes and atmospheric pollution. The instruments used are generally portable and robust, consisting only of a radiation source, reference and sample cells and a detector filled with the gas which is to be monitored.

9.3 Nuclear Magnetic Resonance Spectrometry (nmr)

SUMMARY

Principles

Absorption of electromagnetic radiation in the radio-frequency region of the spectrum resulting in changes in the orientation of spinning nuclei in a magnetic field.

Instrumentation

Powerful and highly homogeneous electromagnet, radio-frequency signal generator and detector circuit, electronic integrator, glass sample tubes.

Applications

Identification and structural analysis of organic materials and study of kinetic effects, mainly from proton and carbon-13 spectra. Useful for quantitative analysis but not widely applied.

Disadvantages

Expensive and complex instrumentation. Moderate to poor sensitivity with continuous wave (scanning) instruments, but greatly enhanced by Fourier transform instruments. Limited range of solvents for studying proton spectra unless they are deuterated.

Absorption of radiation in the radio-frequency, rf, region of the electromagnetic spectrum can be observed for those nuclei which are considered to spin about their own axes. The energy changes are associated with the orientation of the nuclear axis in space relative to an external applied magnetic field and are of the order of $0.1 \, \mathrm{J \, mol^{-1}}$, 10 to 600 MHz (50 cm to 30 m or 3×10^{-4} to $2 \times 10^{-2} \, \mathrm{cm^{-1}}$). This is considerably smaller than the energy changes associated with vibrational and electronic transitions (pp. 361, 377).

All nuclei are assigned a *spin quantum number I* which may be zero, half-integral or integral; only those with a non-zero value can give rise to an nmr spectrum. As the nucleus carries a charge, spinning about its own axis produces

Table 9.9   Nuclear spin quantum numbers and magnetic properties of selected nuclei

NUCLEUS	NUCLEAR SPIN QUANTUM NUMBER I	MAGNETIC MOMENT, μ (AMPERE SQUARE METRE $\times 10^{27}$)	RESONANCE FREQUENCY IN MHz AT 1.409 2 TESLA	RELATIVE SENSITIVITY AT THE NATURAL ISOTOPIC ABUNDANCE
^{1}H	$\frac{1}{2}$	14.09	60.000	1.00
^{2}H	1	4.34	9.211	1.5×10^{-4}
^{12}C	0	—	—	—
^{13}C	$\frac{1}{2}$	3.53	15.085	1.8×10^{-4}
^{14}N	1	2.02	4.335	1×10^{-3}
^{16}O	0	—	—	—
^{17}O	5/2	−9.55	8.134	1×10^{-5}
^{19}F	$\frac{1}{2}$	13.28	56.446	0.834
^{31}P	$\frac{1}{2}$	5.71	24.288	0.066

a *magnetic moment* or *dipole* μ along the axis. This is analogous to the magnetic field associated with a current flowing in a loop of wire. The nucleus also possesses *angular momentum* **I** and for each isotope the relative values of μ and **I** determine the frequency at which energy can be absorbed. In addition, the sensitivity of the technique for a particular nucleus is determined by the value of μ. Table 9.9 lists the spin quantum numbers and magnetic properties of a number of nuclei. It should be noted that ^1H has the highest relative sensitivity and that ^{12}C and ^{16}O, having spin quantum numbers of zero, are inactive. For these reasons *proton magnetic resonance*, pmr, has become one of the most useful techniques in the identification and structural analysis of organic compounds. The only serious handicap is an inherent lack of absolute sensitivity because of the very small energy differences between nuclear spin states which leads to almost equal thermal populations of the energy levels at room temperature. The Maxwell–Boltzmann equation (p. 272) can be used to show that there is an excess of only a few nuclei per million in the lower level. Only this small excess of nuclei are in fact detected by nuclear magnetic resonance.

INSTRUMENTATION

A schematic diagram of a typical continuous wave (CW) or scanning nmr spectrometer is shown in figure 9.26. The magnetic field is provided by a permanent magnet or an electromagnet with a field strength of 1 to 14.1 tesla (10^4 to 14.1×10^4 Gauss). The homogeneity of the field between the poles of the magnet should be at least 3 in 10^9 to ensure narrow absorption bands and good resolution. Samples are contained in long

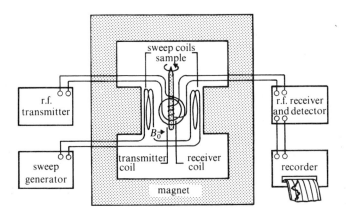

Figure 9.26. Schematic diagram of a continuous wave nmr spectrometer

narrow glass tubes (usually 5 mm o.d.) which are placed in a sample probe located between the pole pieces of the magnet. The apparent homogeneity of the field is improved by rapid spinning of the sample tube by means of an air driven turbine. Around the pole pieces there are sweep coils through which a variable current is passed which enables a spectrum to be 'scanned' by varying the magnetic field over a small range. In this case the rf transmitter operates at a fixed frequency, usually 60 to 200 MHz. Alternatively the field may be fixed and the spectrum scanned by varying the operating frequency of the transmitter. By employing a superconducting magnet, field strengths approaching 10 tesla ($\sim 10^5$ Gauss) and operating frequencies up to 500 MHz can be utilized. The magnetic field and the axes of the transmitter and receiver coils are all mutually perpendicular. This arrangement ensures that radiation from the transmitter can be detected in the receiver coil only when the sample absorbs energy. Some instruments use a single coil, the absorption of energy being detected by an out-of-balance signal in a bridge circuit which incorporates the coil and a capacitor. The spectrum may be recorded on a chart or viewed on an oscilloscope screen by repetitive scanning.

The effect of long-term field instability or drift is counteracted by means of an internal or external field-frequency locking system. This consists of a secondary rf circuit which generates a signal at the resonance frequency of a selected reference nucleus (usually ^1H in tetramethylsilane (TMS) or ^2D in a deuterated compound). The reference nucleus is continuously irradiated and the corresponding absorbance signal monitored. If this diminishes due to field drift, a voltage is generated in a servo-loop circuit which in turn adjusts the current flowing in coils of the electromagnet in such a way as to re-establish the original field. An external locking system employs a second probe, located close to the sample probe, and which houses the reference compound tube. With an internal locking system, the reference compound is added directly to the sample or sealed in a thin tube which is put into the sample tube. Adequate sensitivity can usually be obtained only from solutions of 0.1 M or more, although by computer processing of the data accumulated by multiple scans spectra from 0.01 M solutions or less can be recorded.

A considerable improvement in speed and sensitivity can be achieved with a *pulsed Fourier transform* (FT) spectrometer. Here the sample is subjected to a series of short duration high intensity rf pulses (1–100 μs) covering a wide frequency range (500 – 1 500 Hz). Following each pulse, excited nuclei return to the ground state producing a decaying *emission* signal which is monitored by the receiver coil of a double coil spectrometer and computer-processed to produce a conventional nmr spectrum (p. 413). The time interval between pulses is of the order of a second so that data can be accumulated much more rapidly than with a CW instrument and sensitivity enhancements of one to three orders of magnitude can be achieved quickly by signal averaging.

THE NMR PROCESS

The principle of nmr can be explained in quantum mechanical terms. The angular momentum of a spinning nucleus, quantized both in magnitude and direction, is given by the equation

$$\mathbf{I} = [I(I + 1)]^{\frac{1}{2}} \frac{h}{2\pi} \tag{9.19}$$

where h is Planck's constant and $h/2\pi$ is defined as an angular momentum unit. In the presence of an applied magnetic field, the momentum vector can assume only those orientations in space which result in its component in the direction of the field being an integral or half-integral number of angular momentum units, i.e.

$$\mathbf{I}_z = m_I \frac{h}{2\pi} \tag{9.20}$$

where $\mathbf{I}_z$ represents the magnitude of the component of the angular momentum in the direction of the field, usually designated the z-direction, and $m_I = 0$, $\pm \frac{1}{2}, \pm 1, \pm 3/2. \ldots$ The total number of values for m_I, the magnetic quantum number, is $2I + 1$, as clearly the value of $\mathbf{I}_z$ cannot exceed the value of $\mathbf{I}$ given in equation (9.19).

For a given nucleus, the ratio of the magnetic moment to angular momentum is a constant, i.e.

$$\frac{\mu}{\mathbf{I}} = \gamma$$

where γ is known as the magnetogyric (gyromagnetic) ratio. If μ_z is the component of the dipole in the z-direction

$$\frac{\mu_z}{\mathbf{I}_z} = \gamma$$

and from equation (9.20)

$$\mu_z = m_I \gamma \frac{h}{2\pi} \tag{9.21}$$

The energy of interaction, between a nucleus and the magnetic field it experiences B is given by

$$E = -\mu_z B = -m_I \gamma \frac{h}{2\pi} B \tag{9.22}$$

For a proton or carbon-13 nucleus I is $\frac{1}{2}$, hence m_I can be only $\pm \frac{1}{2}$ and two energy levels, or spin states, are produced, figure 9.27:

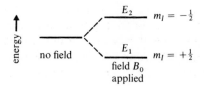

Figure 9.27.   Nuclear spin energy levels for a proton or carbon-13, $I = \frac{1}{2}$

$$E_1 = -\frac{1}{2}\frac{\gamma h}{2\pi} B$$

and

$$E_2 = +\frac{1}{2}\frac{\gamma h}{2\pi} B$$

For energy to be absorbed, radiation of a frequency equivalent to $E_2 - E_1$ must be supplied. Thus, using the Planck relation, the condition for absorption is defined as

$$h\nu = E_2 - E_1 = \frac{\gamma h}{2\pi} B \tag{9.23}$$

or

$$\nu = \frac{\gamma}{2\pi} B \tag{9.24}$$

Equation (9.24) shows that the absorption frequency is directly proportional to the strength of the applied magnetic field at the nucleus.

In practice, using a CW instrument the absorption of energy may be detected by subjecting the sample to radiation of varying frequency at a fixed value of the applied field or vice versa until the conditions required by equation (9.24) are met. At this point, the system is said to be in resonance, both upward and downward transitions occur, and a net absorption of energy is observed because of the small excess of nuclei in the lower level.

The width of an absorption band in nmr spectrometry is determined by the rate at which nuclei absorbing energy return to the lower level. This *relaxation* process is complex, involving dissipation of the excess energy throughout the whole sample, and giving a decaying *emission* signal which FT spectrometers (p. 412) can detect and computer-process. The rate of relaxation in liquid samples is such that absorption bands are narrow ($\sim 0.5\,\text{Hz}$) thus enabling well-resolved spectra to be readily obtained. Spectra of solid samples, however, consist of broad lines and are less easily interpreted.

The amount of structural information given by an nmr spectrum is greatly enhanced by two factors. Firstly, the exact position of resonance is determined by the *chemical environment* of the nucleus. Both the pmr and carbon-13 spectra of an organic compound may therefore show several absorption bands, each corresponding to a particular nucleus or group of nuclei. Secondly, a given band may be split into several peaks as a result of interactions between neighbouring nuclei. The two effects give rise to what are termed the *chemical shift* and *spin–spin coupling* or *splitting*. A third useful feature of the spectrum is that the integrated area of an absorption peak is directly proportional to the number of nuclei responsible for the signal. This facilitates structural correlations and provides a means of quantitative analysis. These factors will now be discussed in the context of pmr.

CHEMICAL SHIFT

According to equation (9.24) at a given frequency all protons will absorb energy at the same value of the magnetic field B. However, the field experienced by a particular nucleus differs in magnitude from that of the applied field because of *shielding* effects by neighbouring electrons. It is because of varying degrees of shielding that protons in different chemical environments absorb at different values of the *applied* field. Differences between such absorptions are referred to as *chemical shifts*.

All protons are subjected to *diamagnetic shielding* by the electrons bonding them to another nucleus. Under the influence of the applied field, these electrons circulate around the field direction and in so doing they create a small localized magnetic field in opposition to the applied field. The magnitude of the opposing field is directly related to the electron density around the proton which in turn is determined by the electronegativities of neighbouring nuclei. The more electronegative the atom to which a proton is bonded, the less the shielding and the smaller is the applied field required to achieve the resonance condition. For example, figure 9.28 shows the relative absorption positions for protons attached to C, Si, N and O. It is conventional to display the spectrum with increasing applied field from left to right. Quantitatively, chemical shifts are measured in frequency (Hz) relative to a standard, the universally accepted reference compound being tetramethylsilane, $(CH_3)_4Si$ (TMS). The desirable properties of this compound are that it has twelve highly shielded protons in identical chemical and magnetic environments

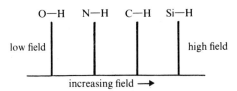

Figure 9.28.   Chemical shift positions for protons attached to different atoms

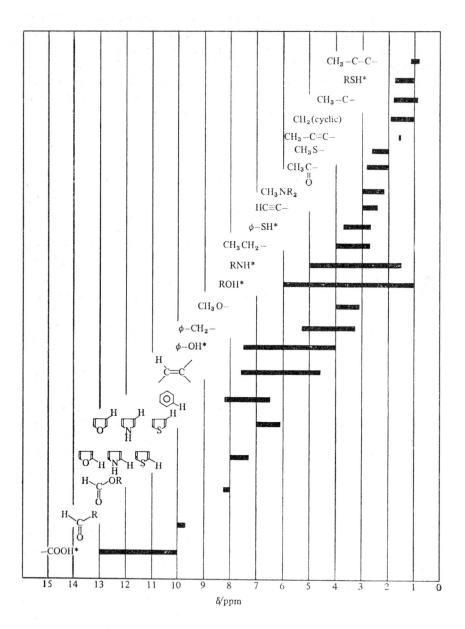

Figure 9.29.   Approximate chemical shifts for protons in various functional groups relative to TMS.

* Chemical shift is highly dependent on hydrogen bonding and exchange effects thereby producing a single sharp resonance peak at a higher field than most other organic protons. Furthermore it is chemically inert, soluble in organic solvents and boils at 27 °C. For aqueous samples, sodium 2,2'-dimethyl-2-silapentane-5-sulphonate, $(CH_3)_3SiCH_2CH_2CH_2SO_3Na$ (DSS) can be used. In practice chemical shifts are expressed in dimensionless units δ by dividing them by the operating frequency of the instrument, thus facilitating comparisons between spectra run on different instruments

$$\delta = \frac{\nu(\text{sample}) - \nu(\text{TMS})}{\nu(\text{instrument})} \times 10^6 \text{ ppm} \qquad (9.25)$$

By including a factor of 10^6, δ values fall within the range 0 to 15 for most organic protons, the values then being expressed in parts per million (ppm). An alternative system, not now officially recognized, is the τ scale in which the TMS peak is assigned the value 10, and τ is defined as $10 - \delta$.

The approximate chemical shift positions for organic protons in different chemical environments are shown in figure 9.29. It will be noted that olefinic protons absorb at a lower field ($> \delta 4.5$) than acetylenic protons ($\sim \delta 2.7$), although the latter are more acidic and therefore experience less diamagnetic shielding. This, and other anomalous effects in unsaturated molecules can be explained in terms of shielding and deshielding zones in space caused by the circulation of π-electrons. Specific examples of the effect, known as *diamagnetic anisotropy* are illustrated in figure 9.30.

For acetylene, which is a linear molecule, the most strongly induced circulation of π-electrons occurs when the molecular axis is parallel to the applied field. The resulting field due to electron circulation can be represented by a cone within which there is a net shielding and outside of which there is a net deshielding, figure 9.30(a). As the acetylenic protons lie within the shielded zones, they absorb at a relatively high value of the applied field. For olefins and aldehydes, the greatest effect is when the double bond is orientated perpendicular to the applied field. For example, all the protons of ethylene and the aldehydic proton of acetaldehyde lie in deshielded zones, figure 9.30(b) and (c) and hence absorbed at a relatively low value of the applied field. The most pronounced effect occurs in aromatic compounds ($\delta 6$–9) where π-electrons can circulate around the ring producing a so-called *ring current*. Aromatic protons lie in deshielded zones (figure 9.30(d)) and absorb at comparatively low values of the applied field.

Lastly, the effect of hydrogen-bonding should be noted. Protons which are hydrogen-bonded are deshielded relative to the non-bonded situation and their chemical shifts can vary over a wide range. The effect is observed in alcohols, phenols, amines and carboxylic acids and, as in the case of infrared spectra it is temperature, concentration and solvent dependent (p. 382).

SPIN–SPIN COUPLING

Absorption bands arising from adjacent protons are split into *multiplet* peaks
by a mutual interaction of the spins. The effect is due to small variations in the

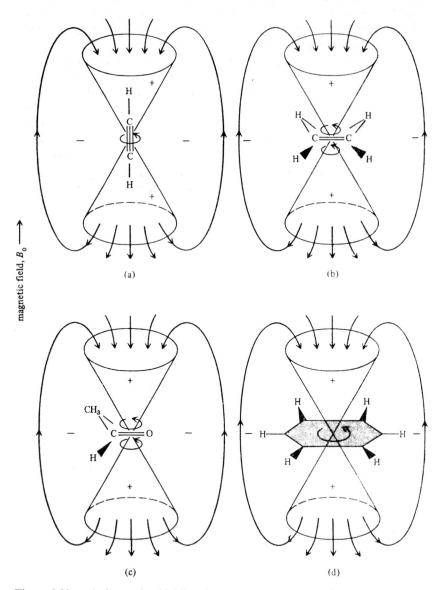

Figure 9.30.   Anisotropic shielding in unsaturated groups. (Shielding zones +;
deshielding zones −). (a) Acetylene. (b) Ethylene. (c) Acetaldehyde. (d) Benzene

effective field experienced by a proton when neighbouring nuclei can occupy two or more energy levels or spin states. It is transmitted through the intervening bonds by a tendency for electron and nuclear spins to be paired.

Consider the case of two single (methine) protons H_A and H_X attached to adjacent carbon atoms and with quite different chemical shifts, figure 9.31(a). The field experienced by H_A is increased or decreased slightly by the two allowed spin states of H_X, designated ↑ and ↓, and which in the gross sample are virtually equally populated. This results in the absorption band for H_A splitting into a *doublet* whose peak intensities are in the ratio 1:1. The effect is mutual in that the two almost equally populated spin states of H_A cause the H_X absorption to split into an identical doublet. The spacing between the component peaks of each doublet is the same and is known as the *coupling constant* J_{AX} usually measured in Hz. It should be noted that J is independent of the applied field whereas the chemical shift difference between H_A and H_X (Δv), measured between the mid-points of the doublets, is directly proportional to the strength of the applied field. A shorthand notation for the appearance

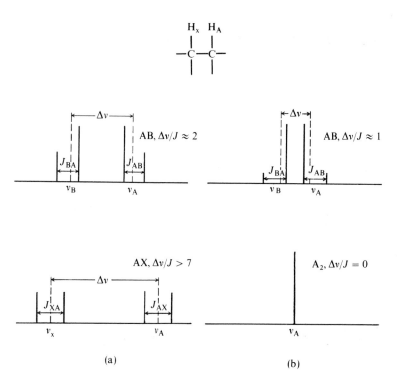

Figure 9.31.   Coupling between adjacent single protons. (a) First order coupling, AX. (b) Second order coupling, AB → A₂

of this spectrum is 'AX' and it is classed as a *first-order* spectrum. This treatment is readily extended to adjacent groups with two or more protons (i.e. methylene and methyl groups). The number of peaks in a multiplet is determined by the number of protons in the *adjacent* group n and is simply $n + 1$. Furthermore, protons on the same carbon atom do not 'split each other' unless free rotation is hindered or there is an adjacent centre of asymmetry. Such protons, and those which are adjacent but have the same chemical shift, are said to be *equivalent*. (N.B. in some instances this may not be the case, but further discussion is beyond the scope of this book.) The intensity ratios within a multiplet follow the coefficients of the binomial expansion of $(a + b)^n$, or Pascal's triangle, and are a result of the statistical weighting given to the various possible combinations of the two spin states for the protons in the adjacent group. The total area of a multiplet is proportional to the number of protons producing the absorption, thus the *triplet* and *quartet* arising from an ethyl group have areas in the ratio 3:2. Figure 9.32 summarizes this information for some common molecular structures.

A first order spectrum is observed only if the chemical shift difference between the coupled groups is large compared to the coupling constant, i.e. $\Delta v / J > \sim 7$. As this ratio becomes smaller, the inner peaks of multiplets grow at the expense of outer peaks and additional or *second-order*, splitting may

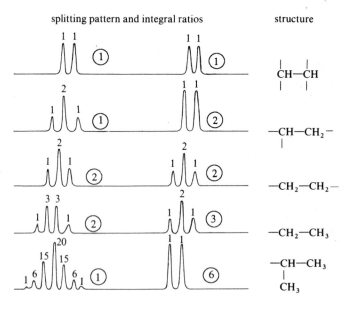

Figure 9.32.   First order splitting patterns for some common molecular structures. (Numbers in circles represent relative total areas of the multiplets)

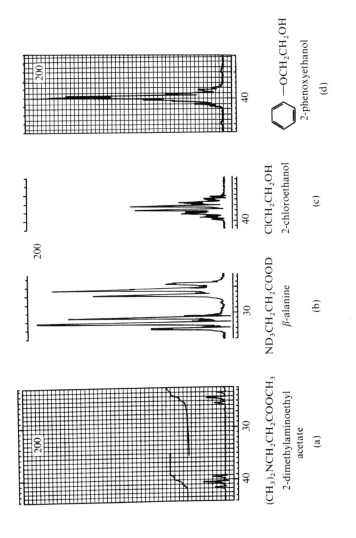

Figure 9.33.   Progressive distortions as $A_2X_2 \rightarrow A_2B_2$ in the configuration ZCH_2CH_2Y, 60 MHz

occur. Chemical shift differences Δv are now measured between the 'centre of gravity' of the multiplets. In the limiting case, where $\Delta v = 0$, the protons are equivalent and the multiplet collapses to a single peak or *singlet*, figure 9.31(b). Progressive distortion of this type is exemplified by compounds having adjacent methylene groups, figure 9.33. The spectrum is described as A_2X_2 or A_2B_2 according to the value of $\Delta v/J$ and the degree of distortion. A significant feature, which is an aid to interpretation, is the symmetrical appearance, even in the presence of considerable second-order splitting. Slight distortion of multiplet intensities from first-order ('roofing' or 'leaning') is nearly always observed but this can be useful in relating coupled groups in complex spectra.

With saturated compounds, the effect of spin–spin coupling is rarely observed for protons more than three bonds apart. Aromatic, olefinic and other unsaturated compounds, however, show more complex splitting patterns because of longer range coupling. The ring protons of substituted aromatic compounds are mutually coupled and mostly give second-order spectra which are difficult to interpret rigorously. Those with similar or identical ortho or para substituents often have a symmetrical appearance similar to an A_2B_2 system. Vinyl groups, mono substituted furans and related compounds with three adjacent single protons give spectra characterized by three groups of four peaks if the chemical shift differences are sufficiently large. Described as an AMX spectrum, the twelve-line pattern results from three unequal coupling constants J_{AM}, J_{MX} and J_{AX}, and is best understood with the aid of a *splitting diagram*, figure 9.34. This shows the individual coupling of each

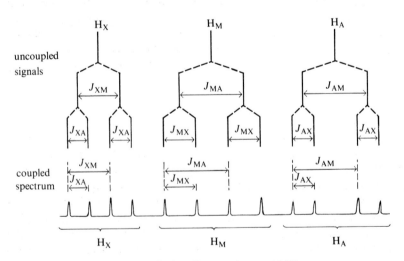

Figure 9.34.   Splitting diagram for an AMX system

proton to the other two and how the three coupling constants can be measured from the spectrum. As the chemical shifts of the three protons become closer, the pattern distorts to ABX and finally, and in a complex manner, to ABC in which the groups overlap. A substantial degree of distortion and second-order splitting can occur before the pattern is unrecognizable.

Chemical Shift Reagents

In the spectra of some compounds the resonances from several groups of protons with similar chemical shifts may overlap, thereby rendering it difficult if not impossible to interpret them satisfactorily. Simpler spectra can be obtained by using an instrument operating at a higher frequency (field strength) which results in increased chemical shift differences without affecting the coupling constants (see below). However, a more powerful spectrometer may not be available and the use of *chemical shift reagents* can provide a very much cheaper alternative. These reagents are co-ordinately unsaturated lanthanide β-diketone complexes, in particular tris(dipivalo-methanato)europium(III) and tris(6,6,7,7,8,8,8-heptafluoro-2,2-dimethyl-3,5-octanedionato)europium(III) (Eu(dpm)$_3$ and Eu(fod)$_3$ respectively). When added to solutions of molecules containing a heteroatom with an unshared electron pair, e.g. aldehydes, ketones, alcohols, esters, ethers, amines etc., further co-ordination by these molecules to the europium complex occurs. Resulting changes in electron densities around protons in the co-ordinating molecules lead to downfield or sometimes upfield shifts of their resonances. The shift is greatest for the group nearest to the co-ordinating heteroatom and becomes progressively less with distance from it. The spectrum is therefore spread out, enabling the originally overlapping signals to be observed clearly. An example of this technique is shown in figure 9.35 for a sample of *n*-hexanol. Although the shift reagent, Eu(dpm)$_3$, causes a degree of line broadening, the expected multiplicity of each group can easily be discerned.

Exchangeable Protons

Protons attached to oxygen, nitrogen or sulphur (acids, alcohols, phenols, amines, etc) are labile in solution and are essentially decoupled from neighbouring protons if the rate of exchange is fast. The resulting spectrum is simplified because the exchanging proton senses only an averaged environment and appears as a sharp singlet. At intermediate and slow rates of exchange, the absorption may be broad or resolved into the component peaks of a multiplet. Traces of acid or alkali in the sample catalyse the exchange process while treatment with D_2O removes the resonance altogether. The latter is often used as a diagnostic test for an exchangeable proton.

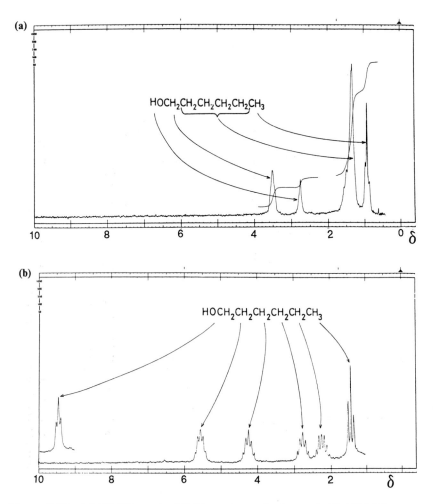

Figure 9.35. (a) 60 MHz proton NMR spectrum of *n*-hexanol. (b) 100 MHz proton
NMR spectrum of *n*-hexanol with 0.29 mole equivalents of Eū(dpm)$_3$ added

CARBON-13 NMR

Carbon-13 spectra are inherently less complex than proton spectra for two
reasons, viz.: (a) chemical shifts between ^{13}C nuclei in different chemical
environments can differ by as much as 200 ppm whereas proton shift differences are seldom more than 10 ppm; (b) as the natural isotopic abundance
of ^{13}C is only 1.1%, coupling between ^{13}C nuclei themselves is not observed
(but see 2-D INADEQUATE spectra, p. 418).

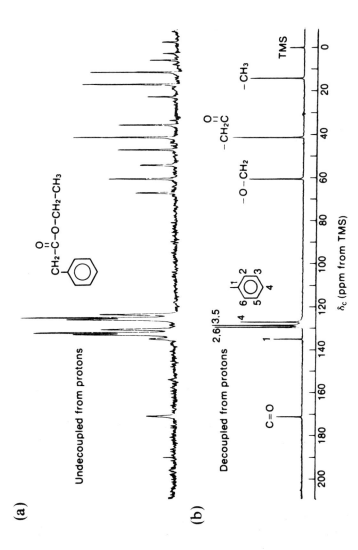

Figure 9.36. ^{13}C nmr spectra of ethyl phenylacetate (from J. A. Moore and D. L. Dalrymple, *Experimental Methods in Organic Chemistry*, W. B. Saunders Co., 1976)

Coupling between ^{13}C nuclei and protons, however, does occur but in practice this can be eliminated by *noise decoupling* which leads to even simpler spectra. Noise decoupling consists of irradiating the sample over a wide frequency range covering all the proton resonances in the sample (white noise irradiation) whilst observing the ^{13}C signals. In some cases ^{13}C-proton coupling can provide useful structural information and this can be done using a partial decoupling technique known as *off-resonance decoupling* which allows only the coupling between ^{13}C nuclei and protons directly bonded to them to be observed.

Examples of coupled and noise decoupled spectra for ethyl phenylacetate are shown in figure 9.36 and the approximate chemical shift positions for ^{13}C nuclei in different chemical environments in figure 9.37. The shift positions of ^{13}C resonances are determined by electronegativity and anisotropy effects as are proton resonances but in a more complex manner. In figure 9.36(a) the methyl quartet ($\delta_c 14$) is due to coupling to 3 equivalent protons and the triplets at $\delta_c 41$ and $\delta_c 60$ are due to the 2 equivalent protons in each methylene group. The ^{13}C attached to the oxygen ($\delta_c 60$) is deshielded more than that attached to the carbonyl ($\delta_c 41$) whilst the ^{13}C of the carbonyl group itself appears at $\delta_c 171$. Longer range ^{13}C-proton coupling (^{13}C–C–H) is just

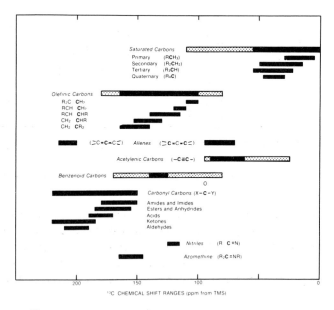

Figure 9.37. ^{13}C chemical shift ranges (ppm from TMS). Extended ranges sometimes occur when polar substituents are attached to the carbons. These extended ranges are indicated by the lightly shaded areas. (From J. A. Moore and D. L. Dalrymple, *Experimental Methods in Organic Chemistry*, W. B. Saunders Co., 1976)

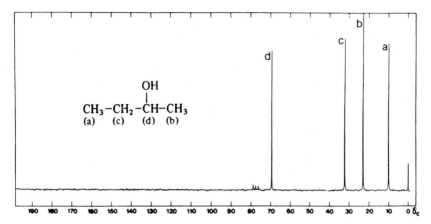

Figure 9.38. ^{13}C spectrum of butan-2-ol (noise decoupled)

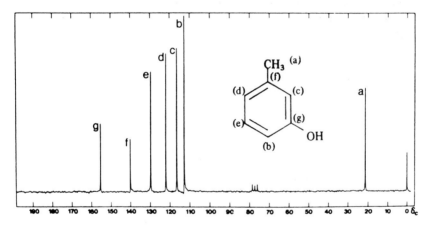

Figure 9.39. ^{13}C spectrum of 3-hydroxymethylbenzene (noise decoupled)

discernible by broadening of the multiplets and the carbonyl singlet. Coupling constants for protons directly bonded to ^{13}C nuclei are of the order of 100–200 Hz whilst longer range coupling constants are typically between 0 and 2 Hz. For more complex molecules, the noise decoupled spectrum is the more readily interpreted. Further examples are shown in figures 9.38 and 9.39.

PULSED FOURIER TRANSFORM NMR (FT-NMR)

Conventional scanning (CW) nmr spectrometers take several minutes to scan a full-range spectrum which is said to be in the *frequency domain* (absorption vs. frequency). With a pulsed or Fourier transform (FT) spectrometer the sample is subjected to a series of single (several microseconds duration), high-power rf pulses of wide frequency-range which excite all the nuclei of a particular element (isotope). Between each pulse, the excited nuclei emit radiation in the form of a rapidly decaying signal as they relax to re-establish the equilibrium population of spin states. This is known as a *free induction decay* signal, or *fid*, and it contains spectral information from all of the nuclei excited by the pulse, although it is not in the form of a CW spectrum. It is said to be in the *time domain* as the signal is monitored as a function of time. Time and frequency domain spectra are mathematically related by a set of Fourier transform equations. A dedicated computer can sample and digitize the fid and transform it into the corresponding frequency domain spectrum in a second or less, compared to the several minutes required to record a CW spectrum. This considerable saving in time can be used to enhance the sensitivity of NMR by repeating the pulse sequence many times and co-adding the fids to improve the signal-to-noise (S/N) ratio. For example, the accumulated data from one hundred pulses, which can be collected and transformed in about two minutes, will increase the S/N ratio by a factor of 10, or in general by the square root of the number of pulses or fids accumulated.

Some examples of fid signals and their corresponding frequency domain spectra are shown in figure 9.40. The fid signal related to a single resonance peak (figure 9.40(a)) is seen to consist of a decaying sinusoidal wave whose frequency corresponds to that of the resonance frequency. Two resonance peaks (figure 9.40(b)) arise from a more complex sinusoidal fid signal whilst the complete ^{13}C spectrum shown in figure 9.40(c) is derived from an fid signal which is far too complex to be understood by visual inspection and can be unravelled only by computer processing. The complexity of such fid signals is due to the superimposition of and mutual interference between the decay signals from all excited nuclei.

Pulsed FT nmr has facilitated the study of nuclei other than ^{1}H where the sensitivity obtainable from a CW instrument is totally inadequate. In particular, ^{13}C nmr, the sensitivity of which is nearly 10^{-4} less than that of the proton (table 9.9), is now a well-established technique that yields information on the skeletal structure of complex molecules. The pulsed technique also enables proton spectra to be obtained from samples as small as a few micrograms.

Multiple Pulse Techniques: 1-D and 2-D nmr
A variety of computer-controlled *pulse sequences* consisting of two or more pulses of appropriate length, frequency-range, power and phase, and separated by variable time intervals, has been developed, giving rise to families of

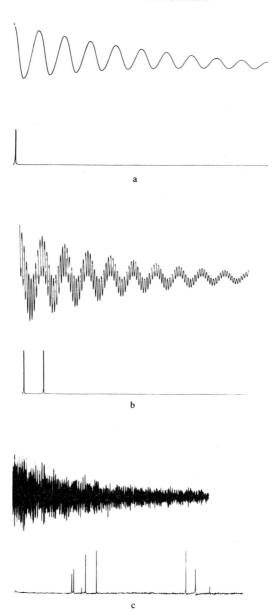

Figure 9.40.   (a) Single decaying sine wave and a plot of the single frequency it represents. (b) Two decaying sine waves and a plot of the two frequencies they represent. (c) Many co-added sine waves in a free induction decay of the ^{13}C nmr spectrum of 3-ethylpyridine and the frequency domain spectrum they represent

1-D (one-dimensional) and *2*-D (two-dimensional) techniques. These techniques provide additional or more easily interpreted data on coupled nuclei, facilitating the identification of signals from chemically different groups of nuclei and correlations between spectra from different elements in the same compound.

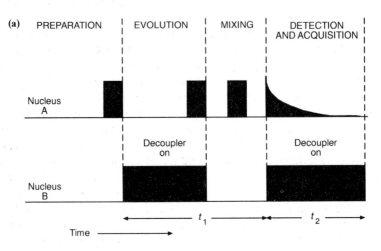

Figure 9.41(a).   Generalized multipulse sequence for 1-D or 2-D nmr

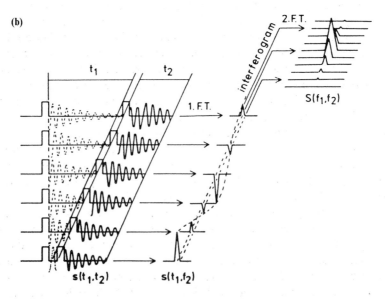

Figure 9.41(b).   Generation of a 2-D nmr spectrum $S(f_1, f_2)$

It is confusing that the term '1-D' is used to describe a conventional absorption vs. frequency (expressed in ppm) spectrum but which is in reality a two-dimensional representation. Likewise, the term '2-D' describes a spectrum where two frequency axes (F1 and F2) are shown at right angles and peak intensities are plotted perpendicular to them. Although this is actually three-dimensional, a 2-D *contour plot*, which is analogous to a topographical map is the simplest and most convenient way of displaying the spectral data. One frequency axis may represent chemical shift in Hz or ppm of one nucleus, usually ^1H or ^{13}C, while the other is either chemical shift of the same or another nucleus or a coupling constant scale in Hz. As in a map, the 'contour lines' depict the intensities or heights of the peaks (figure 9.44). A diagram of a typical pulse sequence is shown in figure 9.41(a). A *preparation period*, if required, is followed by an initial or magnetization rf pulse, then by one or more further pulses with controlled time intervals (*evolution and mixing period*, t_1) between them. The fid is acquired at the end of the sequence (*detection and acquisition period*, t_2) and this is followed by one or two Fourier transformations to give the final 1-D or 2-D frequency domain spectrum. The pulse sequence may be applied to protons and/or carbon-13 nuclei in the sample (or any other nuclei of interest) and spin-decoupling may be used simultaneously. For 2-D techniques, the sequence is repeated many times, the time interval during the evolution period, t_1, being increased by equal increments with each repetition. Fourier transformation of the fids produces a two-dimensional matrix of stored data. By subjecting sets of data points in corresponding positions in the matrix to further Fourier transformations, a 2-D frequency domain spectrum is generated. In effect, this is a Fourier transform of a Fourier transform. The sequence is shown in figure 9.41(b).

The most useful 1-D pulse sequence applied to ^{13}C nuclei is known as *Distortionless Enhancement by Polarization Transfer* (DEPT). A decoupled ^{13}C spectrum (all signals are singlets) and a set of three sub-spectra are generated by variations in the pulse sequence. Each sub-spectrum shows signals from carbons with only one particular proton multiplicity, i.e. methyls, methylenes or methines, which enables signals from each of these three to be identified in the complete spectrum. Quaternary carbons give no sub-spectrum but are easily identified in the main spectrum once the signals due to the methyls, methylenes and methines have been assigned. A set of DEPT spectra for ethyl benzoate is shown in figure 9.42.

Although many pulse sequences have been investigated for 2-D nmr, only a very limited number are of practical interest. One deterrent to their more widespread use is the time required and hence the cost of collecting the data. The time factor arises because the pulse sequences have to be repeated hundreds of times whilst varying the time interval during the evolution period, t_1, in order to provide sufficient data for the final 2-D contour plot. Three of the more valuable 2-D techniques are described below.

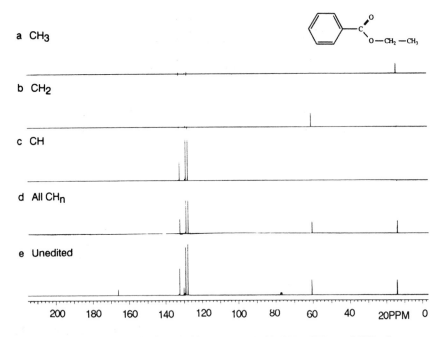

Figure 9.42. DEPT results for ethyl benzoate. (a)–(c): CH_3, CH_2, and CH subspectra. (d) Subspectrum of all protonated carbons. (e) Unedited spectrum. 16 fids were accumulated for each θ value

(i) *Correlated Spectroscopy* (COSY) or *Homonuclear Correlation Spectroscopy* (HOMCOR-2D) provides proton–proton coupling information (^{1}H–^{1}H *connectivities*). This information can also be obtained from a series of 1-D spin decoupled spectra but difficulties arise with strongly coupled protons, overlapping multiplets and in optimizing conditions.

A theoretical COSY spectrum for two coupled methine protons (AX system) is shown in figure 9.43, the chemical shifts of the two protons being δ_A and δ_X ppm. The theoretical 1-D proton spectrum is shown alongside each frequency axis, F1 and F2, the two scales being equal. Two types of multiplet, or arrays of peaks, are observed, viz. those centred on the diagonal F1 = F2 and those where F1 $\neq$ F2 which are termed *cross-peak* multiplets. The *diagonal-peak* multiplets are of no direct interest other than correlating them with both the 1-D spectra plotted alongside. However, the fact that horizontal and vertical lines (not shown) drawn from each cross-peak intersect the diagonal line at the positions of the diagonal peaks confirms that protons A and X are indeed coupled to each other. It should be noted that for each pair of coupled protons or groups of protons the cross-peaks *always* occur in pairs,

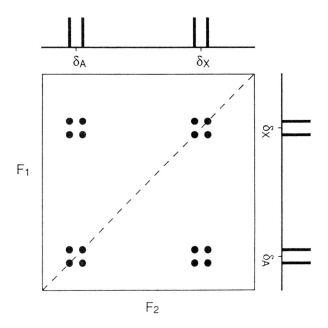

Figure 9.43. Schematic COSY spectrum of a two coupled spins, denoted A and X. For convenience, the normal one-dimensional spectrum is plotted alongside the F_1 and F_2 axes and the diagonal ($F_1 = F_2$) is indicated by a dashed line. This spectrum shows two types of multiplets: those centred at the same F_1 and F_2 frequencies, called diagonal-peak multiplets, and those centred at different frequencies in the two dimensions, called cross-peak multiplets. Each multiplet has four component peaks. The appearance of a cross-peaked multiplet centred at $F_1 = \delta_A$, $F_2 = \delta_X$ indicates that the proton with shift δ_A is coupled to the proton with shift δ_X. This observation is all that is required to interpret a COSY spectrum

equidistant on either side of the diagonal, and that the overall pattern is symmetrical. Furthermore, the co-ordinates of the centre of each cross-peak are the frequencies (chemical shifts) of the A proton and the X proton measured along the F2 and F1 axes respectively. These multiplets arise because of 'mixing' of the signals derived from the coupled nuclei during the evolution period of the pulse sequence.

Figure 9.44 shows a labelled COSY spectrum of geraniol where confirmation of the ^{1}H-^{1}H connectivity (coupling) between protons 5 and 6 has been shown by the dotted lines. Other connectivities can be confirmed in the same way.

(ii) *Heteronuclear Chemical Shift Correlation* (HETCOR) or *Heteronuclear Correlated Spectroscopy* (HETEROCOSY) identifies the specific protons attached to each carbon-13 (^{1}H–^{13}C *connectivities*). The contour plot is related

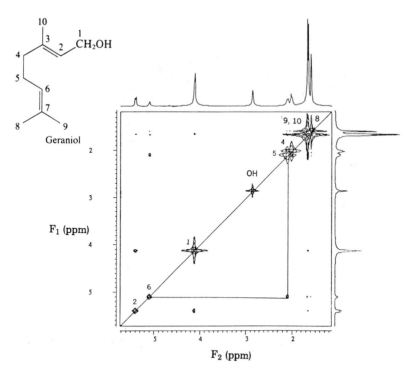

Figure 9.44. Basic COSY spectrum of geraniol, in CDCl$_3$ at 500 MHz. The H-5, H-6 coupling is shown

to the 1-D proton spectrum along the F1 axis at the side and to the broadband decoupled ^{13}C spectrum along the F2 axis at the top. The cross-peaks, which confirm specific ^{1}H–^{13}C attachments, are located at the intersections of horizontal lines drawn from proton peaks or multiplets and vertical lines drawn from ^{13}C peaks. Compared to DEPT, more specific information on ^{1}H–^{13}C connectivities is obtained as illustrated by the HETCOR spectrum of geraniol (figure 9.45).

(iii) *Incredible Natural Abundance Double Quantum Transfer Experiment* (2-D INADEQUATE) identifies directly bonded ^{13}C nuclei (^{13}C–^{13}C *connectivities*) which allows the carbon skeleton of a compound to be elucidated. Because the natural abundance of ^{13}C is 1.1 percent, only about 1 in 10 000 molecules will have adjacent ^{13}C nuclei. The sensitivity of this technique is therefore correspondingly low, and it is very time consuming to collect the data. The contour plot shows a series of widely-separated pairs of doublets (cross-peaks) due to ^{13}C–^{13}C coupling, i.e. all are AX systems. The much more intense singlets of the isolated ^{13}C nuclei (diagonal peaks) which are present in

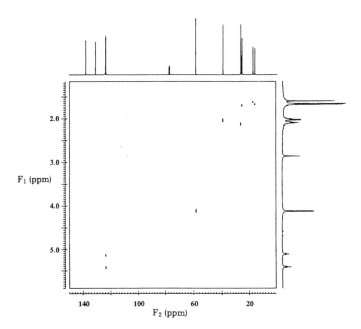

Figure 9.45. The HETCOR spectrum of geraniol, in $CDCl_3$ at 500 MHz for ^1H and 125.7 for ^{13}C

99.99 percent of the sample are suppressed during the pulse sequence. By making the scale of the F1 axis, along which the 1-D ^{13}C spectrum is plotted at the top, half that of the F2 axis, the pairs of doublets separated in the F1 direction, and which link specific ^{13}C nuclei in the 1-D spectrum, appear symmetrically on either side of the diagonal F1 = 2F2. This aids the interpretation of the contour plot, especially for larger molecules with a complex carbon skeleton. The number of doublets corresponding to a particular ^{13}C line in the 1-D spectrum, which are aligned vertically, is determined by how many other ^{13}C nuclei are directly bonded to it. Thus, for a terminal methyl group only one doublet will be observed while methylene, methine and quaternary carbons can have a maximum of two, three or four doublets respectively depending on the structure. The 2-D INADEQUATE spectrum of geraniol is shown in figure 9.46.

QUALITATIVE ANALYSIS—THE IDENTIFICATION OF STRUCTURAL FEATURES

Pmr spectrometry is an extremely useful technique for the identification and structural analysis of organic compounds in solution, especially when used in conjunction with infrared, ultraviolet, visible and mass spectrometry.

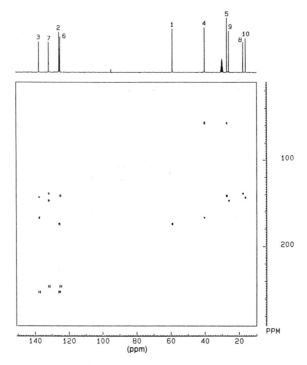

Figure 9.46.   The INADEQUATE spectrum of geraniol in CD_3COCD_3 at 75.6 MHz for ^{13}C

Interpretation of pmr spectra is accomplished by comparison with reference spectra and reference to chemical shift tables. In contrast to infrared spectra, it is usually possible to identify all the peaks in a pmr spectrum, although the complete identification of an unknown compound is often not possible without other data. Some examples of pmr spectra are discussed below.

Figure 9.47 shows the spectrum of 4-ethoxyacetanilide in $CDCl_3$. The triplet and quartet indicate the presence of an ethyl group, and tables show that the chemical shift of the methylene quartet ($\delta4.0$), appearing well downfield, is due to —OPh. The aromatic ring protons are seen around $\delta7$, the somewhat symmetrical pattern suggesting a 1,4-substituted compound, and the upfield group at $\delta6.8$ corresponds to ring protons ortho to the ethoxy group. The sharp singlet at $\delta2.1$, integrating to 3 protons, is shown from tables to be due to an isolated methyl group attached to a carbonyl. The broad resonance at $\delta7.9$, which can be removed by treatment with D_2O, integrates to one proton and is due to the secondary amide group.

The spectrum of divinyl-β,β'–thiodipropionate is shown in figure 9.48

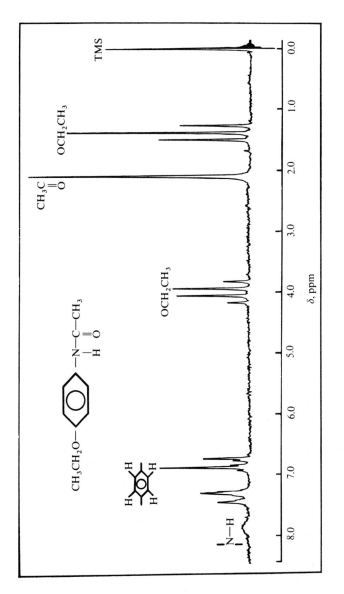

Figure 9.47. Pmr spectrum of *p*-ethoxyacetanilide in CDCl$_3$. Spectrum obtained at a magnetic field of 1.4092 Tesla and a radiofrequency of 60 MHz. The chemical shift scale is the delta scale relative to the methyl protons of tetramethylsilane (TMS)

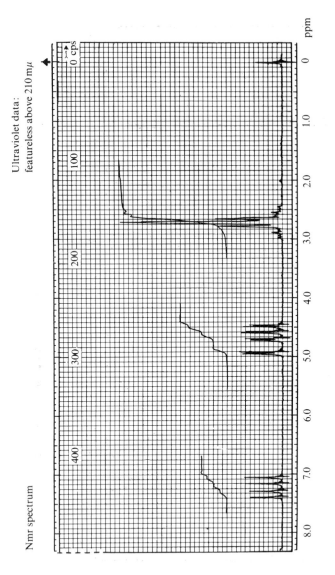

Figure 9.48.   Pmr spectrum of divinyl-β-β′-thiodipropionate, conditions as figure 9.47

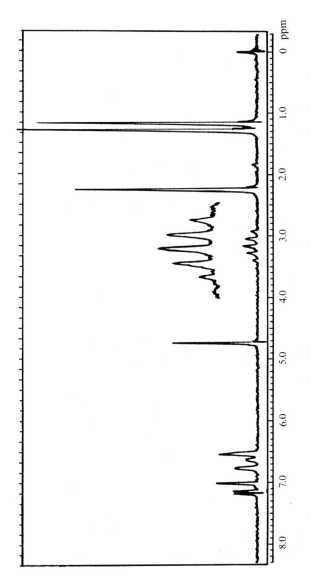

Figure 9.49. Pmr spectrum of thymol, conditions as in figure 9.47

The typical ABX pattern at lowfield is related to the vinyl group, the four peaks at $\delta7.2$ corresponding to the X proton. The geminal A and B protons are well separated between $\delta4.5$ and $\delta5.0$ and the symmetrical but complex A_2B_2 pattern around $\delta2.7$, together with an integral corresponding to four protons, indicates two adjacent methylenes. Tables confirm that the ABX pattern has chemical shifts corresponding to a vinyl group adjacent to an ester group.

Figure 9.49 shows the spectrum of thymol. The doublet at $\delta1.3$ together with the septet at $\delta3.2$ (integral ratio 6:1) are the methyls and methine respectively of an isopropyl group. The outer peaks of the septet are often lost in baseline noise and are better seen by increasing the sensitivity. The sharp resonance at $\delta4.75$, disappearing on treatment with D_2O, is that of the phenolic OH, and the complex pattern between $\delta6.5$ and $\delta7.5$ is typically aromatic. Tri-substitution is confirmed by an integral corresponding to three protons. The sharp singlet at $\delta2.3$, also corresponding to three protons, is in the expected position for a methyl group attached to an aromatic ring. Additional examples of pmr spectra are given in the section on the combined use of spectral data (p. 438).

QUANTITATIVE ANALYSIS

Nmr has been used comparatively little for quantitative analysis although peak areas are directly proportional to concentration. The principle drawbacks are the expensive instrumentation and a lack of sensitivity. The latter can be improved with the aid of computers to accumulate signals from multiple scans or by using a pulsed (Fourier transform) technique. Relative precision lies in the range 3 to 8 %.

APPLICATIONS OF NMR SPECTROMETRY

The technique is currently not used as widely as uv, visible and infrared spectrometry partly due to the high cost of instrumentation. However, it is a powerful technique for the characterization of a wide range of natural products, raw materials, intermediates and manufactured items especially if used in conjunction with other spectrometric methods. Its ability to identify major molecular structural features is useful in following synthetic routes and to help establish the nature of competitive products, especially for manufacturers of polymers, paints, organic chemicals and pharmaceuticals. An important clinical application is *nmr imaging* where a 3-dimensional picture of the whole or parts of a patient's body can be built up through the accumulation of proton spectra recorded over many different angles. The technique involves costly instrumentation but is preferable to X-ray tomography as risks to the health of the patient are virtually eliminated and additional information on their medical condition can be obtained.

9.4 Mass Spectrometry

SUMMARY

Principles
>   Ionization and fragmentation of materials by bombardment with a high energy electron beam, by chemical ionization, field ionization or fast atom bombardment. Analysis of the range of mass fragments produced. Elemental composition of non-volatile materials by application of an rf spark.

Instrumentation
>   Source of high-energy electrons, ion accelerator, magnetic/electrostatic analyser, detector and recorder, high vacuum pumping system.

Application
>   Identification and structural analysis of organic compounds. Determination of trace impurities in a wide range of inorganic materials (spark source mass spectrometry).

Disadvantages
>   Complex and costly instrumentation, difficult to maintain.

Mass spectrometry is a technique for characterizing molecules according to the manner in which they fragment when bombarded with high-energy electrons, and for elemental analysis at trace levels. It is not strictly a spectrometric method as electromagnetic radiation is neither absorbed nor emitted. However, the data obtained are in a spectral form in that the relative abundance of mass fragments from a sample is recorded as a series of lines or peaks. The bombardment process produces many fragments carrying a charge, and this facilitates their separation and detection by electrical and magnetic means. Spectra must be recorded under conditions of high vacuum (10^{-4} to 10^{-6} N m^{-2}) to prevent loss of the charged fragments by collision with molecules of atmospheric gases or swamping of the sample spectrum.

INSTRUMENTATION

The essential components of a mass spectrometer include a *sample inlet system*, an *ionization source* and *acceleration chamber* where sample molecules are ionized, fragmented and accelerated into an *analyser* or *separator*, and an *ion detection* and *recording system*.

Sample Inlet System. Volatile or volatilizable compounds may be introduced into the spectrometer via a pinhole aperture or *molecular leak* which allows a steady stream of sample molecules into the ionization area. Non-volatile or thermally labile samples are introduced directly by means of an electrically heated probe inserted through a vacuum lock. Numerous methods of sample ionization are available of which the most important are *electron impact (EI) chemical ionization (CI), field ionization (FI), field desorption (FD), fast atom bombardment (FAB)*, and *rf spark discharge*.

Electron impact ionization (EI) involves the bombardment of sample molecules with a high-energy electron beam ($\sim 70\,\text{eV}$) which results in a considerable degree of fragmentation into positive ions of varying mass. Chemical ionization (CI) is achieved by first introducing a reagent gas such as methane or isobutane into the electron beam where it is ionized to form several species including CH_5^+ and $C_2H_5^+$. On subsequent introduction of the sample at a level of about 10^3 times less than that of the reagent gas, these species react with sample molecules to yield $(M+1)^+$ and $(M-1)^+$ ions by hydrogen addition or abstraction, where M represents the mass of the sample molecule. There is much less fragmentation into ions of smaller mass. Field ionization (FI) employs a pointed or knife-edged anode and a slit cathode between which a very large potential difference is applied (5 to $20\,\text{kV}$). Sample molecules entering this high potential gradient or electrical force field are ionized to form M^+ and $(M+1)^+$ ions with very little accompanying fragmentation. Both CI and FI are much gentler methods of ionization than EI and are particularly useful where the molecular weight of a compound is sought. The more complex fragmentation patterns produced by EI are useful for the identification of structural features.

Fixed desorption (FD) is similar in principle to FI. It enables ions to be produced directly from solid samples which are deposited from solution onto an anode fitted to a probe that can be inserted into the instrument via a vacuum lock. It is even more gentle than CI and FI, producing molecular ions and virtually no fragmentation. However, the ionization process decays very rapidly so spectra must be scanned quickly and cannot be re-recorded without introducing more sample.

Fast atom bombardment (FAB) involves the production of a high-speed stream of energetic neutral argon atoms ($\sim 5\,\text{KeV}$) which are directed on to the sample that is introduced into the spectrometer on the end of a probe. The samples are first dissolved in a relatively non-volatile matrix, such as glycerol, monothioglycerol or carbowax, which aids the ionization process yielding $(M+H)^+$ and $(M-H)^+$ sample ions together with some fragmentation. FAB is proving particularly useful for non-volatile compounds, including those of a biochemical nature such as peptides, nucleotides and their salts, which hitherto have yielded mass spectra only with difficulty.

Spark source mass spectrometry is used for the examination of non-volatile inorganic samples and residues to determine elemental composition. An rf spark of about 30 kV is passed between two electrodes, one of which may be the sample itself, causing vaporization and ionization. Powdered samples or residues from ashed organic materials can be formed into an electrode after mixing with pure graphite powder.

Analyser or Separator. After ionization and fragmentation of the sample, positive fragment ions are accelerated into the analyser chamber by means of a potential gradient and slit system. The *fragment ions* are then separated in space by allowing them to move through a magnetic and/or an electrostatic field or by measuring the time taken for them to drift through space to a detector (time-of-flight spectrometer). Single *magnetic analysers* are the simplest and a schematic diagram of this type of instrument is shown in figure 9.50(a). *Double-focusing* instruments use an electrostatic followed by a magnetic analyser to improve the resolution. Some mass spectrometers employ a *quadrupole analyser* which consists of a set of four metal rods placed symmetrically around and parallel to the direction of travel of the fragment ions (figure 9.50(b)). Application of a dc potential and an oscillating rf field across the rods causes all ions except those with a particular mass to charge ratio (*m/e*) to follow an unstable path leading to collisions with the rods. Ions with the appropriate *m/e* value follow a stable path to the detector. By progressive alteration of dc potential and rf field, ions with different *m/e* values can be made to follow a stable path to the detector which enables a mass spectrum to be 'scanned'.

Ion Detection and Recording System. Ions from the analyser pass through a slit and impinge either on an earthed electrode called a *collector plate* or on a type of photomultiplier tube known as an *electron multiplier.* The currents produced are amplified and monitored with a high speed recorder employing mirror galvanometers and ultraviolet sensitized chart-paper which operates at several different sensitivities simultaneously to avoid the need for multiple scanning of each sample. The spectra may also be viewed on a cathode ray oscilloscope screen (CRO). Many mass spectrometers are interfaced with a dedicated mini- or microcomputer to facilitate the processing and presentation of the very rapidly accumulated data and to make comparisons with library spectra. This is virtually essential where the spectrometer is also to be interfaced with a gas chromatograph (p. 108) as rapid scanning of eluted peaks can generate hundreds of spectra in just a few minutes. The reliability of mass spectral data is ensured by frequent calibration of the instrument with a standard of known fragmentation pattern, a long chain fluorinated hydrocarbon such as a perfluorokerosene being a common choice.

(a)

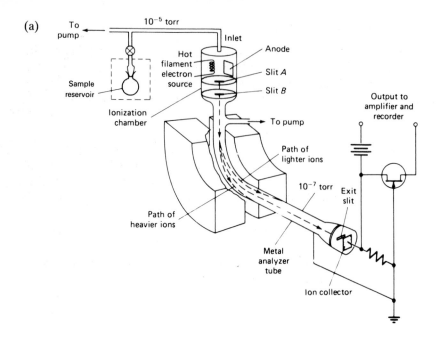

(b)

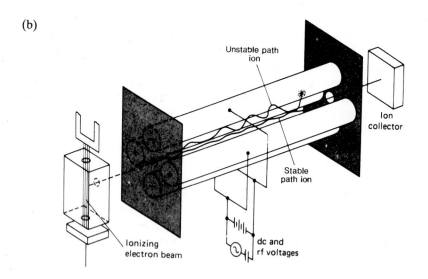

Figure 9.50.   (a) Schematic diagram of a mass spectrometer using a single magnetic
analyser. (b) A quadrupole mass spectrometer.
(From D. Lichtman, *Res. Dev.*, **15** (2), 52 (1964). With permission.)

PRINCIPLE OF MASS SPECTROMETRY

The principle of separation by a magnetic analyser can be rationalized in terms of the kinetic energy of the fragment ions, the accelerating voltage and the magnetic field. The kinetic energy of fragments of mass m and charge e accelerated in a potential gradient V may be equated with the electrical force acting on them,

$$\tfrac{1}{2}mv^2 = eV \tag{9.26}$$

where v is the velocity of the fragments. The magnetic field, acting at right angles to the direction of motion, causes them to be deflected into a circular trajectory in which the centripetal force due to the magnet is balanced by the centrifugal force due to the kinetic energy. Thus,

$$Bev = \frac{mv^2}{r} \tag{9.27}$$

where B is the magnetic field and r the radius of curvature of the trajectory. Eliminating v from equations (9.26) and (9.27) gives:

$$\frac{m}{e} = \frac{B^2 r^2}{2V} \tag{9.28}$$

Equation (9.28) shows that for ions carrying a single positive charge, the mass is directly proportional to the square of the radius of curvature. At a particular value of the accelerating voltage V fragment ions of a given mass will pass along the analyser tube and through an exit slit to impinge on a *collector plate*. A continuous stream of ions of this mass will be registered as an ion current which can be amplified and fed to a recording system. By continuously varying the accelerating voltage, fragment ions of different masses can be 'focused' through the exit slit in turn and an entire mass spectrum recorded. The resolution, or ability to discriminate between ions of a similar mass, can be improved considerably by 'double focusing', an arrangement incorporating separate electrostatic and magnetic fields in the analyser tube. Resolution is defined as $m/\Delta m$ where m and $(m + \Delta m)$ are the m/e values of two adjacent peaks of equal height and separated by a valley which is 10% of the peak height. Resolving power decreases with increasing mass but spectrometers having a resolution of 1 000, i.e. the ability to discriminate between m/e values of 1 000 and 1 001, or between 100 and 99.9, are adequate for many applications. Double-focusing instruments may be capable of resolutions of 20 000 to 50 000 or more.

CHARACTERISTICS AND INTERPRETATION OF MOLECULAR MASS SPECTRA

The recorded mass spectrum of a material is usually represented graphically in the form of a line diagram by expressing the abundance of each fragment ion

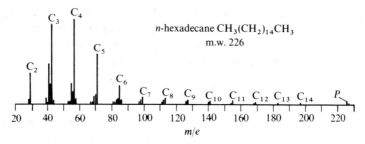

Figure 9.51. Mass spectrum of n-hexadecane

as a percentage of the most abundant one. The peak derived from the latter is called the *base peak*. The *fragmentation* or *cracking pattern* for a single substance is uniquely characteristic and may be used for qualitative identification purposes. The mass spectrum of n-hexadecane is shown in figure 9.51.

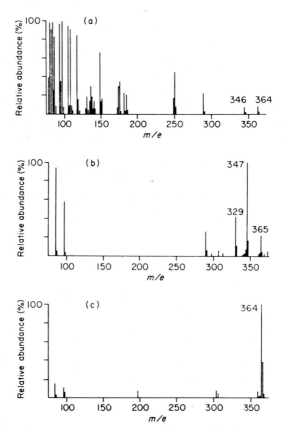

Figure 9.52. Mass specta of tetrahydrocortisone. (a) E.I.; (b) C.I.; (c) F.I.

An aid to identification is the ability to determine the molecular weight with a high degree of accuracy and to establish the empirical molecular formula. The former depends on recognition of the *parent* or *molecular ion peak* M produced when an electron is ejected from the molecule

$$M + e^- \rightarrow M^+ + 2e^-$$

Because of the high electron beam energies used, fragmentation processes predominate in many compounds and the parent peak is often weak or missing from the spectrum. It can sometimes be identified by re-scanning the spectrum at a much lower electron beam energy when its intensity should grow at the expense of all other peaks. Alternatively, increasing the sample size induces bimolecular collisions resulting in proton transfer and a larger $M + 1$ peak. However, the use of CI or FI is being increasingly preferred as much larger M or $(M + 1)$ peaks are obtained. A comparison of EI, CI and FI spectra for the same compound is shown in figure 9.52.

The empirical formula of a compound can be established by evaluating the intensities of isotope peaks $M + 1$, $M + 2$, etc relative to the parent peak. Isotope peaks arise because of the natural abundance of certain heavier isotopes in some molecules of the sample. Thus a compound containing a single carbon atom will produce an $M + 1$ peak 1.08% of the intensity of the parent peak because 1.08% of the molecules contain a ^{13}C atom. If the molecule contains two carbon atoms the $M + 1$ peak will be 2.16% of the parent peak and an $M + 2$ peak due to molecules containing two ^{13}C atoms will be produced of relative intensity 0.01%. Table 9.10 lists the principal

Table 9.10   Principle stable isotopes and relative abundances of some common elements

ELEMENT	PERCENTAGE ABUNDANCE*					
	ISOTOPE		ISOTOPE		ISOTOPE	
carbon	^{12}C	100	^{13}C	1.08		
hydrogen	^{1}H	100	^{2}H	0.016		
nitrogen	^{14}N	100	^{15}N	0.38		
oxygen	^{16}O	100	^{17}O	0·04	^{18}O	0.20
fluorine	^{19}F	100				
silicon	^{28}Si	100	^{29}Si	5.10	^{30}Si	3.35
phosphorus	^{31}P	100				
sulphur	^{32}S	100	^{33}S	0.78	^{34}S	4.40
chlorine	^{35}Cl	100	^{37}Cl	32.5		
bromine	^{79}Br	100	^{81}Br	98.0		
iodine	^{127}I	100				

* calculated using 100 for the most common isotope

Table 9.11   Extract from Beynon's tables of empirical formulae and isotope ratios

	M+1	M+2		M+1	M+2
99			**101**		
$C_2HN_3O_2$	3.40	0.44	CHN_4O_2	2.70	0.43
$C_2H_3N_4O$	3.77	0.26	$C_2HN_2O_3$	3.06	0.64
C_3HNO_3	3.76	0.65	$C_2H_3N_3O_2$	3.43	0.45
$C_3H_3N_2O_2$	4.13	0.47	$C_2H_5N_4O$	3.81	0.26
$C_3H_5N_3O$	4.51	0.28	C_3HO_4	3.41	0.84
$C_3H_7N_4$	4.88	0.10	$C_3H_3NO_3$	3.79	0.65
$C_4H_3O_3$	4.49	0.68	$C_3H_5N_2O_2$	4.16	0.47
$C_4H_5NO_2$	4.86	0.50	$C_3H_7N_3O$	4.54	0.28
$C_4H_7N_2O$	5.24	0.31	$C_3H_9N_4$	4.91	0.10
$C_4H_9N_3$	5.61	0.13	$C_4H_5O_3$	4.52	0.68
$C_5H_7O_2$	5.59	0.53	$C_4H_7NO_2$	4.89	0.50
C_5H_9NO	5.97	0.35	$C_4H_9N_2O$	5.27	0.31
$C_5H_{11}N_2$	6.34	0.17	$C_4H_{11}N_3$	5.64	0.13
$C_6H_{11}O$	6.70	0.39	$C_5H_9O_2$	5.63	0.53
$C_6H_{13}N$	7.07	0.21	$C_5H_{11}NO$	6.00	0.35
C_7HN	7.96	0.28	$C_5H_{13}N_2$	6.37	0.17
C_7H_{15}	7.80	0.26	C_6HN_2	7.26	0.23
C_8H_3	8.69	0.33	$C_6H_{13}O$	6.73	0.39
			$C_6H_{15}N$	7.11	0.22
100			C_7HO	7.62	0.45
$C_2H_2N_3O_2$	3.42	0.45	C_7H_3N	7.99	0.28
$C_2H_4N_4O$	3.79	0.26	C_8H_5	8.73	0.33
$C_3H_2NO_3$	3.77	0.65			
$C_3H_4N_2O_2$	4.15	0.47	**102**		
$C_3H_6N_3O$	4.52	0.28	$CH_2N_4O_2$	2.72	0.43
$C_3H_8N_4$	4.90	0.10	$C_2H_2N_2O_3$	3.07	0.64
$C_4H_4O_3$	4.50	0.68	$C_2H_4N_3O_2$	3.45	0.45
$C_4H_6NO_2$	4.88	0.50	$C_2H_6N_4O$	3.82	0.26
$C_4H_8N_2O$	5.25	0.31	$C_3H_2O_4$	3.43	0.84
$C_4H_{10}N_3$	5.63	0.13	$C_3H_4NO_3$	3.81	0.66
$C_5H_8O_2$	5.61	0.53	$C_3H_6N_2O_2$	4.18	0.47
$C_5H_{10}NO$	5.98	0.35	$C_3H_8N_3O$	4.55	0.28
$C_5H_{12}N_2$	6.36	0.17	$C_3H_{10}N_4$	4.93	0.10
$C_6H_{12}O$	6.72	0.39	$C_4H_6O_3$	4.54	0.68
$C_6H_{14}N$	7.09	0.22	$C_4H_8NO_2$	4.91	0.50
C_7H_2N	7.98	0.28	$C_4H_{10}N_2O$	5.28	0.32
C_7H_{16}	7.82	0.26	$C_4H_{12}N_3$	5.66	0.13
C_8H_4	8.71	0.33	$C_5H_{10}O_2$	5.64	0.53
			$C_5H_{12}NO$	6.02	0.35
			$C_5H_{14}N_2$	6.39	0.17
			$C_6H_2N_2$	7.28	0.23
			$C_6H_{14}O$	6.75	0.39
			C_7H_2O	7.64	0.45
			C_7H_4N	8.01	0.28
			C_8H_6	8.74	0.34

stable isotopes and their relative abundances for a number of elements. The selection of likely molecular formulae which correlate with particular parent masses and their isotope peak ratios is facilitated by a set of tables constructed by Beynon.[3] All possible molecular formulae for compounds containing carbon, hydrogen, oxygen and nitrogen up to a molecular weight of 500, together with the calculated isotope peak ratios, are listed. The presence of chlorine, bromine or sulphur can be inferred from an abnormally large $M + 2$ peak whilst an abnormally small $M + 1$ peak suggests the presence of the monoisotopic elements iodine, fluorine or phosphorus. Beynon's tables also include calculations of the relative intensities of $M + 2$, $M + 4$, $M + 6$, etc peaks for compounds containing one or more atoms of chlorine or bromine. Extracts from the tables are illustrated in tables 9.11 and 9.12. Some of the formulae listed can be automatically eliminated on the basis of the *nitrogen rule*. This states that a molecule of even-numbered molecular weight must contain no nitrogen or an even-number of nitrogen atoms.

Additional information that can be derived from the molecular formula is the number of *unsaturated sites*. The total number in a molecule includes *rings*, *double bonds* and *triple bonds*, the latter counting as two double bonds. The formula for calculating the number of unsaturated sites is:

$$\text{No of sites} = \text{carbons} + 1 - \left(\frac{\text{hydrogens} - \text{nitrogens}}{2}\right) - \frac{\text{halogens}}{2}$$

Thus, for C_7H_7NO, the number of sites is 5 and $C_6H_5CNH_2$ would be a

$$\overset{\parallel}{O}$$

possible structure.

A complete structural analysis can sometimes be accomplished by a study

Table 9.12 Intensities of isotope peaks (relative to the parent peak) for combinations of bromine and chlorine

HALOGEN PRESENT	% M+2	% M+4	% M+6	% M+8	% M+10	% M+12
Br	97.7					
Br$_2$	195.0	95.5				
Br$_3$	293.0	286.0	93.4			
Cl	32.6					
Cl$_2$	65.3	10.6				
Cl$_3$	99.8	31.9	3.47			
Cl$_4$	131.0	63.9	14.0	1.15		
Cl$_5$	163.0	106.0	34.7	5.66	0.37	
Cl$_6$	196.0	161.0	69.4	17.0	2.23	0.11
BrCl	130.0	31.9				
Br$_2$Cl	228.0	159.0	31.2			
Cl$_2$Br	163.0	74.4	10.4			

of the fragmentation pattern. Some general rules for the modes of fragmentation expected from particular types of compound are formulated on the basis
of the concepts of physical and organic chemistry. These are:

1. The relative intensity of the parent peak decreases with increasing
   molecular weight in a homologous series and as the degree of chain-
   branching increases, i.e. it is greatest for short, straight-chain
   compounds. Large parent peaks are observed for aromatic and other
   cyclic structures and for molecules containing double bonds.
2. Cleavage of bonds is more likely to occur at branches in a chain, the
   largest substituent at a branch being eliminated as a radical. Saturated
   rings with side chains tend to cleave at the α-bond leaving the positive
   charge on the ring fragment. Alkyl substituted aromatics cleave at the
   β-bond to give a prominent peak at m/e 91 from the tropylium ion:

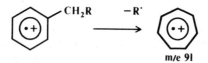

3. Heteroatoms, such as nitrogen, oxygen and sulphur, promote cleavage
   of an adjacent carbon–carbon single bond by forming an ion in which the
   lone-pair electrons participate in resonance stabilization:

$$R-C-CH_2R'-R' \quad C-CH_2R' \longleftrightarrow {}^+C-CH_2R'$$

This produces prominent peaks at m/e 43, 57, 71 etc.

4. Fragmentation may result in prominent peaks with masses not apparently
   related to the original molecule. Rearrangements resulting in the
   elimination of neutral molecules are common as are those due to the
   migration of hydrogen atoms in molecules containing a heteroatom. The
   peaks may sometimes be recognized from the fact that if derived from
   an even-weight parent ion, they are themselves even-numbered.

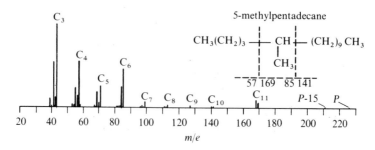

Figure 9.53.   Mass spectrum of 5-methylpentadecane

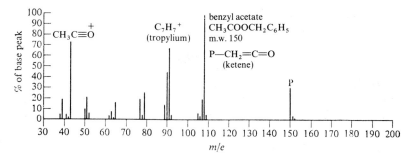

Figure 9.54. Mass spectrum of benzyl acetate

The interpretation of mass spectra and the validity of the above rules may be demonstrated by examining the spectra of several compounds. Returning to the spectrum of n-hexadecane shown in figure 9.51, it will be seen that it is characterized by a small parent peak (Rule 1) and clusters of peaks 14 mass units (CH_2) apart. The largest peak in each cluster is C_nH_{2n+1}, the most intense groups are those at C_3 and C_4, and the decrease in intensity of successive groups is smooth. Contrast this with a spectrum of the isomeric 5-methylpentadecane, figure 9.53 which shows similar clusters but a prominent group at C_6 because of branching (Rule 2). The lack of a smooth decrease in the intensity of successive groups differentiates branched chain from straight chain compounds.

The spectrum of benzyl acetate is shown in figure 9.54. The parent peak at m/e 150 is prominent (Rule 1) as is the tropylium ion peak at m/e 91 (Rule 2). The base peak at m/e 108 is due to a rearrangement (Rule 4) after cleavage of the acetyl group which itself gives a prominent peak at m/e 43 (Rule 3). The size of the $M+2$ peak indicates the absence of sulphur and halogens and the empirical formula $C_9H_{10}O_2$ given in Beynon's tables best fits the isotope peak ratios. The number of saturated sites is $9 + 1 - 5 = 5$, i.e. one ring and four double bonds.

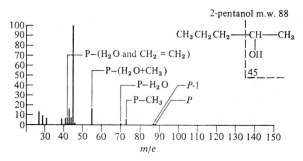

Figure 9.55. Mass spectrum of 2-pentanol

Lastly, figure 9.55 shows a spectrum of 2-pentanol. The base peak at m/e 45 is due to cleavage at the C—C bond adjacent to the OH group (Rule 3) and several peaks appear corresponding to the loss of H_2O and C_2H_2 (Rule 4).

APPLICATIONS OF MASS SPECTROMETRY

Used in conjunction with infrared, nmr, uv and visible spectral data, mass spectrometry is an extremely valuable aid in the identification and structural analysis of organic compounds, and, independently, as a method of determining molecular weights. The analysis of mixtures can be accomplished by coupling the technique to glc (p. 108) although until recently it was achieved by using sets of simultaneous equations and matrix calculations based on mass spectra of the pure components. It is well suited to gas analysis and is widely used by the petroleum industry for both the qualitative and quantitative analysis of hydrocarbon distillates and other petrochemicals. The electronics and metallurgical industries are finding increasing use for spark-source mass spectrometry in the detection and determination of impurity elements at very low levels (ppb) in alloys and semi-conductor materials. Mass spectrometers have proved to be invaluable in the analysis of both terrestrial and extra-terrestrial atmospheres, the latter being accomplished by instruments with light-weight miniaturized components.

9.5 Spectrometric Identification of Organic Compounds

The application of individual spectrometric techniques to the identification and structural analysis of organic compounds has been examined in the preceding sections of this chapter. In most cases the desired information can be gained much more easily and rapidly by studying infrared, uv and visible, nmr and mass spectrometric data in conjunction with one another. Firstly, this is because the information they provide is often complementary in nature and it may not be possible to make a positive identification or a complete structural analysis from the data provided by one technique alone. Secondly, unless an analyst specializes in one of these techniques, only the salient features of a spectrum are readily interpreted and it is thus more profitable to assemble such information from all four techniques before attempting a more detailed study of any one spectrum. The exercise is somewhat analogous to piecing together a jigsaw puzzle. To gain experience in this approach, a number of sets of spectra of varying complexity is presented below.

References
1. WOODWARD, R. B., *J. American Chemical Society*, **63**, 1123, 1941; **64**, 72, 1942

2. RINGBOM, A., *Z. Anal. Chem.*, **115**, 332, 1939

3. BEYNON, J. H. and WILLIAMS, A. E., *Mass and Abundance Tables for Use in Mass Spectrometry*, Elsevier, Amsterdam, 1963

Further Reading

BOLTZ, D. F. and HOWELL, J. A. *Colorimetric Determination of Non-metals*, 2nd Ed., Wiley, New York

DAVIS, R. and FREARSON, M., *Mass Spectrometry*, Wiley, Chichester, 1987.

DENNY, R. C. and SINCLAIR, R., *Visible and Ultraviolet Spectroscopy*, Wiley, Chichester, 1987.

GEORGE, W. O. and MCINTYRE, P., *Infrared Spectroscopy*, Wiley, Chichester, 1987.

PAVIA, D. L., LAMPMAN, G. M. and KRIZ, G. S. JR., *Introduction to Spectroscopy*, Saunders, Philadelphia, 1979.

SANDELL, E. B. and ONISHI, H., *Photometric Analysis of Traces of Metals: General Aspects*, 4th Ed., Wiley, New York, 1978.

SILVERSTEIN, R. M., BASSLER, G. C. and MORRILL, T. C., *Spectrometric Identification of Organic Compounds*, 4th Ed., Wiley, New York, 1981.

WILLIAMS, D. A. R., *Nuclear Magnetic Resonance Spectroscopy*, Wiley, Chichester, 1986.

WILLIAMS, D. H. and FLEMING, I., *Spectroscopic Methods in Organic Chemistry*, 4th Ed., McGraw-Hill, London, 1989.

Sample 9.1
Found: C, 49.4; H, 9.8; N, 19.1 %

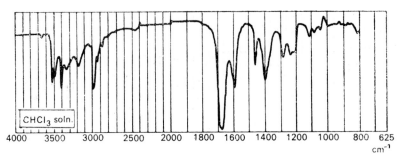

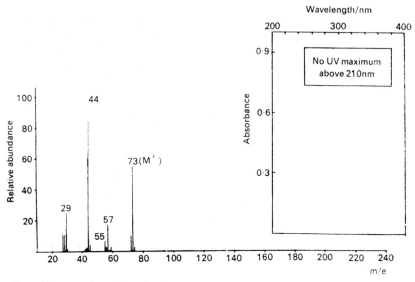

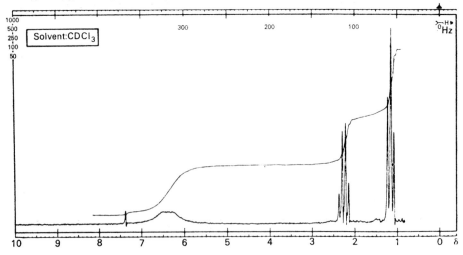

Sample 9.2

Found: C, 70.7; H, 6.0%

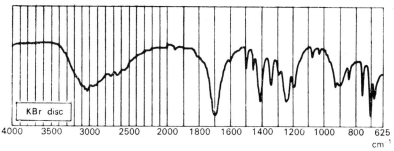

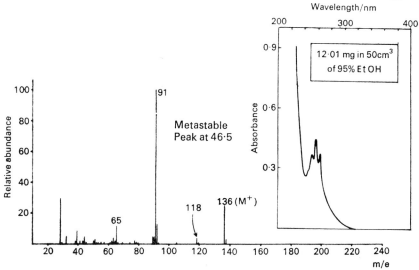

Metastable
Peak at 46·5

12·01 mg in 50cm³
of 95% Et OH

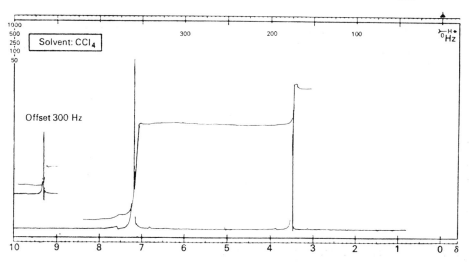

Solvent: CCl₄

Offset 300 Hz

Sample 9.3
Found: C, 61.4; H, 5.1; N, 10.2 %
nmr sweep offset by 30 Hz

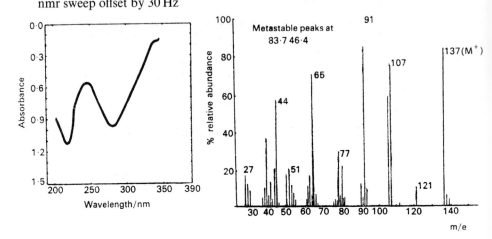

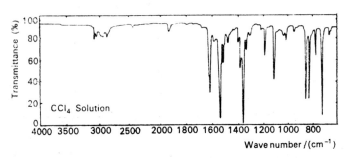

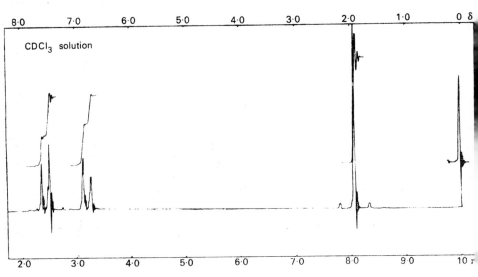

Sample 9.4
Found: C, 33.2; H, 5.0 %
uv transparent above 210 nm

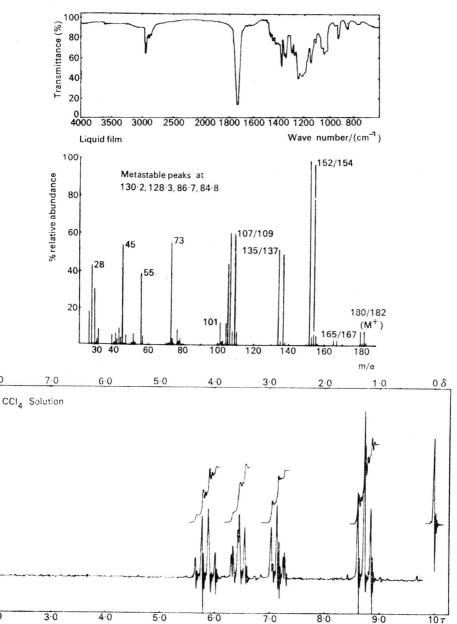

Sample 9.5

uv spectrum A run in ethanol
uv spectrum B run in ethanol/NaOH
nmr spectrum A sweep offset by 120 Hz
nmr spectrum B sweep offset by 210 Hz
after deuteration

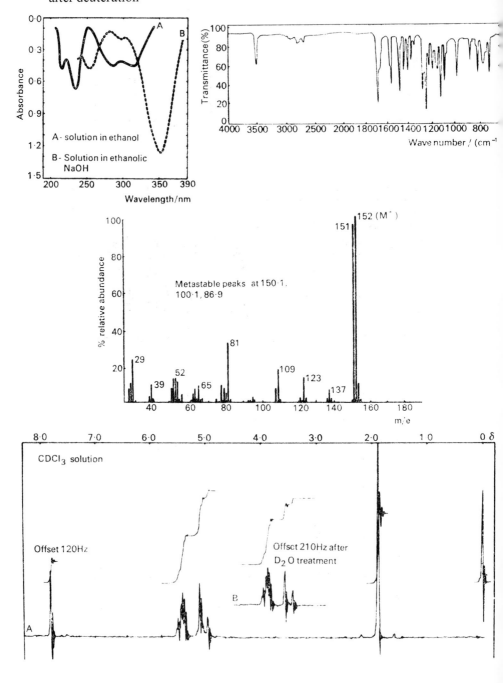

A- solution in ethanol

B- Solution in ethanolic NaOH

Metastable peaks at 150·1, 100·1, 86·9

152 (M⁺)

CDCl₃ solution

Offset 120Hz

Offset 210Hz after D₂O treatment

Sample 9.6
The uv spectrum of this compound shows no maximum above 205 nm. The ir spectrum is obtained on a neat liquid sample.

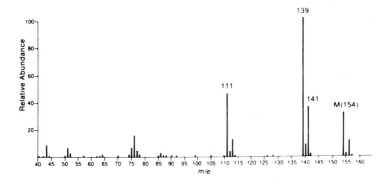

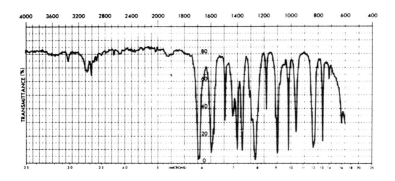

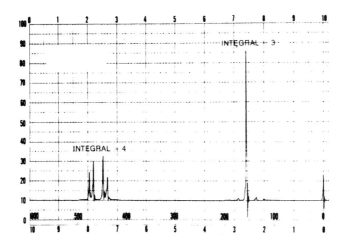

Sample 9.7

The uv spectrum of this compound shows no maximum above 205 nm. The ir spectrum is obtained on a neat liquid sample.

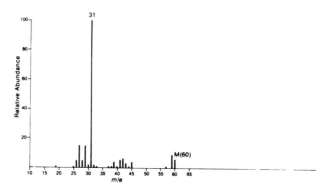

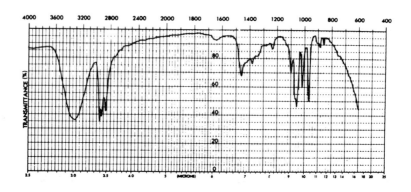

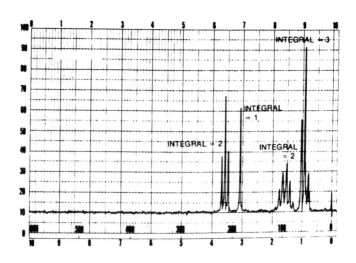

Sample 9.8

The uv spectrum of this compound is determined in 95% ethanol: λ_{max} 250 nm (log ε 3.2). The ir spectrum is obtained on a neat liquid sample.

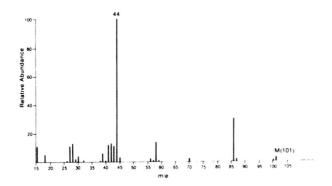

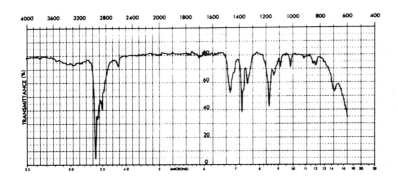

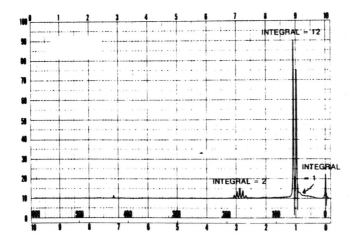

Problems

1. (a) Convert the following percent transmittances to absorbance:

$$90, 50, 10, 5, 1$$

(b) Convert the following absorbances to transmittance:

$$0.434, 0.868, 0.05, 1.0, 2.0$$

2. The concentration of a liquid sample contained in an infrared cell of thickness $0.15 \, \mu m$ is 0.5 M. Calculate the molar absorptivity of the sample if the absorbance at a specified wavenumber is 0.300.

3. A solution of concentration x molar absorbs 70% of the radiation passing through it. If Beer's Law is obeyed, what percentage of radiation would be transmitted at a concentration of $x/3$ molar?

4. A solution containing 3.0 ppm of an analyte (MW = 155) has a transmittance of 65.0% in a 1 cm cell. Calculate:

   (a) the absorbance of the solution;
   (b) the absorbance and percent transmittance for a solution containing 5.2 ppm of the analyte;
   (c) the molar absorptivity of the analyte.

5. Ti and V can be determined simultaneously in steel as their coloured peroxide complexes. 1 mg of each metal in 50 cm^3 of solution gave the following absorbances:

$$\text{Ti} \quad 0.269 @ 400 \, nm; \quad 0.134 @ 460 \, nm$$
$$\text{V} \quad 0.057 @ 400 \, nm; \quad 0.091 @ 460 \, nm$$

Two steel samples, dissolved in 50 cm^3, each 1.000 g, gave the following absorbances:

$$(1) \quad 0.172 @ 400 \, nm; \quad 0.116 @ 460 \, nm$$
$$(2) \quad 0.600 @ 400 \, nm; \quad 0.660 @ 460 \, nm$$

Calculate the percentage of Ti and V in each sample.

6. The absorption characteristics of solutions of the acid-base indicator methyl red at two wavelengths were:

$$1.22 \times 10^{-3} \, \text{M in 0.1 M HCl:} \quad A_{400} = 0.077; \quad A_{528} = 1.738$$
$$1.09 \times 10^{-3} \, \text{M in 0.1 M NaHCO}_3: \quad A_{400} = 0.753; \quad A_{528} = 0.000$$
$$\text{At pH 4.31 (0.1 M ionic strength):} \quad A_{400} = 0.166; \quad A_{528} = 1.401$$

What is the acid dissociation constant of methyl red?

7. To take account of stray light in spectrophotometric measurements, the true absorbance values should be calculated using the formula,

$$A = \log_{10}\left(\frac{I_0 + I_s}{I + I_s}\right)$$

where I_s is the fraction of the incident radiation defined as stray light. For a stray light level of 1%, calculate the true absorbances for solutions whose apparent absorbances are 1, 1.5 and 2.0.

8. If the wavenumber of a C–H stretching band in the infrared spectrum of a certain compound is $2960 \, cm^{-1}$, calculate the wavenumber of the corresponding C–D stretching band in the deuterated homologue.

Chapter 10

Radiochemical Methods in Analysis

SUMMARY

Principles

Emission of ionizing radiations in radioactive decay, nuclear particle and γ-ray spectrometry. Quantitative and qualitative analysis by intensity and spectrometric measurements respectively.

Instrumentation

Gas ionization, solid scintillation, liquid scintillation and semiconductor detectors, autoradiography. Single and multichannel pulse height analysers. Coincidence and anticoincidence circuits.

Applications

Study of chemical pathways in method development. Isotope dilution methods. Radioimmunoassay very important in biochemistry and medicine. Neutron activation analysis used for trace elements in geochemistry, semiconductor technology, pollution studies and forensic science. Relative precision of counting 1 % if 10^4 counts are recorded.

Disadvantages

Sometimes expensive for tracers or irradiation facilities. Special laboratory and handling facilities required. Needs highly skilled operators, and complex instrumentation.

Three main features account for the usefulness of radiotracers in analysis. Firstly, a chemical species may be 'labelled' with a radioactive atom and thus distinguished from non-labelled species of the same type. Only in exceptional circumstances (notably with tritium labels) does the introduction of the label materially affect the chemical behaviour of the species. In the special case of tritium the variations in behaviour (known as isotope effects) arise from the large mass differences between the nucleides ^1_1H and ^3_1H. Secondly, the radioactive characteristics of an atom are unchanged by its chemical environment, so that a label can always be detected wherever it appears. Thirdly, the particles and radiation emitted during radioactive decay are of high energy and may be detected with extreme sensitivity. Thus in an ideal situation determinations may be made of 10^{-12} g of an analyte. Furthermore the energy spectrum of the emissions may be used to identify and distinguish between a number of different radionucleides.

10.1   Nuclear Structure and Nuclear Reactions

For most purposes a simple picture of the atomic nucleus is adequate. It may be regarded as a collection of *neutrons* ^1_0n and *protons* ^1_1p (collectively known as

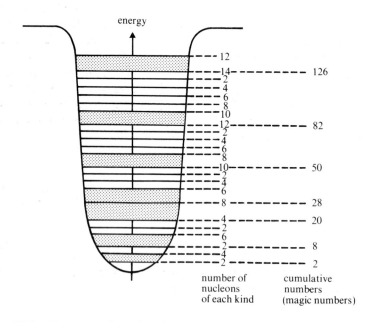

Figure 10.1. Representation of nuclear energy levels showing the energy levels of nucleons and the 'magic numbers' corresponding to filled shells. Shaded areas represent the gaps between the shells

nucleons) arranged in a series of energy levels similar to the arrangement of extranuclear electrons, figure 10.1. The diameter of a nucleus is about 10^{-14} m. Transitions between nuclear levels may occur, accompanied by the emission or absorption of energy which is frequently in the form of electromagnetic radiation (*γ-rays*). The levels are widely spaced, so that the radiation will be of high photon energy and short wavelength (10^{-11} to 10^{-7} nm or 100 keV to 5 MeV). Nuclear transitions and their associated electromagnetic radiations are as characteristic of an individual nucleus as electronic and vibrational transitions are of atoms and molecules.

DECAY REACTIONS

The overall stability of a nucleus is closely related both to total and relative numbers of neutrons and protons within it. If the mass number is large ($A > 209$) a major instability will exist and the nucleus will adjust to a lower energy state, in part, by the emission of large particles, $^4_2$He, *α-particles*. For very heavy nuclei ($A > 238$) spontaneous fission becomes an important mode of decay. In this process, complete fission of the nucleus into two roughly equal parts occurs.

For a given mass number there is a known, stable neutron to proton ratio which varies from 1 to 1.5 as the mass increases, figure 10.2. Any nuclide whose ratio falls outside of these values will be unstable and decay so as to obtain a more stable ratio. In doing so, it will interconvert neutrons and protons and will typically emit *positrons*, β^+ or $^0_1$e, or *negatrons*, β^- or $_{-1}^0$e. Decay processes will involve the emission of a high energy particle in a primary event, closely followed (apart from a few important exceptions) by the emission of γ-rays as the new nucleus adjusts to its ground state. The important modes of radioactive decay are summarized below.

Alpha Decay
This involves the loss of a helium nucleus from a heavy 'overcrowded' nucleus and is accompanied by the emission of γ-rays. The process may be summarized by

$$^A_Z X \rightarrow ^{A-4}_{Z-2} Y + ^4_2 He^{2+} \tag{10.1}$$

An α-particle will have a discrete energy, characteristic of the emitting nuclide, whence α-spectra have sharp peaks.

Negatron Decay
This is a process characteristic of nucleides with high n:p ratios, and involving the loss of an electron from the nucleus, which is usually, but not invariably

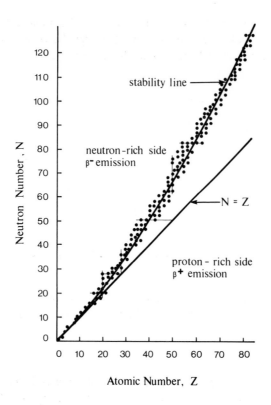

Figure 10.2. Stable nucleides and the stability line

accompanied by the emission of γ-photons. A detailed energy balance reveals that the simple picture cannot account for all the energy lost by the nucleus in the decay and the emission of an additional particle—the antineutrino, $\bar{v}$ is postulated to account for this. The general equation for a negatron emission is

$$\begin{matrix} A \\ Z \end{matrix}X \longrightarrow \begin{matrix} A \\ Z+1 \end{matrix}Y + \beta^- + \bar{v} \qquad (10.2)$$

which implies the decay of a neutron into a proton and an electron (negatron) which is expelled.

$$\begin{matrix} 1 \\ 0 \end{matrix}n \rightarrow \begin{matrix} 1 \\ 1 \end{matrix}p + \begin{matrix} 0 \\ -1 \end{matrix}e \qquad (10.3)$$

The exact energy carried by an emitted negatron will depend upon the angle between its path and that of the antineutrino. As the angle can vary from atom to atom, so will the distribution of energy between the particles. Negatron spectra, figure 10.3 thus do not have sharp peaks.

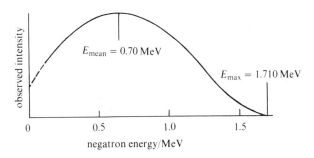

Figure 10.3.   Energy spectrum of ^{32}P negatrons

A low n:p ratio in a nucleus gives a situation which may be stabilized by the conversion of a proton into a neutron. One process which may effect this is positron emission,

$$\,_1^1p \rightarrow \,_0^1n + \beta^+ \qquad (10.4)$$

The positron has a short life and will quickly be annihilated in a reaction with an electron, producing γ-photons of characteristic energy (0.51 MeV). In addition, the basic nuclear process itself is usually accompanied by the emission of gamma radiation. As in the case of negatron decay a complete energy balance reveals a discrepancy which can be accounted for if the emission of a further particle—the neutrino, v is postulated. Overall, positron emission can be summarized in a general equation

$$\,_Z^AX \rightarrow \,_{Z-1}^AY + \beta^+ + v \qquad (10.5)$$

K-electron Capture
An unstable, low n:p ratio may also be adjusted by the capture of an orbital electron, which would naturally involve the nearest shell, K-shell, of electrons

$$\,_1^1p + \,_{-1}^0e \rightarrow \,_0^1n \qquad (10.6)$$

The vacancy left in the low energy K-shell will be filled by an electron from a higher level with the resultant emission of radiations of extra nuclear origin, X-rays, which are distinguished from the accompanying nuclear γ radiation. It should be noted that the n:p criterion is the same for both positron and electron capture processes and it is not unusual to find both occurring with different atoms of the same nucleide.

Internal Conversion

In some limited circumstances, surplus nuclear energy may be transferred directly to an orbital electron. The electron is then expelled from the atom and superficially resembles a negatron. Usually such 'conversion electrons' will have relatively low energies, but the important difference from the negatron is their extra nuclear origin. The process of energy transfer is ill-understood but it should be emphasized that it involves the direct transfer of energy from a nuclear orbital to an extra nuclear electronic orbital and does not involve a γ-photon as was at one time believed. This reaction is commonly observed with unstable isomeric nuclear states which also decay by the emission of low energy γ-photons.

Gamma Decay (Isomeric Transition)

The emission of electromagnetic γ-radiation is the natural accompaniment of most nuclear processes and provides the route for an excited nuclear isomer to decay to its ground state. A typical decay sequence is set out in figure 10.4.

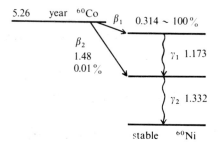

Figure 10.4.   Decay scheme for ^{60}Co

The γ-energies are characteristic of the nuclear energy levels and transitions. Figure 10.12 shows the appearance of a γ-ray spectrum. Such spectra are of immense value in the identification and determination of radionucleides in a radioactive mixture.

THE KINETICS OF DECAY REACTIONS

The rate at which radioactive atoms decay is unaffected by the chemical or physical form of the nucleide and depends only on the number N of atoms present and the *decay constant* λ/s^{-1} for that particular nucleide. In a single step decay where the daughter nucleide is stable, the rate of decay (disintegrations s^{-1}) is simply expressed,

i.e. for
$$X \xrightarrow{\lambda_X} Y \text{ (stable)}$$

$$\text{rate of decay} = -\frac{dN_X}{dt} = \lambda_X N_X \tag{10.7}$$

The number of undecayed nuclei present at time t, is given by the exponential form of the same equation

i.e.
$$N_X = (N_X)_0 \exp(-\lambda_X t) \tag{10.8}$$

A plot of $\log_{10} N_X$ against t is a straight line. A quantity of practical importance is the time taken for the number of parent nuclei to reduce to half the original number. This time is known as the *half life* $t_{\frac{1}{2}}$ and is inversely related to the decay constant.

$$(t_{\frac{1}{2}})_X = \frac{0.693}{\lambda_X} \tag{10.9}$$

Its value is thus characteristic of a particular radionucleide and is a further aid to identification.

When the daughter Y undergoes further decay the mathematical treatment of the system rapidly becomes complex. For example, in a two step decay

$$X \xrightarrow{\lambda_X} Y \xrightarrow{\lambda_Y} Z \text{ (stable)} \tag{10.10}$$

the number of nuclei of Y at time t is given by

$$N_Y = \frac{\lambda_X (N_X)_0}{\lambda_Y - \lambda_X} [\exp(-\lambda_X t) - \exp(-\lambda_Y t)] + (N_Y)_0 \exp(-\lambda_Y t) \tag{10.11}$$

Analytical calculations based upon equations of this complexity are not very fruitful, particularly as they depend upon detecting both radioactive processes with identical efficiencies. Calibration of instruments using suitable standards is thus the normal approach to practical measurements.

BOMBARDMENT REACTIONS AND THE GROWTH OF RADIOACTIVITY

The number of naturally occurring radionucleides is limited and few are of analytical value. For the majority of purposes artificial radionucleides are manufactured. Bombardment reactions are generally used in their production. A suitable target material is exposed to an intense flux of the appropriate particles in a nuclear reactor or particle accelerator such as a cyclotron.

Thermal neutrons in the reactor are efficient in producing (n, γ) neutron capture reactions e.g. $^{58}_{26}$Fe (n, γ) $^{59}_{26}$Fe. The products of these reactions will have an excess of neutrons and generally decay by (β^-, γ) emission. The major disadvantage is that the radioactive atoms will always be diluted with many non-radioactive atoms and chemical separation is not possible. (n, γ) reactions are however usefully exploited in neutron activation analysis (p. 471). With fast neutrons, proton, deuteron or alpha particle bombardment a change in atomic number accompanies the reaction and chemical separation of the 'carrier free' radiotracer becomes possible,

e.g. $\qquad\qquad\qquad\qquad$ ^{14}N (n,p) ^{14}C

$$^{27}\text{Al (d,n) } ^{30}\text{P}$$

The growth and decay of a nucleide in the target during a bombardment reaction may be exemplified by a generalized sequence

$$X \xrightarrow[\sigma_X N_X f]{\text{bombardment}} Y \xrightarrow[\lambda_Y]{\text{decay}} Z \text{ (stable)} \qquad (10.12)$$

The growth of Y will proceed at a rate dependent upon the intensity f of the particle flux, the amount N_X of the target present and the nuclear cross section σ_X for the reaction, figure 10.5. After time t, the activity induced due to Y (i.e. $N_Y \lambda_Y$) will be given by

$$N_Y \lambda_Y = N_X \sigma_X f [1 - \exp(-\lambda_Y t)] \qquad (10.13)$$

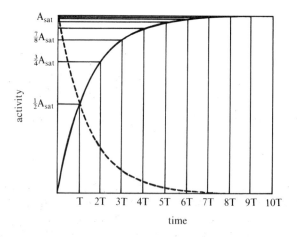

Figure 10.5. Growth and decay of a radionucleide ($T = t_\frac{1}{2}$)

Where σ_X is measured in *barns*, 1 b $= 10^{-28}$ m². If $t > 5\,(t_{\frac{1}{2}})_Y$ the exponential term may be ignored and equation (10.13) becomes

$$N_Y\lambda_Y = N_X\sigma_X f \qquad (10.14)$$

i.e. the *saturation activity*, figure 10.5. Growth and decay considerations are important in activation analysis methods. It should be noted that nuclear reactions are specific for a particular nucleide. Thus N_X represents the number of atoms of the nucleide, which differs from the total number of atoms of the element when the element is not monoisotopic.

10.2 Instrumentation and Measurement of Radioactivity

In all except *autoradiography* which is a special case to be considered separately, radioactive emissions are measured by causing them to activate a *detector* from which a series of electrical pulses are generated. If the sample is maintained in a fixed relationship to the detector, the number of pulses generated within a known time, i.e. *the count*, will be a measure of the radioactivity. With calibration using standard sources, the proportionality between the activity of the source and the count may be determined. In certain circumstances the height (size) of the pulses generated is directly proportional to the energy of the incident particles or photons. *Pulse height analysis* can thus be applied to distinguish between different emissions and radionucleides. The general layout of a counting system is summarized in figure 10.6. α-emitting tracers are rarely used and subsequent attention will focus on β^- and γ detection.

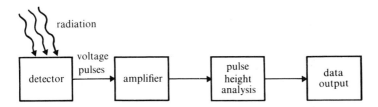

Figure 10.6.   Schematic layout of radiation detection and measurement

RADIATION DETECTORS

Three main types of detector are in widespread use, namely, *gas ionization*, *scintillation* and *solid state*. Although there are a number of common features they are best discussed separately for the sake of clarity.

Gas ionization detectors are based on the interaction between the ionizing radiations and a suitable ionizable gas (e.g. argon), within a closed tube. The

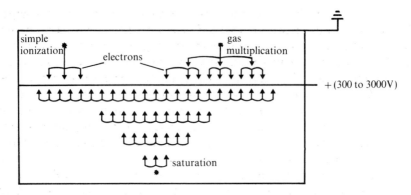

Figure 10.7.   Radiation induced ionization in an argon filled detector

electrons produced in the ionization generate voltage pulses when they are collected by a thin wire anode (300–3 000 V) whilst the argon ions are reduced at an envelope cathode (connected to earth), figure 10.7. Two important types of gas ionization detector (*proportional* and *Geiger–Müller*) have this simple construction. Their characteristics may be appreciated by considering the change in the number of counts recorded as the voltage supplied to the anode is varied, whilst presenting a source of constant activity to the detector. The curve obtained, figure 10.8, is divided into four parts. At low potentials (< 150 V) the potential gradient between the electrodes is shallow. Thus the *primary electrons* produced in the ionizing event move slowly towards the

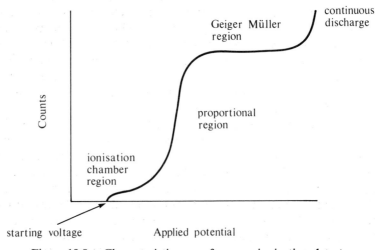

Figure 10.8.   Characteristic curve for a gas ionization detector

anode and they mostly recombine with argon ions before being discharged. As the potential increases so does the efficiency of collection of electrons until all are collected and over a short potential range a plateau is produced. A steadily increasing potential will however cause the electrons to be accelerated towards the anode and produce *secondary ionization* or *gas phase multiplication* in collisions with argon atoms. As a result, the characteristic curve will rise steeply with increase in potential. The size of the pulse produced will be proportional to the energy carried by the incident particle or photon, whence a measure of pulse height analysis may be applied to the signals. Eventually gas ionization reaches its ultimate when a single ionizing event leads to complete ionization of the filling gas, saturating the detector. A further plateau, known as the *Geiger region* results, within which the number of electrons produced is largely independent of the applied potential.

An important feature of all gas ionization detectors is their *dead* or *paralysis time*, which has a direct bearing on their utility. Once initiated a voltage pulse takes several hundred microseconds to die away, and may be prolonged by the discharge of secondary electrons from the cathode as the argon ions are reduced. Until this first pulse is terminated the tube is 'dead' to further radiations and the 'recorded count' C_M will be less than the 'true count' C_T. The relation between the two values will depend upon the length of the dead time t

$$C_T = \frac{C_M}{(1 - C_M t)} \qquad (10.15)$$

The dead time can be shortened by reducing the anode potential immediately the pulse is initiated, and by the incorporation of electron attracting materials which absorb secondary electrons. Typically ethyl alcohol, bromine or carbon dioxide are used for this purpose. *Quenching* a pulse by a combination of the two methods can reduce dead times of Geiger tubes (which generate large pulses) to the range 200 μs to 400 μs and for proportional counters with their smaller pulses to a few microseconds only. In the latter case corrections may generally be ignored, but for a typical Geiger tube dead time losses may amount to 20 % at a count rate of 100 counts s^{-1}.

Gas ionization detectors are widely used in radiochemistry and X-ray spectrometry. They are simple and robust in construction and may be employed as 'static' or 'flow detectors'. In recent years flow studies have received a great deal of attention in the interfacing of radioactive detectors with gas chromatographs. A radio-gas chromatograph, figure 10.9, uses a gas flow proportional counter to monitor the effluent from the gas chromatography column. To achieve satisfactory operation a gas stream of argon containing 5 % carbon dioxide as 'quench gas' is used, and ^{14}C or ^3H labels are monitored by converting the effluents to $^{14}CO_2$, ^3H_2O or organic

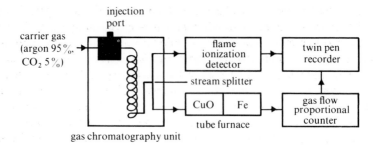

Figure 10.9. Layout of a radio-gas chromatograph

compounds such as benzene or methane, and incorporating them into the gas stream.

Scintillation counters, which constitute an extremely important group, depend upon the absorption of radiation by a scintillator to produce uv light scintillations, which are detected and converted into amplified voltage pulses by a photomultiplier, figure 10.10. *Solid scintillators* are used extensively for the detection and analysis of γ-rays and X-rays, while *liquid scintillators* find widespread employment in the measurement of pure negatron emitters, especially where the particle energy is low (<1 MeV).

Solid scintillators include materials such as sodium iodide, lithium iodide, anthracene, naphthalene and 'loaded' polymers. Sodium iodide detectors are by far the most important, and subsequent discussions will be restricted to

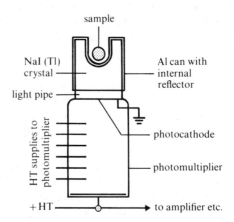

Figure 10.10. A typical scintillation detector. (Well crystal NaI)

these. Crystals of NaI which have been activated by the incorporation of 1 % of TlI were the cornerstone of *γ-ray spectrometry* for many years. On striking a crystal γ-rays interact in a number of different ways—the most important are *photoelectric absorption* and *Compton scattering*. In the former the γ-photon is completely absorbed, and excites electrons to energies directly related to the photon energy. Thus, ultimately the intensity of the light scintillation, and the height of the voltage pulse produced, are proportional to the energy of the incident photon. In this way bases for pulse height analysis and γ-ray spectrometry are provided. When Compton scattering occurs, the γ-photon loses some of its energy in preliminary collisions before being photoelectrically absorbed. It follows that γ-spectra will show distinct 'photopeaks' and a background of lower energy derived from the Compton effect, figure 10.11. Simplicity and high detection efficiencies are the valuable features of NaI(TlI) detectors, which are to be contrasted with the relative complexity, lower efficiencies but exceptional resolution of semiconductor systems, figure 10.11(a).

Liquid scintillators are employed largely for measurements on pure negatron emitters (^{14}C, ^{3}H, ^{32}P, ^{35}S) and are especially valuable when the negatron carries a low energy, e.g. ^{3}H(0.018 MeV), ^{14}C(0.16 MeV). Ideally the scintillator and the sample are dissolved in the same solvent to ensure intimate contact between the radionucleide and the scintillator. Xylene, toluene or dioxan are suitable solvents. The latter will retain the scintillator in solution with up to 20 % of water. Insoluble samples may be measured in suspensions stabilized with silica gel. The energy of the particles is absorbed by the solvent and transferred to the scintillator, finally being re-emitted as a pulse of uv radiation which activates the photomultiplier. A typical 'cocktail' for liquid scintillation counting contains a secondary scintillator, which shifts the scintillations to longer wavelengths. In so doing it will shift the emission peaks into a more sensitive region for the photomultiplier. Typical scintillators are conjugated aromatic molecules whose characteristics are shown in figure 10.12. The intensity of a scintillation is related to the energy of the incident ionizing particle, whence pulse height analysis may be employed to provide a measure of spectrometric discrimination. However, reference back to figure 10.3 will re-emphasize the problems of negatron spectrometry and it will be seen why only two component mixtures can be satisfactorily analysed. Moreover, this can be accomplished only if there is a large difference in the negatron energies (e.g. ten-fold). The important interferences associated with liquid scintillation counting are *chemiluminescence, bioluminescence, chemical quenching* and *colour quenching*. The first arise when a chemical or biochemical reaction within the matrix stimulates the emission of radiation. Refrigeration of the sample to 4°C, removal of the offending chemical or acceleration of the reaction to completion by heating are possible ways of overcoming the phenomenon. Chemical quenching is brought about

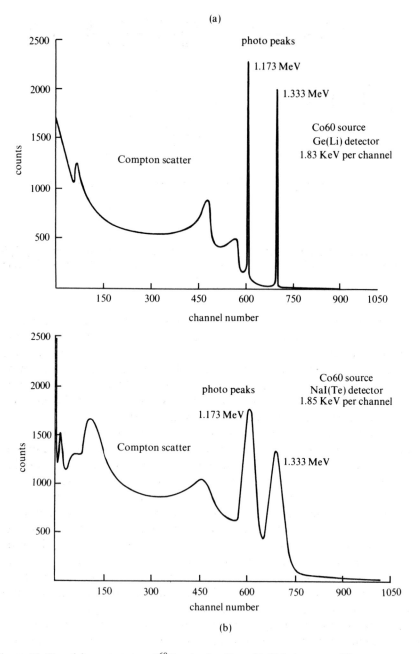

Figure 10.11. (a) γ-spectrum of ^{60}Co obtained by a Ge(Li) detector. (b) γ-spectrum of ^{60}Co obtained by a NaI(T1I) crystal detector

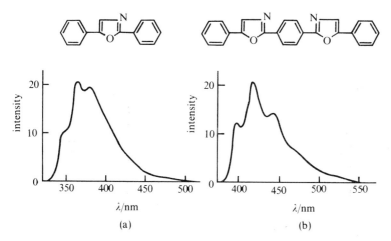

Figure 10.12. Typical primary and secondary scintillators. (a) Primary; 2,5-diphenyloxazole (PPO). (b) Secondary; 1,4-bis-2-(5-phenyloxazolyl)-benzene (POPOP)

by chemical interference with the scintillation energy transitions and is typically a problem with sulphur or oxygen compounds. Removal of the interfering compound is the only answer. Colour quenching can be overcome by bleaching the offending colour or exploiting a secondary scintillator to move the emission peak clear of the absorption peak. Many laboratories now determine ^3H and ^{14}C by combustion to ^3H_2O and $^{14}CO_2$ which are subsequently absorbed and measured in a constant matrix.

Semiconductor detectors depend on the interaction of high energy radiation with crystals of semiconductors such as germanium or silicon. The former generally finds use for medium or high energy γ-radiation and the latter for low energy γ- and X-radiation. When the energy of the radiation is absorbed in the crystal, free electrons and positive holes are produced in numbers proportional to the energies of the incident photons. A thin gold layer is plated on to each end of the crystal and electrodes are attached to this so that a potential of 3–5 kV may be applied across the crystal. Thus the electrons and positive holes will migrate to the electrodes and produce electrical pulses with sizes proportional to the energy of the incident radiation. Detectors based upon these principles provide the basis for modern γ-ray spectrometry and have become of great importance in X-ray spectrometry also.

Although simple in concept, semiconductor detectors are rather complex in construction, because of practical difficulties which have to be overcome. These problems derive from the small numbers of electrons and positive holes produced in the initial interactions. To enable the electrical pulses to be

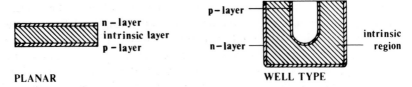

Figure 10.13.   Cross-sections through typical lithium drift radiation detectors

distinguished, it is necessary to reduce thermal randomization to a minimum and to reduce any background currents to a very low level indeed. Refrigeration with liquid nitrogen reduces thermal randomization to a satisfactory level. Background currents can originate from leakage around the outside of the crystal via impurities adsorbed onto the surface. Operation of the detector inside a high-vacuum jacket overcomes this problem. In addition, background currents result from the presence of impurity carriers (*n*-type or *p*-type) within the crystal. To reduce these the semiconductor must be of the highest purity or background currents must be minimized by the use of *p-n* junctions within a lithium drifted crystal. Currently, both types of detector can be found in use.

Lithium drifted detectors are produced by first doping an intrinsic crystal (Ge or Si) with *p*-type impurities such as B, Al, Ga or In. Lithium, a strongly *n*-type element, is then drifted into one end of the crystal. On refrigeration of

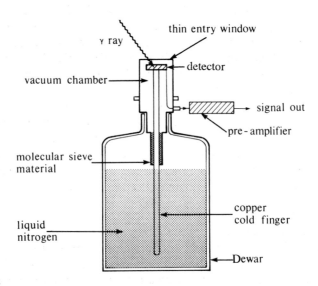

Figure 10.14.   Construction of a 'dipstick' type semiconductor detector system

the crystal a permanent p-n-junction is created within it. The electrical polarity produced leads to n-type impurities migrating to the p-type region and p-type impurities to the n-type region. Thus a depletion layer of highly pure crystal is produced around the junction. Application of a reversed-bias high voltage to the crystal increases the width of the depletion layer. Typical construction of a semiconductor detector is shown in figures 10.13 and 10.14.

The excellent high resolution γ and X-ray spectra which can be obtained from semiconductor detectors make them very important in modern instruments. A typical spectrum is shown in figure 10.11(a) which may be compared with the much broader peaks from the scintillation detector, figure 10.11(b). The spectra are not immune from the problem of Compton scattering (p. 459) but a good quality modern detector will have a photopeak to Compton peak ratio of 40:1 or better. Computer aided spectrum analysis also serves to reduce the interference from the Compton effect.

α-*particle spectrometry* is also carried out using semiconductor detectors whose operation is based on the p-n junction principle.

SOME IMPORTANT ELECTRONIC CIRCUITS

Coincidence and *anti-coincidence* circuits frequently appear in nuclear instruments. Both are concerned with the elimination of background pulses which may arise from electronic 'noise', cosmic radiation or other environmental radiation. The coincidence principle is met frequently in scintillation counters where pulses, thermally generated within the stages of the photomultipliers, can greatly increase the background signal. Two photomultipliers are arranged to view the sample so that scintillations are recorded coincidently by both tubes. On receipt of these signals a coincidence unit will feed a pulse to the display or recording unit whilst ignoring noise pulses, which will not be coincidently received from the two tubes.

Anti-coincidence units are used to eliminate external interferences. The detector is surrounded by a second 'guard' detector so that ionizing radiation from outside (e.g. cosmic rays) must activate the outer detector before penetrating to the inner. A sample placed close to, or inside, the inner detector will activate that only. The anti-coincidence unit is arranged to accept only the latter noncoincident pulses from the measuring detector.

Pulse Height Analysis

Pulse height analysis is a nuclear technique to which frequent reference has been made. It is used to distinguish one radiation from another by means of the proportionality between the radiation energy and the size of the pulses generated. In the simplest case sample radiation may be distinguished from high energy cosmic radiation, or from small 'noise' pulses. On the other hand pulse height analysis can be used to produce a detailed spectrum and thus to resolve a complex mixture of γ- or X-rays.

A *single channel* pulse height analyser utilizes an electronic 'gate' typically 0.1 V wide, which only accepts pulses between the preset upper and lower limits. Scintillation counters frequently employ such devices to remove small noise pulses and large cosmic pulses, as well as to distinguish between particles with different energies. The single channel principle can be extended to circuits in which the gate scans through the pulse height range, and feeds a signal to a ratemeter with an intensity proportional to the number of pulses falling within the gate. In this way a spectrum can be generated on a chart recorder. This process can be likened to the relative movement of a slit and dispersed radiations in an optical spectrometer.

Multi-channel pulse height analysers examine the whole range of pulses coming from a detector simultaneously by using a large number (100–8 000) of channels or gates. The analogue signal from the detector is examined by an *analogue to digital converter* (ADC) with a clock frequency of 50 MHz to 100 MHz. A series of pulses is generated by the ADC for each channel and used to magnetize a 'ferrite core store' similar to those used in digital computers. The information thus stored can be displayed on a cathode ray tube as a spectrum, or relayed to a variety of other data output modes.

Modern pulse height analysers essentially contain dedicated digital computers which both store and process data as well as controlling the display and operation of the instrument. The computer will usually provide spectrum smoothing, peak search and peak integration routines. Analysis may be made by reference to a spectrum library and radionucleide listing. Figure 10.15 summarizes such a pulse height analysis system. A more detailed account of a computer-based pulse height analyser appears in chapter 13 (p. 547).

Autoradiography
Special emulsions which are sensitive to nuclear radiations and X-rays

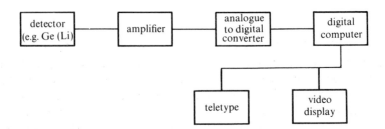

Figure 10.15.   Layout of a nuclear spectrometer with a computer to provide pulse height analysis

are readily available and find considerable use in some fields of analysis. The main use is for locating the position of a label in a specimen, prior to its determination, when a high degree of resolution is needed. Many biological tracer experiments fall into this category as does the assay of thin layer or paper chromatograms. For example, as many as sixteen different amino acids may be located on a single autoradiograph from a thin layer plate. Transmission densitometry may be used to relate the density of 'fogging' to the amount of analyte present. The autoradiograph may also be used as a template to mark off the positions of spots on the TLC plate. Alternative counting procedures may then be used for final assay.

THE STATISTICS OF RADIOACTIVE MEASUREMENTS

At any one instant, only a very small proportion of the total number of unstable nuclei in a radioactive source undergo decay. A *Poisson distribution* which expresses the result of a large number of experiments in which only a small number are successful can thus be used to describe the results obtained from measurements on a constant activity source. In practical terms this means that random fluctuations will always occur and the estimated standard deviation s of a measurement can be related to the total measurement by

$$s = (\text{count})^{\frac{1}{2}} \qquad (10.16)$$

It follows that for the relative standard deviation to be reduced to 1 % or less a minimum of 10^4 counts must be recorded.

10.3 Analytical uses of radionucleides

Although radioactive materials have been used in many different ways to solve analytical problems there are only three groups of methods which are of major importance.

CHEMICAL PATHWAY STUDIES

A method of analysis frequently requires the separation of the analyte from interfering materials present within the sample matrix. Many different procedures are used to effect such separations (chapter 4) and the establishment of quantitative procedures is a singularly important step in method development. Careful study and control of experimental conditions are needed and a large number of experiments may be required to establish the most suitable. Radiotracers are uniquely suited to such studies. The sensitivity of detection means that only very small amounts of tracer need be added to follow the *chemical pathway* of the relevant species. Furthermore it matters little what the physical or chemical state of the tracer is, for measurements may be made on liquids, solids or gases. Chromatography, solvent extraction and precipitation are amongst separation methods widely studied by means of radiotracers. In the individual separation steps the distribution of the species may be studied by simple radioactivity measurements, and subsequently the tracer will serve as a yield indicator for the overall procedure.

RADIOISOTOPE DILUTION METHODS

An important group of analytical methods is based on measurements of the change in isotopic ratio when active and non-active isotopes are mixed. In the simplest case, a known amount w_1 of labelled analyte of known specific activity a_1 is added to the sample. After isotopic mixing has been established sufficient of the analyte is separated (not normally 100%) to allow the new specific activity a_2 to be measured. Measurements of activity and the amount of the analyte separated are thus required. Subsequently the amount w_2 of analyte in the sample may be calculated from equation (10.17).

$$\text{total activity added} = a_1 w_1 = a_2(w_2 + w_1) \tag{10.17}$$

This method has been applied to analyses as different as the determination of vitamin B_{12} in biological tissue (using a ^{60}Co label), and the amount of hydrogen absorbed into the lattice of a transition metal.

As the method does not require quantitative separations, and involves the acknowledged sensitivity accompanying radioactive measurements it is attractive for trace element determination. Limitations are however imposed by the problems of determining the chemical yield. A method known as *substoichiometric yield determination* has been developed to overcome this. Typically, a known (substoichiometric), amount of a complexing agent is used under conditions guaranteeing quantitative reaction with the analyte. The amount of analyte separated is then known from the amount of complexing agent used and does not have to be determined. For example, using ^{59}Fe as a tracer, substoichiometric amounts of Cupferron, and chloroform extraction, iron may be determined at levels of 10^{-9} g cm^{-3} with a relative precision of 2%. *Reverse isotope dilution analysis* employs dilution with a non-active 'carrier' to determine the total amount of a labelled compound in a sample.

RADIOIMMUNOASSAY

A most important technique which has been developed as an extension of the isotope dilution principle is that of *radioimmunoassay* (RIA). Analyses by this method employ substoichiometric amounts of specific binding immuno-

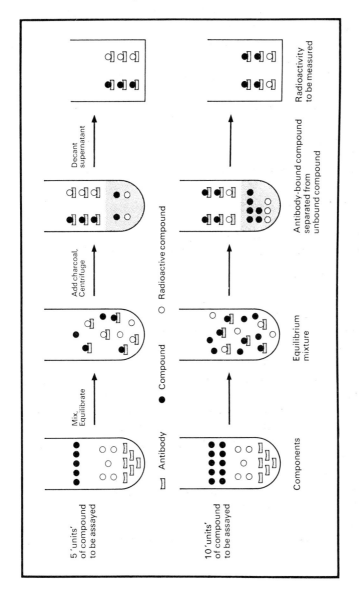

Figure 10.16. The principle of radioimmunoassay. In this example charcoal is used to adsorb the unbound compound and the amount of radioactive compound bound to antibody is measured

468   ANALYTICAL CHEMISTRY

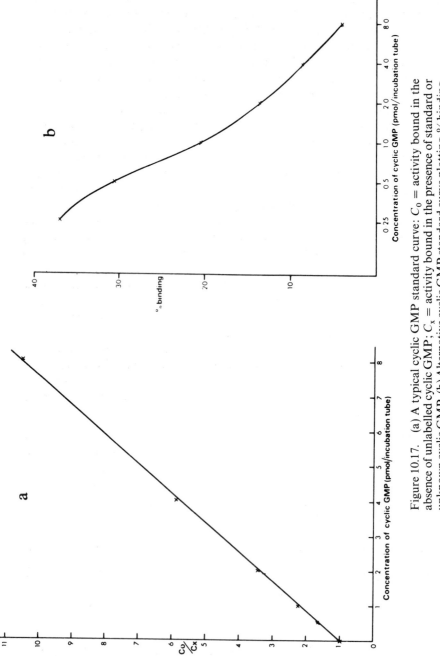

Figure 10.17. (a) A typical cyclic GMP standard curve: C_0 = activity bound in the absence of unlabelled cyclic GMP; C_x = activity bound in the presence of standard or unknown cyclic GMP. (b) Alternative cyclic GMP standard curve plotting % binding

chemical reagents for the determination of a wide range of materials (immunogens) which can be made to produce immunological responses in animals such as sheep or rabbits. It is possible to combine the specificity of an immunochemical reaction with the extreme sensitivity of radiotracer detection. Analytical methods based upon these principles have achieved wide applicability in the determination of organic compounds at trace levels.

In order to establish an RIA method, it is first necessary to obtain the appropriate specific binding agent. This is done by using the analyte to trigger the natural defence mechanism of an animal in which antibodies are produced to combine with, and to neutralize the foreign molecules (immunogens or antigens) entering the bloodstream. These antibodies are specific binding proteins (γ-globulins) which exhibit only a limited reactivity with other immunogens even if they are closely similar in structure. Analytes with relative molecular masses of 6 000 or greater will usually produce the desired response themselves, but smaller molecules (haptens) may need to be attached to a suitable protein before injection into the animal. A binding agent thus produced will normally show sufficient specificity for the original analyte to provide the basis of an RIA method. Once the level of antibody in the blood of the animal is high enough (6–12 weeks), blood is taken from the animal and the serum, together with the antibodies, separated. Further separation is not needed and the binding agent is used in this form. Serum containing antibodies thus produced is often known as *antiserum*. Undoubtedly the production of a suitable antiserum is the most difficult stage in the development of an RIA procedure, requiring specialist biochemical knowledge and skill. Subsequent stages of the analysis are usually straightforward and a range of RIA reagents is now available in kit form. Typical procedures involve the following stages:

1. Mixing samples or standards in solution with a measured amount of the labelled analyte and a measured substoichiometric amount of antiserum.
2. Adjustment of solution conditions followed by incubation (e.g. at 2–4°C for $1\frac{1}{2}$ hours) to ensure maximum and reproducible reaction.
3. Separation of bound and free analyte. This is often done by adsorption of the free analyte on activated charcoal or precipitation of the protein bound fraction with ammonium sulphate.
4. Measurement of the ratio of radioactivity in the bound and free fractions.

The activity ratio A may be related to the amount of analyte X in the sample by

$$A = \frac{Z}{(X + Y - Z)} \tag{10.18}$$

where Y and Z are, respectively, the amount of labelled analyte added and the amount equivalent to the antiserum added (i.e. bound). However, in practice it is advisable to construct a suitable calibration graph because of the danger of cross-reactions and frequent lack of linearity. The steps are summarized diagrammatically in figure 10.16 and typical calibration graphs (dose response curves) in figure 10.17.

Applications of RIA

The determination of small amounts of organic compounds in complex matrices has long been a difficult analytical problem. If separation procedures are used they frequently need to be complex and are time-consuming. By using specific binding reagents, RIA methods avoid such difficulties and facilitate the analysis of many compounds at picogram levels which would otherwise be extremely difficult or impossible to determine. In the many recent studies of biologically active compounds such as hormones and steroids, RIA has been extensively employed. The monitoring of drugs and drug metabolites presents similar analytical problems, and here RIA has been used for both medicinal and forensic investigations. There is also a growing interest in the use of RIA as a method for screening water supplies. Table 10.1 indicates the scope of applications of RIA.

The principal advantages of RIA are its sensitivity, specificity and simplicity of operation. There are three main disadvantages. Firstly, the lengthy development periods for new methods often result from the difficulties of producing the specific binding agent. Secondly, cross-reactions with other molecules similar to the analyte can sometimes interfere. Thirdly, poor precision can result unless careful control of experimental conditions is

Table 10.1   A selection of compounds for which radioimmunoassays have been developed, indicating the wide applicability of the technique

DRUGS	STEROIDS	POLYPEPTIDE HORMONES
Amphetamines	Anabolics	Adrenocorticotrophic hormone
Barbiturates	Androgens	(ACTH)
Diazepam	Corticosteroids	Follicle stimulating hormone
Digoxin	Oestrogens	(FSH)
Gentamicin	Progesterones	Glucagon
Methadone	Steroid glucuronides	Growth hormone
Morphine		Insulin
Nicotine		Luteinizing hormone (LH)
Penicillins		Prolactin
Prostaglandins		
Tetrahydrocannabinol		
Tubocurarine		

employed to ensure reproducible binding reactions. Relative precisions of 1–3 % are typical.

RADIOACTIVATION ANALYSIS

Nuclear bombardment reactions in which the product is radioactive constitute the basis of radioactivation analysis (p. 453). Although in principle any bombardment–decay sequence may be used the analyst is largely concerned with thermal neutron activation. Equation (10.13) relates the induced activity to the amount of the parent nuclide (analyte). However, practical difficulties arise because of flux inhomogeneities. It is common therefore to irradiate a standard with very similar characteristics alongside the sample, e.g. for a silicate rock sample a standard solution would be evaporated on to a similar amount of pure silica. On the assumption that identical specific activities for the analyte are then induced in the sample and standard the amount w_2 of analyte is readily calculated

$$\frac{a_1}{w_1} = \frac{a_2}{w_2} \qquad (10.19)$$

where a_1 and a_2 are the activities measured for standard and analyte respectively under identical conditions, and w_1 and w_2 refer to the respective masses. Neutron activation analysis is an extremely sensitive, selective and precise technique, table 10.2.

Table 10.2 Sensitivity of neutron activation analysis in determination of elements by (n, γ) reactions with a neutron flux of $10^{12}\,\mathrm{n\,cm^{-2}\,s^{-1}}$

GRAMS	ELEMENT
10^{-12}	Eu, Dy
10^{-11}	Mn, Pd, In, Sm, Ho, Re, Ir, Au
10^{-10}	Na, Sc, Cu, Ga, As, Br, Kr, Y, Sb, Pr, Tb, La, Er, Tu, Yb, Ta, W, Th, U
10^{-9}	P, Ar, K, Rb, Co, Ru, Cd, Cs, Ba, Ce, Nd, Gd, Hf, Os, Pt, Hg
10^{-8}	Cl, Si, Ni, Zn, Ge, Se, Mo, Ag, Sn, Te, Xe, Tl
10^{-7}	Ca, Sr, Fe, Zr, Bi
10^{-6}	S, Pd
10^{-5}	Mg, Pb

These valuable characteristics derive from a combination of factors. Firstly, extremely sensitive instrumentation with a facility for spectrometric distinction between radionucleides is available. In particular, a Ge(Li) or intrinsic solid state detector together with a multichannel analyser can be used to make activity measurements by γ-ray spectrometry, often with a bare minimum of sample treatment. Secondly, activation cross sections can be large (up to 10^5 barns) and intense neutron fluxes are available (10^{14} neutrons $cm^{-2} s^{-1}$) whence $N_Y\lambda_Y$ becomes large (equation 10.13)). Thirdly, the reagent blank problem which is so common in trace element analysis is largely eliminated, for contaminants so introduced after activation will not be active and thus not detected. Finally, when sample processing prior to measurement is needed, the problem of working with microgram amounts of material can be simplified by the addition of non-active 'carrier' which does not affect the final activity measurement.

Neutron activation analysis is an attractive method in many trace element problems, or where the total amount of sample is limited. Many geochemical studies of trace constituents and semi-conductor (transistor) developments have used the technique, whilst in recent years pollution investigations have provided a new focus. Interest in the forensic potentialities is also growing. Small flakes of paint, single hairs and a variety of other small samples have been analysed and identified by activation analysis. In recent years activation analysis has lost further ground to ICP-MS which provides more comprehensive information and is more readily operated. Sensitivity is also comparable in many cases.

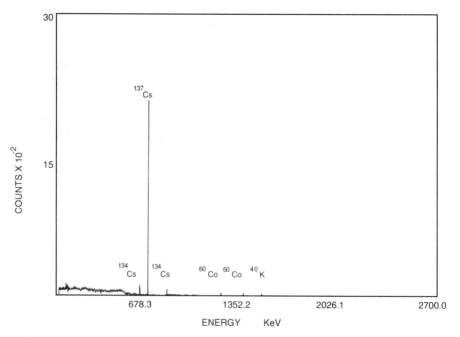

Figure 10.18. γ-Ray spectrum of a soil sample from Belarus showing the identification of ^{134}Cs and ^{137}Cs fallout from the Chernobyl accident

ENVIRONMENTAL MONITORING

One aspect of environmental monitoring which has been highlighted in recent years concerns the detection, identification and determination of radioactive substances. As a high proportion of radionucleides emit γ-radiation, high resolution γ-ray spectrometry is a valuable tool for detection and measurement. Typically, a semiconductor detector with a multichannel pulse height analyser will be used. Identification will be by peak energy and quantitative analysis by peak area. Figure 10.18 illustrates the identification of ^{134}Cs and ^{137}Cs in fallout from the Chernobyl accident. α and β^- measurements may also be needed. Separation of the radioactive material is then required prior to measurement. This is accomplished by the use of techniques such as precipitation, solvent extraction, ion-exchange, and where α-counting is used for transuranic elements, electrodeposition.

Problems

1. ^{51}Ti decays to ^{51}V by the emission of β^- particles and γ-rays. A spectroscopic examination gave the following energies for the radiations.

$$\beta^- \qquad 1.50\,\text{MeV}, 2.105\,\text{MeV}$$
$$\gamma \qquad 0.928, 0.605, 0.323\,\text{MeV}$$

Construct a reasonable decay scheme for the decay.

2. Calculate the disintegration rate of $1\,\text{cm}^3$ of tritium at standard temperature and pressure. What will the activity be after 2.5 years? The half life of tritium is 12.26 years. Express your answers in curies and becquerels.

3. A sample was irradiated in a neutron flux and the induced radioactivity monitored after it was removed from the flux. The fully-corrected results are presented below.

Plot a log/linear curve of activity v. time and analyse it into its components. Deduce the half-lives and the initial activities of the components.

Time (hours)	Activity (counts/min.)	Time (hours)	Activity (counts/min.)
0	7 300	4.0	481
0.5	4 680	5.0	371
1.0	2 982	6.0	317
1.5	1 958	7.0	280
2.0	1 341	8.0	254
2.5	965	10.0	214
3.0	729	12.0	181
3.5	580	14.0	153

4. $10\,\mu g$ of natural zinc was irradiated in a thermal neutron flux of 5.10^{12} neutrons $\text{cm}^{-2}\,\text{s}^{-1}$ for one week. Calculate the induced activity due to ^{65}Zn which is produced from an (n, γ) reaction on ^{64}Zn.

The natural abundance of ^{64}Zn is 48.88%, and its activation cross-section to thermal neutrons is 0.50b. The half life of ^{65}Zn is 245d.

5. Natural sodium is monoisotopic and undergoes activation by thermal neutrons as follows:

$$^{23}Na(n,\gamma)^{24}Na \xrightarrow[t_{\frac{1}{2}} = 14.9\,hrs]{\beta^-/\gamma} {}^{24}Mg\,(stable)$$

The activation cross-section is 0.53 barns.

If 100 μg of sodium is irradiated for 1 day in a flux of 5.10^{12} neutrons cm^{-2} s^{-1}, what will its activity be when it has been removed from the reactor and transferred to a laboratory 3 hours' journey time away?

6. A mixture is being assayed by radioisotope dilution analysis. 10 mg of the labelled analyte (0.51 μCi mg^{-1}) was added. 1.5 mg of the pure analyte was separated and its specific activity measured and found to be 0.042 μCi mg^{-1}. What was the amount of analyte in the original sample?

7. In a radioimmunoassay procedure, 1 pmol of labelled analyte was added to the sample. Antiserum equivalent to 75 pmol of analyte was then added, and after incubation the bound and free fractions separated by precipitation with ammonium sulphate solution. The activity ratio, bound:free, was measured to be 2.88. Calculate the number of moles of analyte in the original sample.

Further Reading

ANON, *Sampling and Measurement of Radionuclides in the Environment*, HMSO 1989.
CHAPMAN, D. I., 'Radioimmunoassay', *Chemistry in Britain*, **15**, No. 9, 439, 1979.
KNOLL, G. F., *Radiation Detection and Measurement* (2nd edn), Wiley, New York, 1989.
McKAY, H. C., *Principles of Radiochemistry*, Butterworth, London, 1971.
MALCOLME-LAWES, D. J., *Introduction to Radiochemistry*, Macmillan, London, 1979.
PASTERNAK, C. A. (Ed.), *Radioimmunoassay in Clinical Biochemistry*, Heyden, London, 1975.
SALMON, L., *Instrumental Neutron Activation Analysis in Environmental Studies of Trace Elements*, AERE-R 7859, HMSO 1975.

Chapter 11

Thermal Techniques

Thermal methods of analysis may be broadly defined as methods of analysis in which the effect of heat on a sample is studied to provide qualitative or quantitative analytical information. Such studies have a long history, but it is only in the last 30 years that instrumental improvements have led to methods which are both simple and reliable to operate. This has been accompanied by a steadily widening applicability. Such techniques are now applied across a wide range of areas in which analytical chemistry is used. Thermal analysis may be defined as a group of techniques in which a property of the sample is monitored against time, or temperature while the temperature of the sample, in a specified atmosphere, is programmed. There are a substantial number of sample properties on which the effect of heat has been studied, which has led to the development of a number of recognized techniques for which the appropriate instrumentation is commercially available. The most important of these are summarized in table 11.1. In this text, only *thermogravimetry (TG)*, *differential thermal analysis (DTA)*, *differential scanning calorimetry (DSC)*, *pyrolysis – gas chromatography (pyrolysis-GC)* and *thermomechanical analysis (TMA)* will be discussed in any detail.

Thermal events are usually studied by recording the change in thermal property as the temperature is varied to give a *thermal analysis curve* or *thermogram*. The main thermal events are summarized in table 11.2. Such curves are characteristic of a sample in both qualitative and quantitative senses. However, they can be complex, leading to difficulties in assigning the detail of the curve to particular thermal events. Nevertheless, complex curves can still be used as a 'fingerprint' for identification of the sample. Derivatives may often be of help in interpretation. The best results may be obtained by using a combination of thermal techniques, e.g. TG with DSC or DTA. It should be recognized that results in thermal analysis are very dependent upon the conditions and parameters surrounding their measure-

475

Table 11.1　The main thermal analysis techniques. (From *Introduction to Thermal Analysis*, M. E. Brown, Chapman and Hall.)

PROPERTY	TECHNIQUE	ABBREVIATION
Mass	Thermogravimetry	TG, TGA
	Derivative thermogravimetry	DTG
Temperature	Differential thermal analysis	DTA
Enthalpy	Differential scanning calorimetry	DSC
Dimensions	Thermodilatometry	
Mechanical properties	Thermomechanical analysis (Thermomechanometry)	TMA
	Dynamic mechanical analysis	DMA
Optical properties	Thermoptometry or thermomicroscopy	
Magnetic properties	Thermomagnetometry	TM
Electrical properties	Thermoelectrometry	
Acoustic properties	Thermosonimetry and thermoacoustimetry	TS
Evolution of radioactive gas	Emanation thermal analysis	ETA
Evolution of particles	Thermoparticulate analysis	TPA

Table 11.2　Thermal Events. (From *Introduction to Thermal Analysis*, M. E. Brown, Chapman and Hall.)

$A(s_1) \longrightarrow$	$A(s_2)$	phase transition
	$A(l)$	melting
	$A(g)$	sublimation
	$B(s)+\text{gases}$	decomposition $\begin{cases} \text{thermal} \\ \text{radiolytic} \end{cases}$
	gases	
$A(\text{glass}) \to A(\text{rubber})$		glass transition
$A(s)+B(g) \to C(s)$		oxidation / tarnishing
$A(s)+B(g) \to \text{gases}$		combustion / volatilization
$A(s)+(\text{gases})_1 \to A(s)+(\text{gases})_2$		heterogeneous catalysis
$A(s)+B(s) \to AB(s)$		addition
$AB(s)+CD(s) \to AD(s)+CB(s)$		double decomposition

ment. For example, the history of the sample and its mass, the nature of the sample holder, the rate of heating or cooling, the surrounding atmosphere and its flow rate, will all have an effect.

11.1 Thermogravimetry

SUMMARY

Principles
Study of the change in mass of a sample as the temperature is varied.

Instrumentation
Sensitive balance with the sample pan inside a furnace whose temperature can be accurately controlled and programmed for change. Facilities for controlling the atmosphere of the sample. Electronic integration and derivative curve presentation.

Applications
Qualitative and quantitative analysis for a wide range of sample types, especially for inorganic materials and polymers. Kinetic studies where weight changes can be clearly attributed to a particular reaction. Chemical reactions, volatilization, adsorption and desorption may be studied. Relative precision at best ca.1% but very variable.

Disadvantages
Limited to samples which undergo weight changes, thus melting, crystal phase changes etc. cannot be studied. Complex thermal traces are often difficult to interpret.

Thermogravimetry (TG), formerly known as *thermogravimetric analysis (TGA)*, is based on the very simple principle of monitoring the change in weight of a sample as the temperature is varied. By controlling the atmosphere e.g. with O_2 or N_2 it may be possible to encourage or suppress oxidation reactions, thus controlling to some extent the nature of the thermal events occurring. When heated over the range of temperatures, ambient to approximately 1000°C, many materials undergo weight changes giving characteristic curves. Where the changes can be linked to a particular thermal event, such as oxidation, or loss of water of crystallization, the size of the step in the curve can be used for quantitative analysis. Where thermograms are complex, or where changes are subtle, derivative curves

(DTG) can be valuable in interpretation. Figure 11.1 shows an example of a straight forward thermogram and Figure 11.2 a more complex one showing the value of DTG.

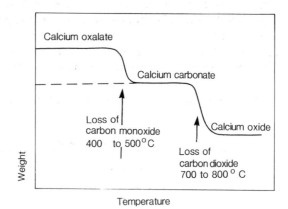

Figure 11.1.   TG curve for calcium oxalate

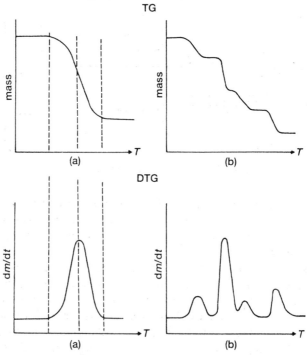

Figure 11.2.   Comparison of TG and DTG curves

INSTRUMENTATION

Early designs of thermobalances consisted of little more than a good quality chemical balance with one pan suspended in an electric furnace. Linked chart recorders provided a trace of temperature and mass changes. In some senses this experimental arrangement can still be seen in modern instruments. However, the evolving requirements of greater sensitivity, both in terms of sample size and mass change, mean that current instruments are based on the use of electronic micro-balances. The schematic arrangement of such a thermobalance is shown in figure 11.3. For use in

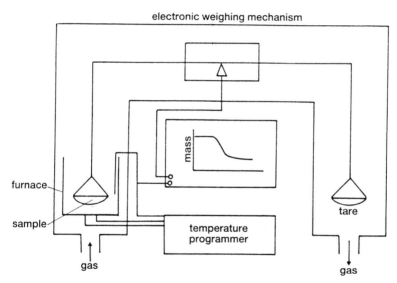

Figure 11.3.   A schematic thermobalance

TG, a null-point mechanism is to be preferred, as this ensures that the sample remains in the same part of the furnace, even when the mass changes. Thus any effects due to thermal gradients within the furnace tube remain constant.

The design and operation of the furnace are of critical importance in obtaining good quality, reproducible thermograms. Most instruments use electrical resistance heaters, although heating techniques using infrared, laser irradiation or microwave induction heating have been investigated. The latter has a particular attraction as it would enable uniform heating throughout the sample to be employed. The arrangement of the sample pan may be of horizontal, toploading, or suspended design. The effects of convection currents within the furnace tube, which can lead to uneven heating can be minimized by a series of baffles. Even so, there will still be a thermal gradient within the furnace, and careful temperature calibration is needed. One method of calibration which has been used effectively is the *Curie-point* method. Ferromagnetic materials lose their magnetism on heating at exactly reproducible temperatures or Curie points. A range of metals or alloys with Curie points between 150°C and 1000°C is available. If suitable ferromagnetic calibration standards are placed in the sample pan of the balance, and a large permanent magnet is placed below the pan, the sample will experience a downward attraction leading to an apparent increase in weight (figure 11.4). At the Curie point the loss of the ferro-magnetism will be reflected by an apparent loss of weight, enabling the temperature experienced by the balance pan to be accurately known. Using a range of standard materials, an accurate calibration curve of the furnace can

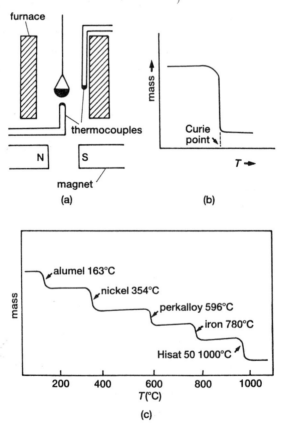

Figure 11.4.   Curie-point method of temperature calibration

be produced. However, this does not completely solve the problem of accurate assessment of the temperature actually experienced by the sample. With the exception of microwave induction heating, the heat is absorbed by the exterior of the sample and transferred to the interior by conduction. Thus a temperature gradient will exist within the sample. Many materials that are the subject of TG investigation, will have low thermal conductivities whence this effect will be pronounced. The only remedy is to reduce the sample size to the absolute minimum, e.g. 10 mg, and avoid rapid heating programmes.

APPLICATIONS OF TG

TG may be used to characterize and compare samples using the thermogram as a fingerprint. Where the thermal processes taking place are known or where a step in the thermogram of a mixture may be clearly related to one component, quantitative analysis may be possible. Weight changes in chemical reactions may be monitored in order to follow the kinetics of the reaction. The earlier developments of modern TG methods in the 1950s and 1960s were exploited largely in the field of inorganic chemistry and produced a revolution in inorganic gravimetric analysis. Some analyses have already been exemplified in figures 11.1 and 11.2. Subsequently however the use of TG has spread into more diverse fields. Noteworthy are applications to polymers for thermal stability studies, as well as qualitative and quantitative analysis, whilst in pharmaceutical preparations both active ingredients and excipients can be analysed. Figures 11.5 and 11.6 show examples of these applications.

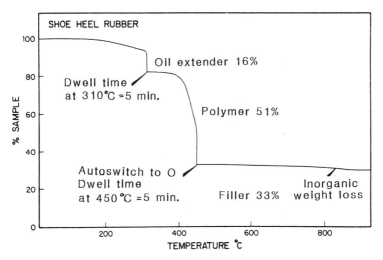

Figure 11.5.   Determination of extender, polymer and filler in an elastomer

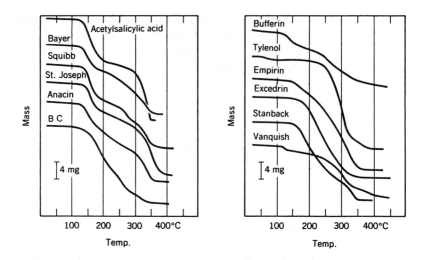

Figure 11.6.   TG curves of commercially available analgesics as determined by Wendlandt and Collins

11.2   Differential Thermal Analysis (DTA)

SUMMARY

Principles

Monitoring of the temperature difference between a sample and an inert reference as they are heated uniformly. Endothermic or exothermic changes in the sample lead to characteristic deviations in temperature, which can be used for qualitative and quantitative analysis.

Instrumentation

Programmed electric furnace containing sample and reference. Sensitive thermocouples and recorders to produce a ΔT vs T plot. Facility to control atmosphere of sample.

Applications

Fingerprinting by the pattern of thermograms, qualitative and quantitative analysis by peak area or height for a wide range of materials.

Study of thermal characteristics, stability, degradation and reaction kinetics on small samples, over a temperature range of $-175°C$ to $1000°C$ and above. In some cases precision may be good (1%) but it is variable and may be much poorer.

Disadvantages

Small sample sizes are often required to minimize thermal conductivity problems. Less satisfactory than DSC with regard to resolution of thermal traces and quantitative data.

Differential thermal analysis (DTA) is based upon the measurement of the temperature difference (ΔT) between the sample and an inert reference such as glass or Al_2O_3 as they are both subjected to the same heating programme. The temperature of the reference will thus rise at a steady rate determined by its specific heat, and the programmed rate of heating. Similarly with the sample, except that when an exothermic or endothermic process occurs a peak or trough will be observed. Typical behaviour is shown schematically in Figure 11.7.

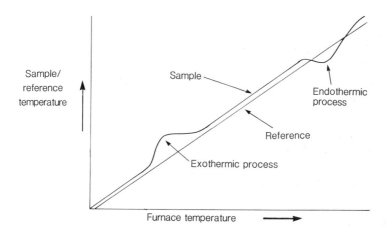

Figure 11.7. Schematic representation of the variation of sample and reference temperature in DTA

In practice ΔT vs furnace temperature is plotted giving a thermogram of the type illustrated in figure 11.8. Figure 11.9 shows a curve for the DTA examination of calcium oxalate monohydrate exemplifying also the effect of changing the atmosphere from nitrogen to air.

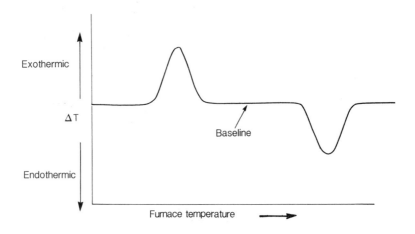

Figure 11.8. ΔT vs T plot to give a schematic DTA trace

Mostly, the convention with exothermic peaks (*exotherms*) shown as positive and endothermic peaks (*endotherms*) as negative is accepted. However, there are exceptions and care should always be exercised in the interpretation of recorded data.

INSTRUMENTATION

The essential instrumentaion for DTA involves a single furnace containing both the sample and reference, which are heated at the same rate. The temperatures of the sample and reference are measured separately by individual thermocouples. A thermogram is then plotted of the difference in temperature between sample and reference against overall furnace temperature. Two differing conformations for sample and reference heating

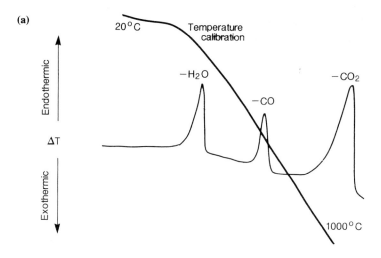

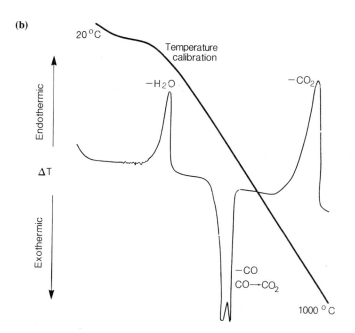

Figure 11.9.   DTA study of calcium oxalate $CaC_2O_4.H_2O$ showing the effect of changing the atmosphere from (a) nitrogen to (b) air

may be found in use. In the first the sample and reference are heated in recesses in the same heating block, figure 11.10(a). This arrangement is known as a *classical DTA* instrument. In the second arrangement, the sample and reference pans are placed on separated heating blocks or heat sinks, with the temperature of the block being measured rather than the sample itself, figure 11.10(b). Such an arrangement comprises a *calorimetric* or *Boersma* DTA instrument. The two variations provide essentially the same data, but the latter, whilst less dependent on the thermal properties of the sample, has a slower response to thermal changes. The temperature range of DTA instruments is obviously dependent upon the nature of the furnace and thermocouples and the limitations imposed by them. Typically instruments are designed to operate in the range from room temperature to ca. 1000°C. However, considerable extension to both higher and lower temperatures is possible with suitably modified equipment. Calibration is by heating known reference materials of accurately known external characteristics.

APPLICATIONS OF DTA

Typical DTA curves have been illustrated in figures 11.8 and 11.9. The combination of exotherms and endotherms will be unique to a particular sample composition. Thus the pattern of the thermogram can be used as a finger print for qualitative analysis, whilst the areas under the curve may be used for quantitative analysis. These principles can be widely applied to samples of very different types, e.g. minerals, inorganic compounds, pharmaceuticals, polymers, foodstuffs and biological specimens. Typical samples sizes are from 1 mg upwards allowing, where necessary, for measurements to be made on small samples. DTA may be used effectively in a simple characterization or a purity estimate by studying the melting

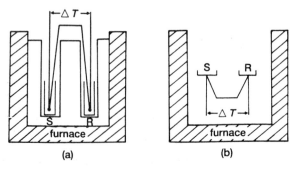

Figure 11.10.   Differential thermal analysis (DTA). (a) Classical apparatus
(S = sample; R = reference). (b) Calorimetric

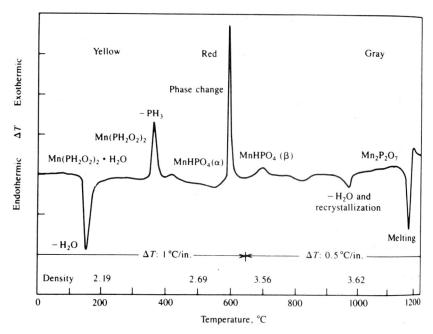

Figure 11.11.   DTA curve for $Mn(PH_2O_2)H_2O$

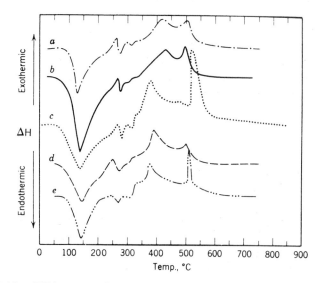

Figure 11.12.   DTA curves of potato and corn starch. (a) Potato starch; (b) potato starch, duplicate run; (c) corn starch; (d) methanol-extracted corn starch; (e) ammonia-pregelatinized corn starch

characteristics of a specimen. On the other hand much more complex systems may be usefully studied. Figure 11.11 illustrates a complex behaviour pattern for an inorganic compound and figure 11.12 the characterization of starch in foodstuffs. Reaction kinetics may be studied and measurements of enthalpy changes made.

Applications in polymer analysis are illustrated in figures 11.13 and 11.14. The former shows schematically the typical thermal processes which can occur on heating a polymer, and the latter an analysis of a seven component mixture based upon melting points.

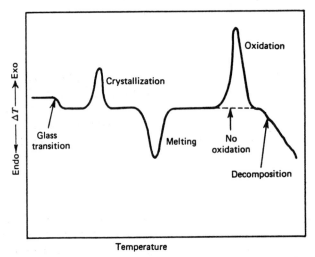

Figure 11.13.   Schematic DTA curve of a typical polymer. Note: glass transitions reflect a heat capacity change but none H = 0. Thus they are seen as 'steps' on the baseline rather than peaks

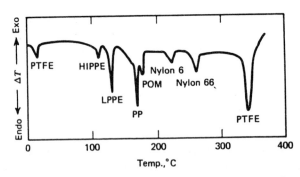

Figure 11.14.   DTA curve of a seven-component polymer mixture showing endothermic peaks associated with melting

Probably the main weakness of DTA as a method of analysis remains the difficulty of linking the thermal changes shown on the thermogram, with the actual thermal processes taking place. It should be noted that data obtained by DTA are often similar to that available for differential scanning calorimetry. Indeed the two techniques overlap extensively and may be seen as complementary. A comparison of the two techniques is made at the end of the next section.

11.3   Differential Scanning Calorimetry (DSC)

SUMMARY

Principles
   Sample and an inert reference heated separately, with the power supply to the sample heater variable so that the temperature difference can be maintained at zero even when endothermic or exothermic changes occur. The difference in power supplied to the two heaters is monitored as the analytical signal (ΔE). Alternatively, the differential heat flow to sample and standard when they are heated from the same source is monitored.

Instrumentation
   Sample and reference crucibles with separate heaters. Thermocouples with feedback to sample heater so that the power is varied to maintain $\Delta T = 0$. Data output equipment to provide ΔE vs temperature curves, derivative curves and peak integration. Facility to vary atmosphere of sample.

Applications
   Widespread study of thermal properties on an extensive range of sample types. Qualitative and quantitative analysis. Relative precision is very variable at best ca 1% but can be much poorer.

Disadvantages
   Usually limited to small sample sizes. Thermograms are often complex and thus difficult to interpret fully.

Differential scanning calorimetry (*DSC*) is a technique which aims to study the same thermal phenomena as DTA, but does so on a rather different principle. Hence, although the data obtained are very similar, they may differ in detail. Typical DSC equipment will operate over the temperature

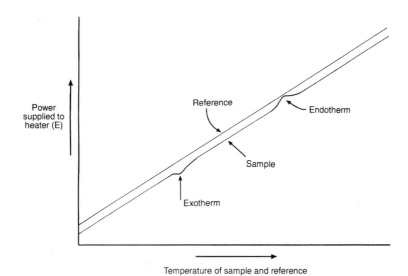

Figure 11.15.   Schematic representation of the variation of power supply to the
sample and reference in DSC

range from ambient to ca. 700°C. However, as with DTA, specially
modified equipment can extend this substantially in both directions.

In principle, like DTA, DSC involves the heating of the sample and an inert
reference in parallel. However, for *power compensated DSC*, the two are
heated quite separately with separate electrical heaters. The heaters are
programmed to ensure that the temperatures of both sample and reference
advance at exactly the same rate. It follows that when endotherms or
exotherms occur in the sample, the power to the heater will need to be varied
in order to maintain $\Delta T = 0$, figure 11.15. Thus by monitoring the difference
in power supplied to the heaters (ΔE) the thermal changes in the sample
may be followed. Figure 11.16 illustrates a typical DSC thermogram
schematically. It is worthy of note that the measurement of ΔE is effectively
a direct measurement of the energy change in the sample. This makes DSC
particularly appropriate as a technique for the measurement of ΔH values
which can be derived from the areas of the peaks obtained. *Heat flux DSC*
attains similar results by heating sample and reference from the same source.
Thermocouples are used to sense the differential heat flow (supply) to the
sample and standard.

A comparison of figures 11.15 and 11.16 with 11.7 and 11.8 will show the
peaks for endotherms and exotherms reversed in direction for DTA relative
to DSC. This is a commonly met presentation but as there is no agreed

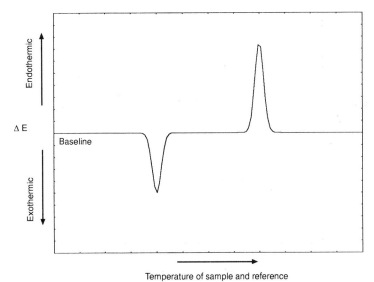

Figure 11.16.   Schematic DSC curve

convention, different authors may use different presentations. Confusion can arise unless care is exercised in the interpretation of the thermograms.

INSTRUMENTATION

The equipment for power compensated DSC involves two parallel temperature measurement systems. Sample (ca. 50 mg) and reference in small pans are placed on the separate blocks (figure 11.17). Each block is provided with a

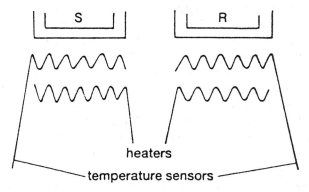

Figure 11.17.   Power-compensated differential scanning calorimetry (DSC) apparatus (S = sample; R = reference)

separate heater and thermocouple with feedback to the heaters so that the power supply can be varied as necessary to ensure that $\Delta T = 0$ at all times as the temperature is raised overall. A thermogram is then generated by monitoring the difference in power supplied to the two heaters (ΔE) and plotting this against the overall temperature (figure 11.16). Data presentation at its simplest may be on a chart recorder, but increasingly microcomputers or microprocessors are used to give added flexibility to the presentation. Built into the system will be a facility to control the atmosphere of the sample. For heat flux DSC the sample and reference in separate containers are placed on

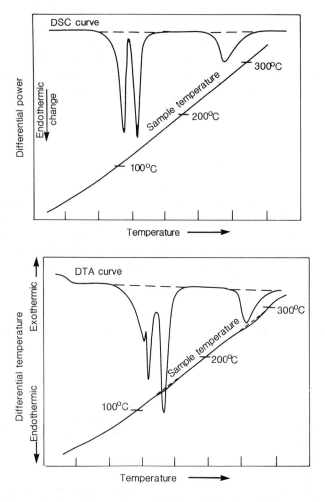

Figure 11.18.   Comparison of DSC and DTA for $CuSO_4.5H_2O$

separate platforms which sit on a heated metal (Cu/Ni alloy) disc. Thermo-couples are placed to monitor the heat flow from the disc to the sample and standard. The differential heat flow will then reflect the different thermal behaviour of the sample and standard. A plot of this against overall tempera-ture produces a graph analogous to those from power compensated DSC. Calibration is by use of standards (e.g. indium) with accurately known thermal characterization.

APPLICATIONS OF DSC

DSC essentially studies the same thermal phenomena as DTA, albeit using a different principle. Thus DTA and DSC provide very much the same information and their applications are similar. Reference back to the section on the applications of DTA will suffice to indicate the scope of DSC. Some differences in the quality of the information obtained sometimes exist however, leading to a preference for one technique over the other for particular purposes.

DTA AND DSC

The experimental set up for heat-flux DSC is very similar to that for calorimetric or Boersma DTA. Thus heat-flux DSC will have the same freedom from the thermal properties of the sample and slower response times associated with Boersma DTA. DSC will generally have better resolution as illustrated in figure 11.18. Finally, as has been discussed earlier, by measuring the power differential, DSC is making a direct measurement of enthalpy changes and is thus the more satisfactory tool for thermodynamic measurements.

11.4 Thermomechanical Analysis (TMA) and Dynamic Mechanical Analysis (DMA)

SUMMARY

Principles
    Measurement of the effect of heat on the mechanical properties of a sample, e.g. expansion, compression, penetration, extension and resonant frequency of oscillation.

Instrumentation
    Quartz probes fitted with thermocouples to measure the temperature, and follow the movement of the sample. Linked transducer i.e. a linear

variable density transformer to sense the probe movement and produce a related electrical signal. Sample furnace, programmers and various output devices.

Applications

Investigation of the mechanical properties of a range of materials, especially polymers, and their change with heating over the range $-100°C$ to $1000°C$. Quality control of mechanical properties.

Disadvantages

Information is restricted largely to mechanical properties and cannot easily be related to the actual composition of a sample.

One of the more recently exploited forms of thermal analysis is the group of techniques known as *thermomechanical analysis (TMA)*. These techniques

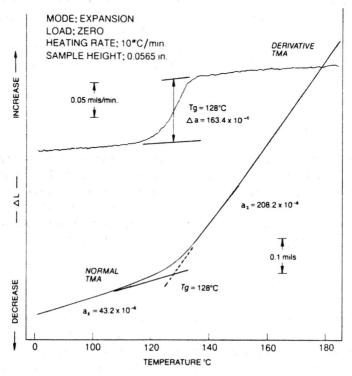

Figure 11.19.   TMA expansion study of an epoxy printed circuit board showing the determination of the glass point (T_g)

are based on the measurement of mechanical properties such as expansion, contraction, extension or penetration of materials as a function of temperature. TMA curves obtained in this way are characteristic of the sample. The technique has obvious practical value in the study and assessment of the mechanical properties of materials. Measurements over the temperature range from $-100°C$ to $1000°C$ may be made. Figure 11.19 shows a study of a polymeric material based upon linear expansion measurements.

INSTRUMENTATION

A TMA analyser will need to accurately measure both the temperature of the sample, and very small movements of a probe in contact with the surface of the sample. A typical analyser, as illustrated in figure 11.20(a) and (b), uses a quartz probe containing a thermocouple for temperature measurement, and is coupled to the core of a linear variable differential transformer (LVDT). Small movements at the sample surface are transmitted to the core of the LVDT and converted into an electrical signal. In this way samples ranging from a few μm to cm thicknesses may be studied with sensitivity to movements of a few μm. For studying different mechanical properties the detailed construction of the probe will vary as is illustated in figure 11.20(c).

APPLICATIONS OF TMA

TMA has mainly been used in the study of polymers. The mechanical properties study may be used to characterize a polymer as well as to assess its mechanical utility. There is an obvious application to quality control. The ability to study small specimens gives the technique a distinct advantage over more traditional methods of mechanical testing if sample size is limited. A typical TMA study has already been exemplified in figure 11.19.

DYNAMIC MECHANICAL ANALYSIS

An associated technique which links thermal properties with mechanical ones is Dynamic Mechanical Analysis (DMA). In this, a bar of the sample is typically fixed into a frame by clamping at both ends. It is then oscillated by means of a ceramic shaft applied at the centre. The resonant frequency and the mechanical damping exhibited by the sample are sensitive measurements of the mechanical properties of a polymer which can be made over a wide range of temperatures. The effects of compositional changes and methods of preparation can be directly assessed. DMA is assuming a position of major importance in the study of the physico-chemical properties of polymers and composites.

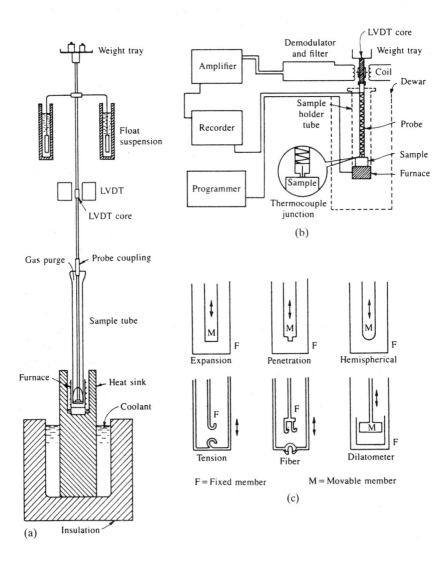

Figure 11.20. Schematic representation of a thermomechanical analyser (a) and (b) with various probe configurations (c)

11.5   Pyrolysis – Gas Chromatography

SUMMARY

Principles

Controlled pyrolysis of small samples followed by GC separation of the products. Identification by MS or FT-IR.

Instrumentation

Furnace, Curie-point or heated filament pyrolysers linked to packed column or capillary column gas chromatograph. GC-MS or GC-FTIR interfaces.

Applications

Potentially applicable to a wide range of organic materials. Most use has been made in the analysis of polymers and oils. Essentially a qualitative technique with poor precision for quantitative measurements.

Disadvantages

Relatively poor reproducibility because of the dependence on accurately reproduced heating profiles and the use of very small samples.

Evolved gas analysis (EGA) is based on the study of gases or volatile breakdown products emitted by a sample on heating. The identity and properties of the volatile materials emitted serve as a basis for the analysis of the sample. One particular technique of EGA which has attracted substantial attention is *Pyrolysis–gas chromatography* (Py–GC). As the name suggests this technique uses gas chromatography to separate the breakdown products of the sample which have been produced by carefully controlled pyrolysis. The pyrogram thus obtained will generally show a complex pattern of peaks which may be used for both qualitative and quantitative analysis in the ways discussed in section 4.2.1. Identification of the peaks present in the pyrogram has been carried out for some time by standard GS–MS procedures. However, more recently, with the development of FT-IR instruments, GC–FTIR is also being exploited in this way. An example of Py–GC–MS analysis is given in figure 11.21 and of Py–GC–FTIR in figure 11.22.

INSTRUMENTATION

The essential instrumentation is divided into three parts: (a) the pyrolyser, (b) the gas chromatograph and (c) the MS or FTIR instruments. In this chapter

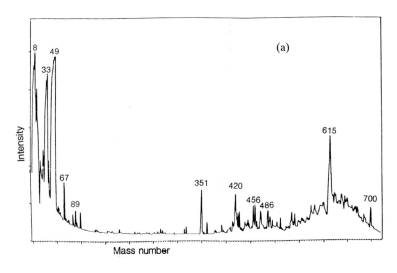

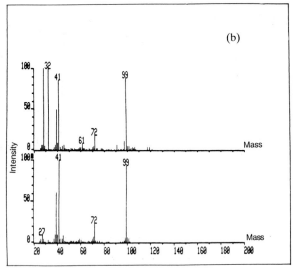

Figure 11.21. (a) Pyrogram of polystyrene cement. (b) Mass spectrum of peak 36 from polystyrene cement and its first library fit: 3-isothiocyanatopropene

interest focuses on pyrolysers as the other instruments are discussed elsewhere.

It has been recognized that best results are obtained when the temperature of the sample is raised rapidly and reproducibly to the pyrolysis temperature

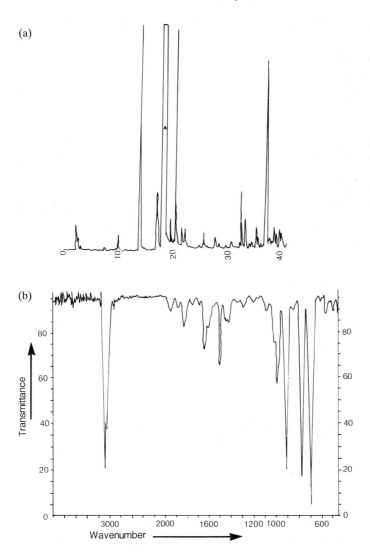

Figure 11.12. (a) Py–GC trace for polystyrene. (b) FTIR spectrum of 'peak A' identifying it as styrene (courtesy of P. J. Haines, Kingston University)

and then held closely at that temperature for the desired pyrolysis time. One obvious way of achieving this aim is by the use of an electrically heated microfurnace. Considerable difficulties were encountered in the development of such furnaces with suitable characteristics, and although pyrolysers of this type are now readily available and in use, they still suffer from the relative disadvantage of rise times of several seconds. A design for a modern furnace is shown in Figure 11.23.

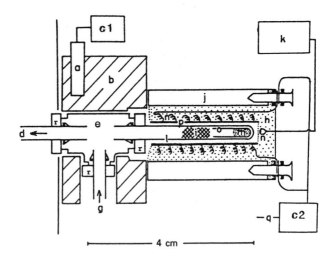

Figure 11.23. Controlled furnace-type pyrolyser: a, heater; b, Al block; c, variable transformer; d, gas outlet to column; e, Swagelok union; f, column oven; g, gas inlet; h, cement; i, glass wool plug; j, insulating block; k, pyrometer; l, stainless steel chamber; m, sample; n, heater thermocouple; o, pyrolysis tube; p, ceramic tube; q, line voltage. Reprinted from Irwin, *Analytical Pyrolysis*, Marcel Dekker Inc. NY (1982)

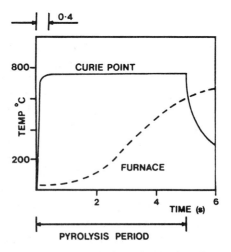

Figure 11.24.   Comparison of pyrolysis times for Curie-point pyrolysis and furnace pyrolysis

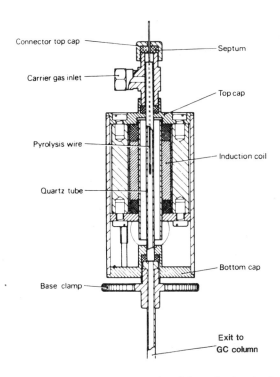

Figure 11.25   A Curie-point pyrolyser. Reprinted from Irwin, *Analytical Pyrolysis*, Marcel Dekker Inc. NY (1982)

An alternative approach has been to use *Curie-point pyrolysers*. The use of the Curie point in accurately reproducing a temperature has already been discussed for the calibration of TG furnaces (p. 480). In a slightly different way the Curie point can be used for accurately reproducing pyrolysis conditions with the added advantage that the rise time is only about 0.4s. The sample, typically 50 μg, is either placed on or encapsulated in the end of a pyrolysis wire composed of the appropriate ferromagnetic alloy. Induction heating can then be used to raise the temperature rapidly to the Curie point where it can be accurately maintained until the induction coil is switched off. The heating profile for a Curie point pyrolyser is contrasted with that for a typical microfurnace in figure 11.24 and the construction of a pyrolysis unit in figure 11.25 with typical pyrolysis wires in figure 11.26.

A third type of pyrolyser sometimes utilizes a filament heated by its own electrical resistance. The most effective pyrolysers of this type use an initial pulse of heating at a high voltage to produce a high current and rapid heating

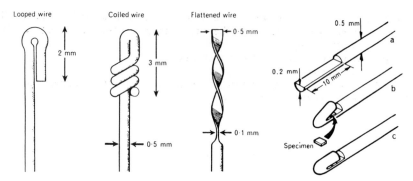

Figure 11.26. Various pyrolysis wire configurations. Reprinted from Irwin, *Analytical Pyrolysis*, Marcel Dekker Inc. NY (1982)

to the pyrolysis temperature, i.e. 700°C in 12 ms, followed by reduction to an accurately controlled maintenance voltage to maintain the pyrolysis temperature.

Problems

1. The decomposition of copper(II) sulphate pentahydrate follows the course shown below. Analyse the graph and calculate the most likely reactions.

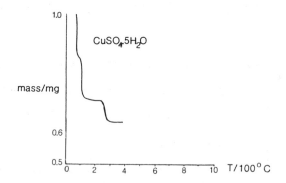

2. A sample of silica packing material with 5 μm particles for use in liquid chromatography has been reacted with various silanizing reagents. A sample of 400 mg, heated at 6°C min^{-1} lost a very small amount by 110°C, but started to lose weight rapidly at 150°C to a plateau corresponding to a loss of 14.0 mg. Above 200°C, further weight was lost and a final plateau obtained at a loss of 45.5 mg by 700 °C.
   Suggest why
   (a) the loss up to 110°C is small
   (b) the loss up to 700°C is in two stages and calculate the percentage by weight of coating material.

3. A sample of white plastic tape was placed in a thermobalance and heated at 10°C min^{-1} in nitrogen to give the thermogravimetric curve below. By careful measurement identify
   (a) the moisture content of the polymer tape;
   (b) the filler content of the tape;
   (c) the identity of the polymer from its decomposition temperature and the stages of the decomposition.

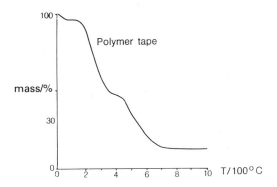

4. Compare and contrast the techniques of DTA and DSC.
5. Polyethylene terephthalate (9.94 mg) gave a peak of area 116.3 cm^2 on melting on a DSC, whereas 5.89 mg of pure indium ($\Delta H_{fus} = 28.45 \, J g^{-1}$) gave a peak of 40.0 cm^2. Calculate the latent heat of fusion of this polyethylene terephthalate, and compare with the pure crystalline value $\Delta H_{fus} = 117.57 \, J g^{-1}$. Comment on the answers.

Further Reading
BROWN, M. E., *Introduction to Thermal Analysis–Techniques and Applications*, Chapman and Hall, 1988.
DODD, J. W. and TONGE, K. H., Thermal Methods. In *Analytical Chemistry by Open Learning*, Wiley, 1987.
HAINES, P. J. *Thermal Methods of Analysis: Principles, Applications and Problems*, Blackie A & P, 1995.
IRWIN, W. J., *Analytical Pyrolysis*, Dekker, 1982.
WENDLANDT, W. W., *Thermal Analysis*, 3rd Ed., Wiley, 1986.
WILLARD, H. H., MERRITT, L. L., Jr., DEAN, J. A. and SETTLE, F. A., Jr., *Instrumental Methods of Analysis*, 7th Ed., Wadsworth Publishing Co, 1988.

Chapter 12

Overall Analytical Procedures and their Automation

In chapter 1 a general pattern for analytical procedures was introduced and the various stages of an analysis identified as: sampling; pretreatment; separation or masking; measurement; interpretation of results. Subsequent chapters have dealt with separation methods, measurement techniques and the interpretation of results in more detail. It remains to examine sampling more closely and to consider, by way of example, some overall analytical schemes.

12.1 Sampling and Sample Pretreatment

REPRESENTATIVE SAMPLES AND SAMPLE STORAGE

It is axiomatic that unless the sample is correctly selected, taken and handled then there is no point in carrying out an analysis. A sample must provide a meaningful measure of the analyte, and only careful consideration of the overall problem by the analyst and his technical colleagues will ensure correct sampling. For a completely homogenous system any part of the whole is suitable for analysis. With a non-homogenous system a more difficult sampling problem exists. For example, a truck load of an ore will contain particles of vastly different sizes and of variable composition and to obtain a sample representing the average composition of the load, a systematic selection of material from various parts and levels must be made. The selected material must then be crushed, ground and mixed to provide a single homogenous sample. Thus the analyst may be presented with several kilograms of crushed material within which there is still some variation in particle size. The final selection of the material on which the analysis is to be carried out may be made by *coning and quartering*. In this process the powdered material is poured into a cone shaped heap, and subsequently divided into four equal parts. Two opposite quarters are combined into a second cone, the remaining two being discarded. The process is repeated until the amount of sample is reduced to that required for analysis. Coning and quartering should ensure that a representative selection of particle sizes is taken at each division.

A similar situation arises with a large metallurgical specimen although crushing and grinding is not possible. Sawing or drilling is used to obtain a systematic set of samples from different parts of the specimen. These may then be analysed individually to provide a profile of the analyte distribution, or mixed, coned and quartered to provide a sample representative of the overall composition. A rather different and more complex sampling situation is discussed in example 3 (p. 514). Having obtained a properly representative sample, care and thought must be given to its storage prior to analysis. Many factors may need to be taken into account, but the prime objects, as in the sample treatment discussed below, must be to avoid losses or contamination and to prevent undesirable changes in chemical form. Use of the right container is important. For instance, solutions can leach trace metals from the wall of a glass container or lose them by adsorption onto it. Biological samples present special problems as they decompose rapidly unless refrigerated or stored in a sterile environment.

In a limited number of cases, direct analysis of a sample is possible and is generally preferable; normally, however, some degree of pretreatment or 'conditioning' is required. There are three important pitfalls to guard against in pretreatment. Firstly, additional analyte(s) may contaminate the sample during chemical or mechanical conditioning. This is a particular problem where trace constituents are concerned. Secondly, losses of a volatile analyte may easily occur and thirdly, the analyte may undergo undesirable changes of chemical form. Hence, it is necessary to consider carefully what pretreatment is suitable in the light of the required analysis. Generally, moisture content should be determined before any pretreatment. Dissolution is by far the most important method of preparing a sample for analysis. Where a solution cannot easily be obtained, chemical attack is necessary. Aqueous acids or alkalis are effective in many cases. Organic samples are generally 'dry oxidized' by heating in an open crucible or sealed 'bomb' or 'wet oxidized' with concentrated HNO_3/H_2SO_4 mixtures. Inorganic samples which resist attack by acids or alkalis are best fused with a suitable flux, e.g. $LiBO_2$, Na_2CO_3, Na_2O_2, KOH. Alternatively, hydrofluoric acid may be used to effect dissolution and simultaneous volatilization of SiO_2 if it is not an analyte. However, this acid attacks glass and is hazardous to use. Acid leaching of the analyte(s) from an insoluble matrix is sometimes preferred so as to simplify the preparation procedure and minimize subsequent interference from matrix

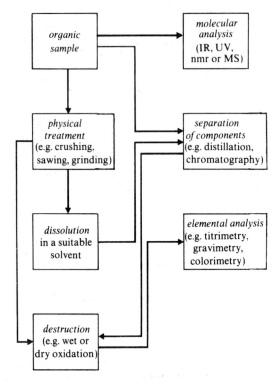

Figure 12.1. Outline procedures for analysis of an organic material

elements. Typical sequences for the analysis of organic and inorganic samples are illustrated in figures 12.1 and 12.2.

SAMPLE CONCENTRATION AND CLEAN-UP: SOLID PHASE EXTRACTION

Concentration of an analyte prior to measurement may be necessary where the level is likely to be close to or below the practical detection limit of the technique to be used. Solvent extraction and ion-exchange may be used for this purpose. Where a complex and/or contaminated ('dirty') sample is to be analysed, a *clean-up* procedure is often employed before determination of the analyte(s). This is designed to avoid interference by other components of the sample, i.e. the matrix, and is particularly desirable prior to a gas or liquid chromatographic separation as the quality of the chromatography can be greatly improved and the working life of the column extended.

Solid phase extraction (spe) has recently become particularly important in this context. Spe, or solid–liquid extraction, is based on the principle of

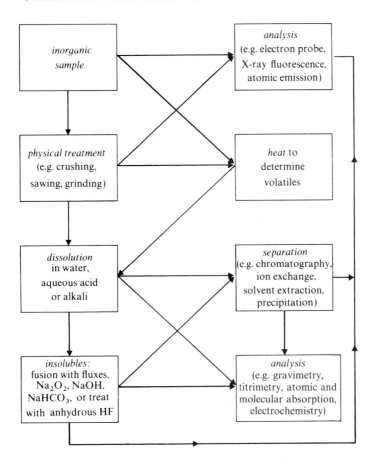

Figure 12.2.   Outline procedures for analysis of inorganic materials

chromatography. It involves passing a solution of the sample through a small plastic cartridge packed with a sorbent consisting of particulate silica or a chemically-modified silica. The sorbents used are similar to the bonded-phases used in hplc but have a coarser particle size (usually about 40 μm). The cartridge is essentially a plastic syringe barrel partially filled with the sorbent but with space above the packed bed to hold the sample solution and washing or eluting solvents. The sample solution and other solvents can be pushed through the sorbent bed under positive pressure using a syringe plunger or drawn through by applying gentle suction at the bottom end as shown in figure 12.3.

By appropriate choice of the combination of sorbent and solvent, either the analyte(s) of interest can be retained (sorbed) whilst the matrix components

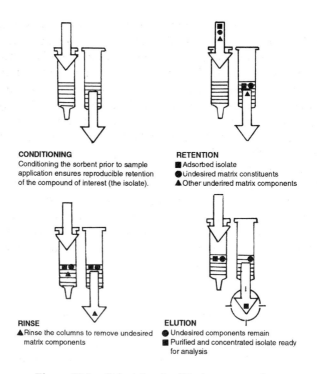

CONDITIONING
Conditioning the sorbent prior to sample
application ensures reproducible retention
of the compound of interest (the isolate).

RETENTION
■ Adsorbed isolate
● Undesired matrix constituents
▲ Other underired matrix components

RINSE
▲ Rinse the columns to remove undesired
matrix components

ELUTION
● Undesired components remain
■ Purified and concentrated isolate ready
for analysis

Figure 12.3.   Principle of solid phase extraction

pass through or the other way round. If the analyte(s) is/are retained, the
sorbent is then rinsed through with some of the solvent to remove any
remaining matrix components, and the analyte(s) subsequently 'eluted' with a
small volume of another solvent so as to provide a matrix-free and more
concentrated solution. Alternatively, if the matrix components are retained
whilst allowing the analyte(s) to pass through the sorbent, the resulting
matrix-free solution is ready for analysis after rinsing the sorbent through
with the same solvent to remove any remaining analyte(s). All four chromatographic sorption mechanisms (p. 75) can be exploited depending on the
nature of the sample. Thus, hydrophobic and dipolar interactions, H-
bonding, ion-exchange and exclusion may contribute or predominate in the
retention of sample components. Sorbents exhibiting mixed retention
mechanisms and new bonded-phases are increasing the range of applications
of spe. It is a versatile, and rapid technique requiring only small volumes of
solvents, and the cartridges are cheap enough to be discarded after use thus
obviating the need for regeneration. Batches of samples can be processed in
parallel by using a vacuum manifold to provide suction for a number of

cartridges simultaneously, and automated extractions are possible. A recent variation of spe allows trace levels of substances in liquid or gaseous samples to be determined by concentrating them onto a fused silica fibre coated with a layer of a gas chromatographic stationary phase, e.g. polydimethylsiloxane. The fibre can be attached to a modified microsyringe so that after concentration of the required sample components it can be inserted directly into a gc injection port for them to be thermally desorbed into the carrier gas stream.

Another concentration technique similar in principle to supercritical fluid chromatography, sfc (p. 143) is *supercritical fluid extraction*, sfe. Requiring more expensive equipment than solid phase extraction, it involves pumping carbon dioxide under supercritical conditions (above 31.1°C and 72.9 bar or 1050 psi) through an extraction chamber containing the sample. The extracted analytes are either collected in a suitable solvent prior to analysis or passed directly into a gas, liquid or supercritical chromatograph. Sfe is fast, efficient and capable of concentrating trace constituents from a wide range of materials and manufactured products.

12.2   Examples of Analytical Problems and Procedures

The choice and development of an analytical procedure will involve several identifiable stages. Firstly, it is essential to define the problem, and to this end the following questions are a guide.

What should be determined?
What is the expected concentration range?
What degree of precision and accuracy is required?
What are the other components of the sample, and will they interfere?
What is the physical condition of the sample?
How many samples are to be analysed, and is there any scope for automation?

Only with these questions answered and a sound knowledge of the available techniques and methods is it possible to devise a satisfactory procedure.

Once the procedure is decided the next step is calibration, which will require suitable standards and determination of the effect of the matrix and reagents on the measurements made. The precision and accuracy of the procedure should then be assessed by the analysis of either synthetic or 'spiked' samples. In the latter method, the recovery of a known amount of analyte added to an actual sample matrix is checked. Finally, the procedure selected should be compared with other procedures available. The following examples will serve to illustrate some aspects of sampling problems, method selection and procedures.

EXAMPLE 1   EVALUATION OF METHODS FOR THE DETERMINATION OF FLUORIDE
IN WATER SAMPLES

In selecting a suitable method for the determination of fluoride at low levels
(0.1 to 5 ppm), reference to analytical texts and the literature reveals that
there are a large number of methods available, most of which are spectrophotometric. A critical appraisal of such methods was required by a laboratory
involved in the routine determination of fluoride in drinking and river waters.
Eight methods selected from the literature are listed in table 12.1. The
distillation, titrimetric and ion-exchange methods (6–8) were rejected,
without further investigation, on the basis of published data, the findings of
which are summarized in the table. The four spectrophotometric methods
included three based on the bleaching effect of fluoride ions on a coloured
metal complex (1–3) and one in which a coloured fluoride complex is formed
directly (4). It was decided to investigate these four and the potentiometric
method on a comparative basis in the laboratory. The particular aspects to be
considered included precision, accuracy, sensitivity, interferences, speed,
convenience and cost. For the spectrophotometric methods, temperature
effects, the stability of the colour formed and of the reagent solutions were
also considered.

All measurements were made using the same set of instruments, and
reagents were of analytical grade wherever possible. Sodium fluoride (dried at
120°C for 2 hours) was used to prepare a 500 ppm standard fluoride solution
which was stored in a polythene container and from which more dilute
solutions were prepared as required. Precision was evaluated statistically,
estimates of standard deviations being based on at least ten replicates at each
of two concentration levels within the range of fluoride expected. One of the
most important considerations was the effect of interferences, particularly
those which reduce the concentration of free fluoride ions by complex
formation or precipitation, e.g. hydrogen ions, aluminium, iron and calcium.
The three metals in particular are frequently present in water samples and
must be removed or complexed to ensure reliable results. Only in the case of
measurements with the fluoride ion electrode was the problem of interferences easily and reliably eliminated. This was achieved by using a 'total
ionic strength adjustment buffer' consisting of an inert electrolyte of high
ionic strength (e.g. 1 M KCl) to which were added one or more of the
following: sodium acetate, sodium citrate, EDTA, sodium hexametaphosphate
(Calgon). The potentiometric method also surpasses the others for speed,
simplicity, precision and accuracy as indicated in table 12.1. Furthermore, it
is particularly suited to the continuous monitoring of fluoride levels in
drinking water. The spectrophotometric methods are lengthy because of the
time required to develop a stable colour (up to 1 hour), the alizarin red-S

Table 12.1 Comparison of methods for the determination of fluoride in water samples

METHOD	REAGENT	PRECISION	ACCURACY (based on recoveries from 'spiked' samples)	INTERFERENCES	COMMENTS
1. spectrophotometric	alizarin red-S	4–5%	<100%	preferably removed by ion exchange or distillation	poor colour stability, sensitive to small temperature changes, lengthy
2. spectrophotometric	eriochrome cyanine R (solochrome)	6–8%	<100%		good colour stability, sensitive to small temperature changes, lengthy
3. spectrophotometric	S P A D N S	4–5%	<100%	aluminium is the most serious	fair colour stability, lengthy
4. spectrophotometric	alizarin–complexone	~3%	<100%	readily complexed to release F^-	good colour stability, lengthy
5. potentiometric (fluoride electrode)	—	~3%	100%		rapid and simple
6. distillation—titration	converted to H_2SiF_6	NOT EVALUATED			lengthy, unreliable
7. titration	$Th(NO_3)_4$ alizarin red–S indicator				poor end point
8. ion-exchange—spectrophotometric	beryllium salt, S P A D N S				lengthy, complicated, poor precision

complex being especially poor in this respect. It was noted, however, that for the three bleaching methods (1–3) the rate of change of absorbance by the blank closely followed that of solutions containing fluoride, i.e. the difference between the blank and a sample absorbance is nearly constant.

Table 12.2   Analytical scheme for a carburettor and combustion chamber cleaner

STEP OR DETERMINATION	COMMENTS	CONCLUSION OR RESULT
1. Karl–Fischer titration for total H_2O in original sample	amperometric end point	30% w/w H_2O
2. NH_3 in original sample by Kjeldahl distillation	titration of excess acid after absorbing NH_3 in standard acid solution	0.5% w/w NH_3
3. total N_2 in original sample by decomposition and Kjeldahl distillation		0.4% w/w N_2
4. original sample extracted with a saturated NaCl solution in a calibrated vessel	removes H_2O quantitatively	30% w/w H_2O
5. infrared spectrum of organic phase from 4 recorded		showed presence of aromatic material, a carboxylic acid salt and butyl cellosolve
6. hydroxyl number for organic phase from 4	acetylation reaction	10% w/w butyl cellosolve
7. acid number for organic phase from 4	potentiometric titration with alcoholic KOH	~0.5% w/w free acid
8. organic phase from 4 extracted with HCl and infrared spectrum of organic residue recorded	removes butyl cellosolve	showed presence of aromatic material and a carboxylic acid
9. aqueous phase from 4 evaporated to dryness and infrared spectrum of residue recorded		no water-soluble organic material detected
10. original sample evaporated on steam-bath under N_2 and infrared spectrum of residue recorded	viscous liquid residue, orange-brown colour	probably contains a dye, spectrum similar to that of oleic acid
11. acid number for residue from 10	see 7	7.5% w/w oleic acid
12. residue from 10 chromatographed on silica column	separated into two fractions	
13. solvent removed from fractions eluted in 12 and infrared spectra of residues recorded	1st *fraction* almost colourless residue 2nd *fraction* orange residue	spectrum typical of a mineral lubricating oil spectrum identical to that of oleic acid (dyestuff <0.5%)

The final laboratory report favoured the use of the fluoride electrode for routine determinations and the alizarin-complexone procedure as an alternative or referee method. These two showed the most consistent agreement during the analysis of a variety of water samples from rivers, wells and reservoirs, the former giving 100% recoveries with 'spiked' samples.

EXAMPLE 2   ANALYSIS OF A COMPETITIVE PRODUCT

A manufacturer of motoring products planned to market a carburettor and combustion chamber cleaner which would remove oily and other deposits. As a preliminary step in formulating the product, a complete analysis of a cleaner already marketed by a competitor was required. The competitive formulation was stated to include aromatic petroleum distillates and butyl cellosolve.

On receiving the sample, its appearance and odour were carefully noted as the inferences drawn at this stage can frequently save the analyst considerable time and effort and may help to establish the qualitative composition. In this case, the sample consisted of an orange–brown free-flowing liquid with an odour both of ammonia and of an organic solvent reminiscent of paint-thinners. Agitation of the sample caused foaming, suggesting the presence of a soap or other surfactant. On the basis of these observations, an analytical scheme was initiated which ultimately involved fourteen steps or determinations. In some instances, decisions to run particular determinations were taken only after the results of earlier analyses were known. The scheme therefore evolved and was modified partly as a result of planning and partly by hindsight. A summary of the completed scheme together with the results is given in table 12.2. The ammonia was considered to be present almost entirely as the ammonium salt of oleic acid which thereby functioned as a detergent, the odour of ammonia suggesting that it was present in slight excess. The proportion of aromatic petroleum distillate, which was classified as a solvent naphtha, was determined by difference. In view of the surprisingly high water content, a duplicate formulation was prepared to confirm that a stable one-phase system could be formed from the constituents. The composition of the competitive product by weight was found to be:

aromatic petroleum distillate	38%
butyl cellosolve	10%
water	30%
mineral lubricating oil	13%
oleic acid	8%
ammonia	0.5%
dyestuff	<0.5%

A feature of this analytical scheme is the marked reliance on infrared spectrometry and titrimetry. The former is particularly applicable to the qualitative characterization of unknown organic materials whilst titrimetry provides a rapid, precise and cheap means of quantitative analysis. The routine titrimetric determination of water, total acid (acid number) and total base (base number) forms a significant proportion of the work load in some analytical laboratories. It is instructive to consider how other techniques might have been applied to the solution of this particular problem, e.g. nmr spectrometry and chromatography.

EXAMPLE 3   THE ASSESSMENT OF THE HEAVY METAL POLLUTION IN A RIVER ESTUARY

Many heavy metals (e.g. Pb, Tl, Zn, Hg) which are known poisons are present in industrial wastes. The effluents from most factories, after some degree of pretreatment, are discharged into rivers. Atmospheric contaminants from smelting works or combustion processes eventually enter the natural drainage system as 'fall out', and are carried into the rivers. It is probable that the deposition of sediments and the higher pH of marine water, which leads to precipitation, results in a build-up of the heavy metal pollutants in the river estuary. An assessment of this build-up is essentially an analytical problem.

The natural levels of heavy metal pollutants in the environment vary considerably, but are generally low (less than 100 ppm). In the cases of lead and mercury very low levels are normal (less than 1 ppm) and concentrations of a few parts per million in food are known to be dangerous. The analytical methods used must, therefore, be able to cope with concentrations of hundreds of ppm and have a potential sensitivity well below 1 ppm. The characteristics of a number of techniques which will meet these requirements are summarized in table 12.3. The practical part of the problem, as always, starts with sampling. A systematic programme of sampling, including the bed and banks of the river as well as the water, and estuarine life would generate an enormous number of samples, especially as the analysis of repetitive samples taken over a year would be essential to give the survey perspective. Although such surveys have been attempted, various 'short cut' expedients have been devised. An important example of such a 'short cut' is based on an assessment by marine biologists, who say that it is the degree to which pollutants enter the food chain which is even more important than the overall level of pollution. Thus, analysis concentrated on 'pollution indicators' such as shellfish which are low down in the food chain will serve as a measure of the overall build-up of pollutants in estuarine life. To this end, extensive analyses of limpets and mussels have been made. Samples were taken from various parts of the

Table 12.3   Characteristics of some trace metal techniques

TECHNIQUE	APPROXIMATE OPERATING RANGE	RELATIVE PRECISION	REMARKS
inductively coupled plasma-atomic emission spectrometry	major–ppb	—	sensitive, precise qualitative and quantitative technique with a wide linear response
atomic absorption	0.1 ppm–100 ppm	0.5–2%	excellent quantitative technique but not suitable for qualitative analysis
flame photometry	1–100 ppm	1–4%	limited in scope because of susceptibility to interference
electron probe	50 ppm–500 ppm	5–10%	useful for localized analysis for parts of samples; analyses surface layer only; costly equipment; qualitative and quantitative applications
X-ray fluorescence	20 ppm–0.1%	5–10%	best suited to heavy elements in light matrices, very simple to operate; costly equipment; qualitative and quantitative applications
neutron activation (γ-ray spectrometry)	10 ppm–10^{-4} ppm	1–5%	expensive and specialized; qualitative and quantitative information at ultra trace levels

estuary at different depths. Careful note was taken of the materials supporting the shellfish. In addition, similar samples were obtained from 'open sea' sites to provide a comparison with the estuarine samples and an indication of the natural levels. Other important pollution indicators for this type of site are vegetation and sediments. The former give additional indications of incorporation of metals into the food chain. The latter, by virtue of the presence of hydrous oxides of Al, Mn, and Fe, concentrate metals by sorption. Samples of these indicators were also collected.

The actual analysis of the samples started with a qualitative survey to identify the principal metals present. ICP-AES was selected as being most suitable for this because it provided a permanent record, a multielement analysis and good sensitivity.

Solutions of the shellfish tissue were prepared for analysis by wet oxidation, and the sediments were digested with 40% v/v HNO_3. ICP-AES showed a wide variety of elements to be present with particular build-ups of cobalt,

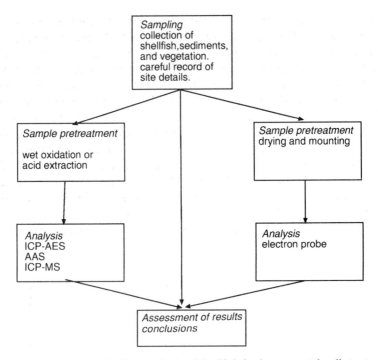

Figure 12.4.   Outline analysis of shellfish for heavy metal pollutants

nickel and zinc (50–100 ppm). The presence of lead, cadmium and mercury at dangerously high levels was not indicated. Additional qualitative studies were made by electron probe analysis and ICP-MS. The levels proved to be too low for effective use of the former method whilst the second confirmed the ICP-AES results. Facilities for X-ray fluorescence were not available and flame photometry was not considered suitable for the elements of interest.

Systematic quantitative analysis was subsequently carried out utilizing the sensitivity, precision and simplicity provided by ICP-AES. Pb and Hg were monitored using the superior sensitivies of AAS for these elements. All measurements were made on the solutions derived from the wet oxidation or extraction of the samples. A limited series of ICP-MS analyses were made to provide a check on the results being obtained by atomic absorption and ICP-AES. Figure 12.4 summarizes the various stages in the solution of this analytical problem.

EXAMPLE 4   THE ANALYSIS OF HYDROCARBON PRODUCTS IN A CATALYTIC REFORMING STUDY

In hydrocarbon reforming processes the vapour of an alkane is passed over a supported metal catalyst such as platinum on silica or alumina. Dehydrocyclization, isomerization and cracking reactions all take place to produce ring compounds, alkane isomers and shorter chain alkanes and alkenes. The octane rating of the product is dependent upon the proportion of aromatic compounds synthesized while the low boiling cracked products are wasted. To establish optimum reaction conditions for a catalyst it is necessary to make an extensive series of studies in which the composition of the product is monitored as the temperature, pressure, flow rate, etc. are varied. Rapid, on line analysis of a complex mixture of hydrocarbons is needed. Gas chromatography offers the best solution to this problem, as it provides for ready separation of the volatile constituents, their identification from retention times and their quantitative analysis from peak areas. However, with complex mixtures it becomes necessary to obtain independent confirmation of the peak identities and to establish that the peaks obtained correspond to single components. Integrated gas chromatographic and mass spectrometric analysis can provide this additional information.

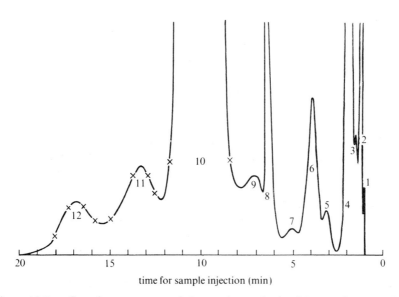

time for sample injection (min)

Figure 12.5.   Gas chromatogram of the products obtained from *n*-hexane when passed over a SiO_2/Pt catalyst at 450°C. (Column: 2 m, Apiezon 'L' on 100 to 120 mesh Diatomite 'C'; Carrier gas, helium, 45 cm^3 min^{-1}; Temperature 30°C)
X = typical sampling points
Identity of peaks

1. propane	7. 2-methyl-1-butene
2. *iso*-butane	8. 2-methyl pentane
3. *n*-butane	9. 3-methyl pentane
4. butene-1, butene-2	10. *n*-hexane
5. *iso*-pentane	11. methyl cyclopentane
6. *n*-pentane	12. benzene

In a series of experiments with silica supported platinum catalysts the feedstock (*n*-hexane) was passed continuously over the catalyst for thirty minutes, and the products collected in a liquid nitrogen trap. Ten samples of the products were injected into a gas chromatograph interfaced with a mass spectrometer. Good separations were obtained on an Apiezon 'L' column at a temperature of 30°C and a typical chromatogram is shown in figure 12.5. The effluent from the column corresponding to each peak on the chromatogram was sampled four times and a mass spectrum obtained in each case. The sampling points are indicated in figure 12.5. A synthetic mixture of the expected components was also analysed under identical conditions. Peak identities were tentatively assigned on the basis of retention times in this standard chromatogram. The uniformity within the spectra from the same peaks established the purity of the components and comparisons with standard spectra confirmed the tentative identifications. Figures 12.6(a) and 12.6(b)

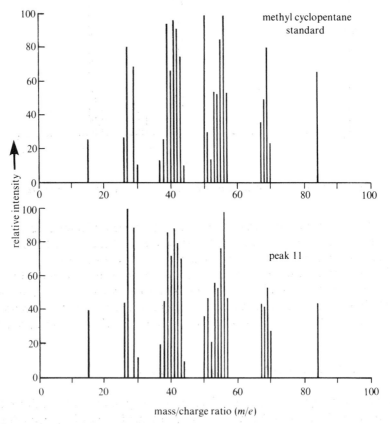

Figure 12.6(a).   Mass spectrometric confirmation of peak 11 as methyl cyclopentane

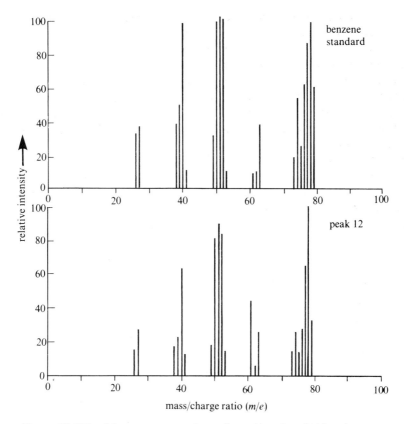

Figure 12.6(b).   Mass spectrometric confirmation of peak 12 as benzene

show typical spectra for peaks 11 and 12 which were identified as methyl cylopentane and benzene. Subsequent assessments of the proportions of the products were made by the comparison of integrated peak areas.

This example illustrates the power of a modern analytical procedure to reduce a complicated problem to a set of relatively straightforward measurements. Furthermore it highlights the effectiveness of interfacing two techniques to produce a method which has a resolution unmatched by either individually.

12.3   The Automation of Analytical Procedures

One area of analytical chemistry which is currently developing rapidly is the automation of methods. Some degree of automation has been used for a number of years in instruments such as automatic burettes coupled to absorptiometric or electrometric end point detectors, and in data output devices which provide continuous pen recording or signal integration facilities. The major features of recent developments include the scope for instrumental improvements provided by solid state electronic circuits and the increasing application of digital computers (chapter 13).

A fully automated analytical method has a number of significant advantages over the manual version. It can be used to carry out a very large number of similar analyses or to provide continuous monitoring of an analyte with minimum operator involvement. Moreover, the results obtained will frequently be more reliable. An instrument neither suffers the tedium of repetitive work, nor does it make subjective judgements of readings. Lastly, an automatic instrument can often be used in an environment where it is impossible for an operator to work. Obvious examples are the inside of a nuclear reactor or the outside of a spacecraft. Automated analysis may thus provide information more cheaply and reliably than manual analysis as well as some data that would otherwise be unobtainable.

It is obvious that the simpler a method of analysis, the easier it will be to automate. Non-destructive methods which involve a minimum of sample treatment are the most attractive. X-ray fluorescence, for example, has been successfully applied to the continuous monitoring and control of process streams. However, the scope of automated analysis is wide and methods have been designed with a basis in non-specific properties (pH, conductance, viscosity, density) as well as those characteristic of the chemical constituents of a sample. The chief areas in which automatic analysis is applied are the quality control of industrial processes, environmental monitoring and the analysis of biological fluids. Each application demands a close and frequent check on the levels of both desirable and undesirable constituents within the system. Such checks generate a need for continuous analytical monitoring or frequent sampling. It is convenient to consider separately the automatic methods designed to meet these two alternatives.

THE AUTOMATION OF REPETITIVE ANALYSIS

Automatic analysis consists essentially of the same steps as the corresponding manual method (p. 4). In some cases this may be simple, the requirements amounting to a mechanical device for presenting the sample to the detector, a timer to control the time of measurement and a data recorder. However, if sample pretreatment and separations are necessary a variety of wet chemical stages needs to be automated. Such automated steps may be included in what remains essentially as an operator procedure. For example, an automatic pipette may be used for sampling; a fraction collector in a chromatographic separation; or an automatic sequence for making a measurement and recording data. By the use of modern computer controlled robots it is becoming possible to fully automate such an analytical sequence. Laboratory robots are discussed a little more fully at the end of the chapter. On the other hand an analysis can be automated at all stages. Pioneers in automated wet analysis were Technicon Ltd with their *Autoanalyser*. Subsequent developments and imitators have extended the scope of this system for several hundred different analytes and matrices.

Figure 12.7 illustrates the outline of such an analysis. An automatic pipette extracts a preset volume of the liquid sample (or solution) from a cup presented to it on a turntable. The measured sample is mixed with the reagents in the

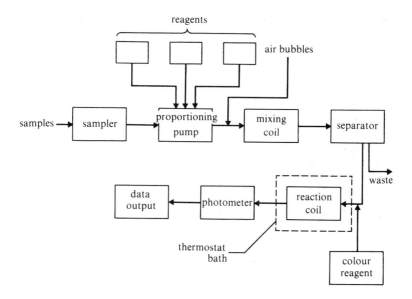

Figure 12.7.   Schematic representation of an automated colorimetric analysis

appropriate proportions, and propelled through the instrument by the *peristaltic proportioning pump*. This pump operates by means of moving bars, attached to a chain drive, which sequentially compress the plastic sample and reagent tubes to drive the liquids forward through the instrument. The incorporation of a succession of air bubbles at regular intervals prevents diffusion and cross-contamination between reagents and samples. A glass coil inserted into the stream provides a delay long enough for mixing to be completed within the separated sections of the solution. If interfering materials are present and cannot be masked, a separator unit is incorporated prior to the measurement stage. For a colorimetric method, the colour reagent is added and a reaction coil to allow for mixing and colour development is inserted. This coil may be immersed in a suitable thermostat bath. Finally the coloured solution is fed to the detector via a 'debubbler' which removes the air. The most convenient detection devices are filter photometers and spectro-photometers, which are readily operated as 'flow instruments', although emission instruments may be used. An Autoanalyser can be concurrently processing a number of samples and analysis rates of 150 samples per hour have been reported. The major disadvantage is the inability to handle solid, or high viscosity samples without a prior manual dissolution step. Figure 12.8 summarizes the automated determination of magnesium in urine by means of its reaction with o,o'-dihydroxyazobenzene, and fluorimetric detection.

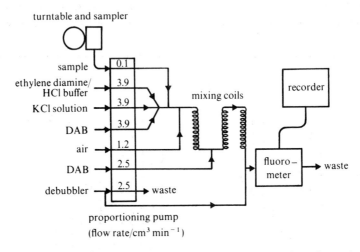

Figure 12.8.   The automated determination of magnesium in urine using o,o'-dihyroxyazobenzene (DAB). The final reagent concentrations obtained are:

| ethylenediamine | 0.28 M | HCl 0.14 M |
| DAB | 0.07 mM | KCl 0.10 M |

A more recent development is a technique known as *flow injection analysis*, in which a discrete volume of a liquid sample is injected into a *carrier stream*. Reagents required for the development of the analytical property of the analyte, e.g. colour developing reagents for spectrophotometry, are already present in the stream. The stream then flows straight to the detector and the technique depends upon the controlled and reproducible dispersion of the sample as it passes through the reaction zone. Thus the reaction does not necessarily need to develop to completion, although delay coils may be added as required for the reaction to proceed far enough for detection.

With the carrier stream unsegmented by air bubbles, dispersion results from two processes, *convective transport* and *diffusional transport*. The former leads to the formation of a parabolic velocity profile in the direction of the flow. In the latter, radial diffusion is most significant which provides for mixing in directions perpendicular to the flow. The extent of dispersion is characterized by the dispersion coefficient D.

$$D = \frac{C_0}{C_{max}}$$

where C_0 is the initial concentration of the analyte and C_{max} is the maximum concentration at the detector. Values of D in the ranges <3, $3\text{–}10$, and $10\text{–}15$ would represent limited, medium and large dispersions in sequence. Figure 12.9 shows a typical output trace for the enzymatic determination of alcohol in beverages and figure 12.10 a schematic representation for the determin-

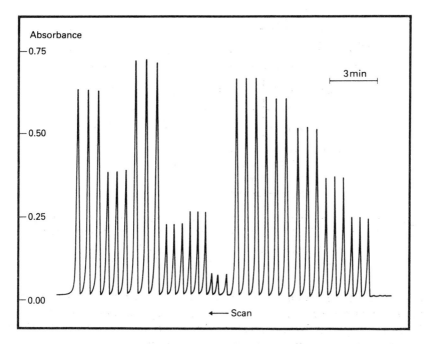

Figure 12.9.   Continuous-flow enzymatic determination of ethanol in beverages. To the right is a series of six standards (0–0.4%) followed by six samples – orange juice (1 and 10% v/v), bottled beer (1 and 10% v/v), red wine (1% v/v) and liqueur (1% v/v), all analysed in triplicate (from P. J. Worsfold, *Chemistry in Britain*, December, 1988)

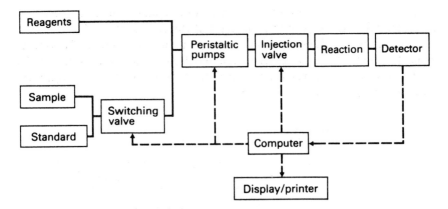

Figure 12.10. Block diagram of a remote spectrophotometric, flow injection based monitor for the determination of nitrate in river water. Reduction to nitrate by copperized cadmium followed by colour development with sulphanilimide and N-I-naphthylmethylenediamine dihydrochloride (from P. J. Worsfold, *Chemistry in Britain*, December, 1988)

ation of nitrate in river water. As well as reagents in the carrier stream reactions may be effected by percolating the stream over beads on which the reagent has been immobilized, a procedure which has been most useful in enzymatic assays.

Flow injection analysis is a fast developing technique with many potentialities. Particular attractions are the relative simplicity of operation and automation, together with sample throughputs which may exceed 100 per hour. Thus routine monitoring of process streams and pollution control are obvious areas for application.

CONSTANT MONITORING AND ON LINE ANALYSIS

The provision of a constant measure of an analyte in the process stream of a chemical plant presents a unique set of problems. Ideally a specific analytical detector such as an ion-selective electrode is inserted directly into the stream, to provide an electrical signal directly related to the amount of analyte present. This ideal is rarely realized and it is more usual to divert some of the stream through a sampling loop and to examine the material diverted. It is then possible to extend the scope and precision of the analysis. Figure 12.11 shows schematically, a method for monitoring the acidity of a process stream. The sample loop flows into a titration cell which contains a glass electrode to measure the pH, with the signal from the electrode being fed to a reagent pump of accurately known stroke. A continuous acid-base titration is carried out with the rate of supply of alkaline titrant (speed of the pump) being controlled by the potential of the glass electrode. Thus, if the acid content of the process stream goes up, so does the speed of the pump and vice versa. The rate at which the pump is operating reflects the pH of the process stream and can be used to generate a control signal to the feedstock valves so as to adjust the feedstock composition.

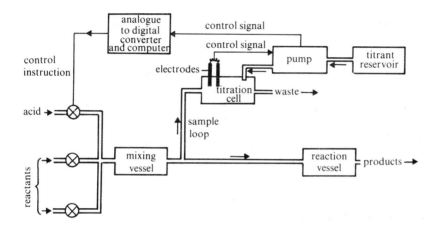

Figure 12.11.   Schematic diagram of an analyser/controller for an acid catalysed reaction

Non-destructive analysis is especially valuable in an on line situation. X-ray fluorescence has above all become of major importance for the analysis of inorganic process streams. Cement production is an example of the successful application of this technique. The X-ray analyser can be used for the simultaneous assay of the various feedstocks (iron ore, clay and limestone) for Fe_2O_3, Al_2O_3, SiO_2 and CaO. In turn the signals from the analyser are used to control the feedstock supplies to the blending mill and to maintain an optimum product composition.

A number of valuable methods of analysis are essentially *batch processes* themselves and cannot be operated on a continuous basis. Their use in an on line mode will then require frequent sampling and rapid presentation of the sample to the instrument followed by rapid analysis. It may be necessary to use a battery of analysers working in parallel to cope with the number of samples presented. A particularly important example of 'batch' analysis is the use of gas–liquid chromatography. The power of this technique to provide rapid analysis of complex mixtures has led to its use for monitoring many organic processes, and its integration into many chemical plants.

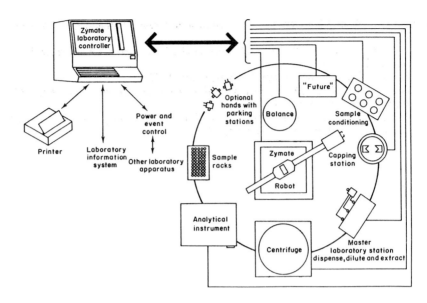

Figure   12.12.   Robot   arm   system   for   sample   preparation   with   different
         workstations (reproduced by permission of Zymark Corporation)

LABORATORY ROBOTICS

The development of reliable robots together with sophisticated, programmable, computer controllers, has begun to make an impact on laboratory
practice in analytical laboratories. For example, where large numbers of
samples need to be routinely processed before measurement, a robot can
provide for sustained and prolonged operation. In so doing skilled operators
can be released for other duties thus achieving a saving in staff costs.

   Robot design has developed along a number of lines. *Cylindrical robots*
use a horizontal arm, which moves up and down and swivels around a central
column. *Cartesian robots* move and operate along the axes of the familiar
cartesian coordinates. *Robot arm* systems aim to parallel the innate
flexibility of the human hand and arm and have perhaps the greatest potential.
The tools fitted to the robots for manipulations can be changed at will and
automatic changes may be built into the program of the controller, thus
automating the whole process. Figure 12.12 is illustrative of the use of a
robot arm.

Further Reading
EWING, G. W. (Ed.), *Environmental Analysis*, Academic Press, 1977.

FOREMAN, J. K. and STOCKWELL, P. B., *Automatic Chemical Analysis*, Wiley, New York, 1975.

FOREMAN, J. K. and STOCKWELL, P. B. (Ed.), *Topics in Automatic Chemical Analysis*, Wiley, New York, 1979.

NICHOLLS, G. D., *On-line Process Analysis*, Wiley, 1988.

SKOOG, D. A. and WEST, D. M., *Fundamentals of Analytical Chemistry*, 4th Ed., CBS Publishing, New York, 1982.

VALCARCEL, M. and LUQUE DE CASTRO, M. D., *Flow injection analysis – principles and applications*, Ellis Harwood, Chichester, 1987.

Chapter 13

The Role of Computers and Microprocessors in Analytical ₍Bio₎Chemistry

13.1 Introduction

The management of an analytical chemistry laboratory involves a number of different but related operations. Analysts will be concerned with the development and routine application of analytical methods under optimum conditions. Instruments have to be set up to operate efficiently, reproducibly and reliably, sometimes over long periods and for a variety of analyses. Results will need to be recorded and presented so that the maximum information may be extracted from them. Repetitive analysis under identical conditions is often required, for instance, in quality assurance programmes. Hence a large number of results will need to be collated and interpreted so that conclusions may be drawn from their overall pattern. The progress of samples through a laboratory needs to be logged and results presented, stored, transmitted and retrieved in an ordered manner. Computers and microprocessors can contribute to these operations in a variety of ways.

INSTRUMENT OPTIMIZATION

It is rare to find an instrumental method, which, when optimized for the determination of one analyte in a particular matrix, can then be directly applied to another analyte and matrix. Thus each analyte and matrix combination will require a different package of instrument settings. The number of parameters which need to be controlled in this way is often large. For example, in flame atomic absorption spectrometry it will include the nature of the fuel/support gas mixture, rate of nebulizer feed, optical alignment, slit-width, monochromator setting, power to lamps and photomultipliers, and integration time. The traditional method of dealing with this is to adjust each individual parameter in turn whilst holding the others constant. A record of the settings is kept and the instrument manually reset when the analysis is to be repeated at a later time. Computers and microprocessors can help in the initial selection of most parameters, which are simply entered from a keyboard or keypad, but more particularly by allowing the storage and recall of groups of settings more quickly and accurately than can be done manually. Furthermore, greater stability of operation results from the continued monitoring of these settings. Modern spectrometers and chromatographs often allow for the analysis of a number of analytes in sequence, with a computer controlling the adjusting of instrument settings according to a predetermined program. The condition of the various components of an instrument can be continuously monitored by inbuilt self-diagnostic software so that any deterioration in performance can be brought to the attention of the operator and malfunctions reported as they arise.

DATA RECORDING AND STORAGE

One feature of modern developments in analysis has been the increasing speed with which measurements can be made. This trend is well exemplified by the changes which have taken place in obtaining and presenting spectra. Manual plotting of a series of wavelengths or frequencies and corresponding detector responses was replaced by the use of x/t and x/y plotters. Currently, fast Fourier transform and other software allow the rapid display on a VDU screen of a complete spectrum in a matter of seconds. In the sequential spectrochemical analysis referred to earlier, ten or more elements may be determined in a few minutes. To provide for the rapid acquisition and storage of this data a computer with a considerable amount of memory is essential. Selected information may then be extracted quickly, or a complete spectrum recalled in a few seconds.

DATA PROCESSING AND DATA ANALYSIS (CHEMOMETRICS)

Rarely will it be possible to draw conclusions directly from the raw data of analytical measurements and it is usual for some refinement of the data to be carried out. In its simplest form this could merely comprise background corrections, but it is often much more complex, requiring corrections for a number of factors as in mass spectrometry, X-ray fluorescence and electron probe microanalysis (p. 332). More complex routines made available by computers include spectrum smoothing, stripping one component from a spectrum or making peak area measurements from chromatograms.

Data reduction and interpretation are much aided by computer methods. The principle of extracting as much information as possible from analytical measurements through the application of statistical and other mathematical methods, usually with the aid of appropriate computer software, is known as *chemometrics* (p. 33).

The subject has been given considerable impetus by the rapid growth in the power of computers and in the sophistication and speed of graphic routines. Among the more useful branches of chemometrics are statistics, signal processing, optimization, modelling, factor analysis, image analysis, pattern recognition, cluster analysis and library searching. Some of the most impressive examples are found in molecular spectrometry where ms, nmr or ir spectra are used extensively as qualitative tools. A computer library can store a large number of reference spectra which are available for rapid recall. The computer will be able to carry out a search for the most likely identity of an unknown sample and present the analyst with a limited number of possibilities. It is important to note that the computer will not replace the skill and experience of the analyst, but carry out routine preliminaries enabling him to concentrate his/her efforts on the final interpretation. In almost all cases quantitative analysis is carried out by calibration with standards. A computer can store such calibration data and automatically evaluate routine measurements immediately so that the instrument functions as a direct reader.

LABORATORY MANAGEMENT

The logging of samples submitted to an analytical laboratory is a complex matter involving the recording of many details from receipt to completion of the required analyses and the generation and use of a great deal of information. If the laboratory is very large, or more than one is involved, the task becomes formidable. The following list of requirements illustrates the complexity of the operation:
Before and during the analysis:
    origin of sample(s), date of receipt and order number
    allocation of sample number(s)
    listing of analyses required
    identity of analyst(s)
    status of each analysis
    abnormalities encountered during the analyses.
Subsequent to the analysis:
    time taken for each analysis
    computation of results
    statistical and other applications of chemometrics to the data

library and data bank searches
generation of report in required format
transmission of report to required location(s)
archiving of results
collation with other results and data
receipts for work done

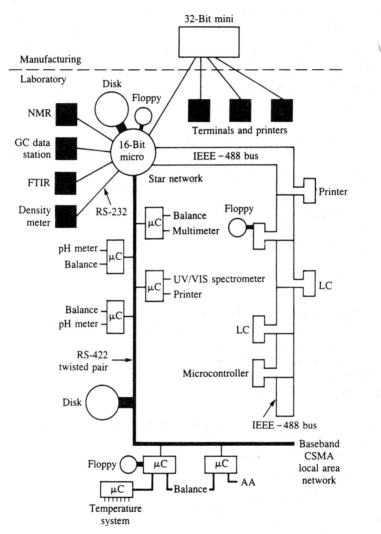

Figure 13.1(a). A local area network (LAN) linking the components of an analytical laboratory. μC = microcomputers (after *Anal. Chem.*, **54**(12), 1296A (1982), courtesy of American Chemical Society)

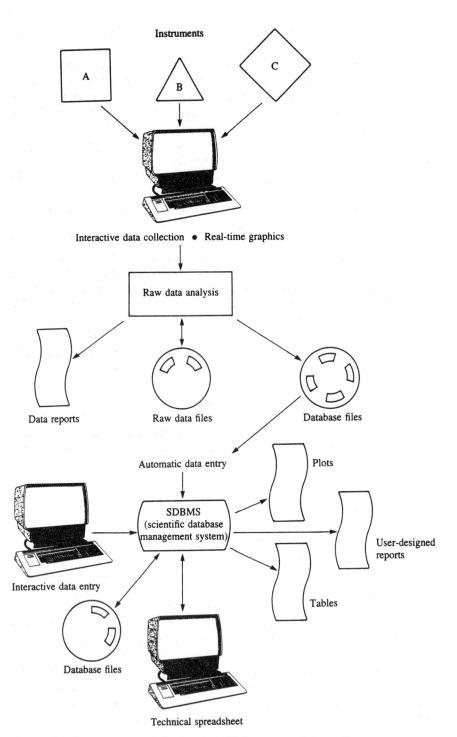

Instruments

A

B

C

Interactive data collection • Real-time graphics

Raw data analysis

Data reports

Raw data files

Database files

Automatic data entry

Plots

SDBMS
(scientific database
management system)

User-designed
reports

Interactive data entry

Tables

Database files

Technical spreadsheet

Figure 13.1(b). Diagrams of the LMS 1100 laboratory information management system (LIMS) (after Harder and Koski, *Am. Lab.*, **15**(9), 28 (1983), with permission)

A computerized laboratory can utilize a software package called a *laboratory information management system* (LIMS) to carry out or control all of these requirements provided that computerized instrumentation, computer terminals, printers and plotters, disk drives etc. are linked together in a *local area network* (LAN) to facilitate the transfer of information between them. A typical LAN is shown in figure 13.1(a). It includes a minicomputer, which runs the LIMS, various analytical instruments and microcomputers, some with high resolution graphics capabilities, and a considerable amount of disk memory.

There are numerous advantages in running a laboratory through a LIMS, *viz* the ability to capture, transfer and process data for a variety of instruments and to report the results in any specified format, the ability to use or construct databases (data banks) from any of the information generated, stored or accessible outside the laboratory, and the ability to log the progress and status of each sample that passes through the laboratory. The saving in time and effort for laboratory personnel and the relative freedom of data and results from human errors are further major benefits. LIMS can be purchased as customized commercial software packages or developed by a company's 'in-house' programmers. Privileged access to various parts of the system can be built in so that, for example, only specified personnel can change the parameters associated with a particular method of analysis or the availability of confidential information can be restricted. An overall picture of sample flow through the laboratory can be used to pinpoint sources of delay so that changes in the operation can be introduced to improve efficiency. A diagrammatic representation of a LIMS is shown in figure 13.1(b).

EXPERT SYSTEMS

These consist of sophisticated software packages and data banks of information that act as sources of knowledge and advice on a particular subject area, e.g. chromatography, atomic emission techniques, the analysis of polyurethane plastics, the quality control of anti-depressant drugs. They can be accessed interactively by the user to extract information and develop ideas. Systems become more 'expert' as additional data and opinions from numerous sources are fed into them. The software currently available, and which is likely to evolve into a form of artificial intelligence, is already capable of making logical deductions after rationalizing the information fed into the computer and will in the future be able to explain the reasoning processes leading to its decisions and advice. A diagrammatic representation of an expert system for an analytical separation is shown in figure 13.2.

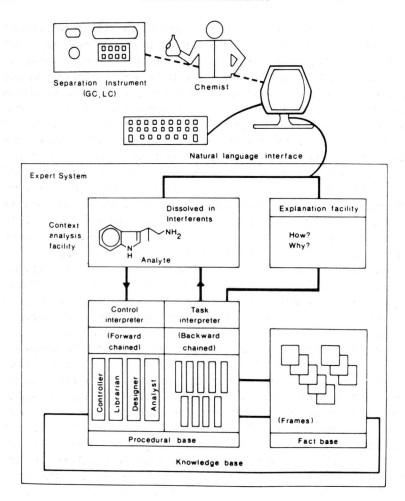

Figure 13.2.   Expert system for separation sciences

A brief description of computers, microprocessors and computer/instrument interfacing in the context of analytical chemistry is given in the following sections.

13.2   Computers and Microprocessors

Modern digital computers contain three main components which are termed the *hardware*. These are the *central processing unit* (CPU), the *memory* and the *input-output* (I/O) systems. The CPU consists of an *arithmetic logic unit*, which performs the computations, *registers* for the temporary storage of information, a *clock* to enable the various operations to be synchronized, and a *control unit*. The latter passes instructions, results of computations and data to and from the arithmetic logic unit, the memory and the I/O devices, generally under the control of a *program* (*software*) stored in the memory. All instructions and data are stored, transmitted and received, and computations executed in a *machine code* consisting of groups of *binary numbers*, i.e. 0 and 1, which are easily represented electrically by one of two *states* (*voltages*) (usually 0 V and 5 V). Groups of signals are passed between the various components along sets of parallel lines known as the *bus system*. Thus the computer operates at extremely high speed, the various hardware components consisting essentially of very large networks of on/off switches corresponding to binary 1 and 0 and connected by printed circuits. The term binary number or *binary digit* is commonly abbreviated to *bit* and computers are designed to handle 8, 12, 16, 24 etc. groups of bits at a time, the number in the group being termed the *word length*; an 8-bit word is called a *byte*. Large or *mainframe* computers may use word lengths of 16, 32 or 64 bits whereas mini- and microcomputers (see below) use 8, 16 or 32 bit words.

There are two types of memory used in computers. The main memory is based on integrated circuit chips and all parts can be *accessed* with great rapidity and with equal ease. In mainframe computers this is known as the *main store* whilst in smaller computers, including microcomputers, it is called the *random access memory* (RAM). Some parts of RAM may be reserved for the storage of programs or data which are to be protected from change or accidental erasure. Such a reserved area of memory is called *read only memory* (ROM). ROM chips, sometimes called *firmware*, are often used in integrators and microcomputers dedicated to particular tasks. The size of the main store or RAM is usually quoted in terms of the number of words which can be accommodated simultaneously, as this is an indication of the power of the computer. A minimum of about four thousand (4K) words is needed for even simple scientific applications, and microcomputers can now be purchased with up to 256 megabytes (256 000K) of RAM, whilst mainframe computers are very much larger. Additional memory space (called *backing store* on mainframe computers) is available in the form of magnetic tapes or disks and is used for long term storage of programs and data. It is virtually unlimited in size but it is relatively slow to access. Disk storage is much more widely used than tape, especially for microcomputers. Rigid (hard) 15″–16″ diameter disks with capacities of 1 000 megabytes or more are used in mainframe computers whilst mini- and microcomputers employ 8″ hard or 8″, $5\frac{1}{4}$″ and $3\frac{1}{2}$″ mini 'floppy' (flexible) disks with capacities ranging from one to several hundred megabytes for programs and data depending on their size and the density of the stored information.

Two of the most recent developments in disk storage are the *compact disc ROM (CD-ROM)*, which is similar to an audio CD and read by a laser beam, and the *floptical disk*, a hybrid magnetic/optical storage medium which can be read or written to (R/W). The current storage capacity of a CD-ROM is 250 megabytes and that of a floptical is 20 to 40 megabytes but will ultimately be considerably more. Both can accommodate photographic quality images and/or complete libraries of data (*data bases*).

Input-output (I/O) devices enable the user to communicate with a computer, and instruments to be *interfaced* with it. Electric typewriters, visual display units (VDU), printers, plotters and magnetic tape and disk units are the principal I/O devices. Although computers use machine code in all their operations, software for the user is written in a more intelligible format known as a *high-level language*, e.g. BASIC, FORTRAN, ALGOL, PASCAL etc. These take the form of abbreviated or shorthand instructions in English which the computer has to translate into machine code. The translation step involves the conversion of each alphanumeric or other character in the user's program or data into a corresponding 8-bit number between 0 and 255 (the ASCII code system). ASCII coded characters are then converted into machine code before execution. High level languages have been devised with particular types of application in mind, and BASIC in its various forms has been the most widely used in the context of analytical applications.

MINI- AND MICROCOMPUTERS

The distinction between mini- and microcomputers is becoming essentially one of size and price. Minicomputers, which use 16- or 32-bit words, had much larger memories than microcomputers and could be used for the control of several laboratory instruments on a *time-sharing* basis. However, microcomputers are becoming ever more powerful. Although some still use 8-bit words, 16-bit and 32-bit word machines are becoming standard, and this, together with clock speeds of 33 to 100 MHz, results in a faster computing capability. Their memory size (above) now exceeds that of earlier mini- and mainframe computers and they can be linked together into *micro-networks* (LANS) based in one or covering several laboratories. The *disk operating system* (DOS) of a microcomputer, e.g. MS-DOS, is now loaded from a hard disk on startup (*booting* the system) rather than being resident on a ROM chip. This enables the system to be upgraded easily or even another DOS altogether to be employed. One further distinguishing feature of the microcomputer is the use of *microprocessor*-based CPU (see below). Many analytical instrument manufacturers now offer their instruments complete with a *dedicated* microcomputer and supporting *software packages* to perform appropriate instrument control tasks, data processing and display and/or printout of results; the provision of *high-resolution* (1280 × 1024 pixels) *colour graphics* software to display such things as chromatograms and spectra on a VDU screen has become an important feature. These and more general applications software packages frequently run in a *windows* environment and multicolour (up to 16.7 million colours) display mode. This gives great flexibility to the user allowing information in the form of option menus, help screens, data tables, spectra, chromatograms editing and file-handling routines, expanded scale graphical displays etc., to be temporarily superimposed on an existing screen within a series of rect-angular frames (windows) that can be enlarged, moved or instantly erased when no longer required. The use of a *computer mouse* as an alternative to entering commands and parameters at the keyboard greatly speeds up the use of this type of software. Maximum flexibility for the user is achieved by offering a high-level programming language (usually BASIC) for editing exist-ing programs or for writing entirely new ones.

MICROPROCESSORS

A *microprocessor* is a miniaturized CPU which is all on one integrated circuit chip and occupies only a few square centimetres in area. Many analytical instruments are now *microprocessor-controlled*, the control programs for setting and monitoring instrumental parameters, for diagnostic routines and for some data processing (sometimes called *data reduction*) being stored on ROM chips. Some RAM may also be available to enable the user to write and store sets of method parameters for particular routine analyses and to perform a degree of additional data processing. Microprocessors are also at the centre of every microcomputer on the market. The layout of a typical microprocessor system is shown in figure 13.3. Clock frequencies, which determine the rate of computation, have risen steadily from below 10 MHz to well over 50 MHz as new processors have been developed. The fastest currently available is a *pentium* processor with a clock speed of 100 MHz enabling *multitasking* and *real-time data processing* to be readily accom-plished.

13.3   Instrument-Computer Interfaces

The purpose of interfacing instruments with computers is to enable raw analytical data to be collected as it is produced, then processed, stored and displayed or printed out. This may be accomplished as it is gathered, i.e. in

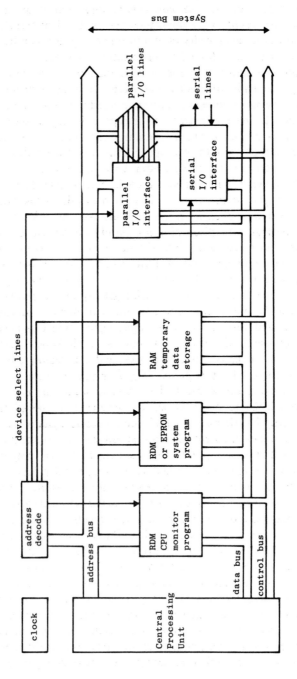

Figure 13.3.   Layout of a typical microprocessor system

real-time, or at some later time, i.e. *post-run*. Complete chromatograms or spectra can easily be stored in the main memory or RAM or transferred to disk. The immense storage capacity of mainframe computers can be used to provide large libraries of data (*data banks*) for future reference.

Interfacing can be achieved in various ways depending on the nature of the detector signal produced by the instrument. Many generate *analogue* signals, i.e. a continuously variable voltage or current related to the concentration or nature of the analyte. As computers can accept only digital information, *analogue-to-digital converters* (ADC) are required to facilitate the direct input of data. These devices convert, for example, an input voltage which typically may vary between 0 and 1 V into integer numbers within a specified range, e.g. 0 to 4 095 on a linear scale, and represent them in binary form. The *resolution* of such an ADC signal would be 1 in 4 096, which allows discrimination between analogue signals differing by as little as 2.44×10^{-4} V for a 0–1 V input range. This degree of discrimination would be described as *12-bit resolution* as the binary representation of numbers up to 4 095 requires 12 bits. The resolution of commercial ADC units is usually 10, 12, 14 or 16 bits.

The frequency at which an ADC needs to *sample* an analogue signal and the total number of samples required varies greatly and depends on the nature of the particular analytical technique and instrument. In all cases, it is vital that the digitized version of an analogue signal, which is necessarily discontinuous, represents the original signal as faithfully as possible. In theory, an infinite number of *samples* of the analogue signal, called *data points*, would be required for a perfect digital representation but this is impractical and a compromise has to be made. A limited number of data points are collected, the precise number and frequency of sampling being determined primarily by the rate at which the analogue signal changes and the time scale of the measurements. It may be sufficient to collect individual data points or small groups of points intermittently and over comparatively long time intervals if the analogue signal does not vary rapidly. This would be the case, for example, where absorbance readings for a quantitative spectro-photometric UV analysis are to be taken. A faster and perhaps variable sampling rate might be required during a potentiometric titration, whilst a rate of 10 to 100 Hz would be necessary where the detector signal is varying widely in a much shorter time, such as during a chromatographic run or the recording of an infrared spectrum. Extremely fast sampling rates, sometimes as high as 50 kHz, are necessary in mass spectrometry where scans lasting only a few seconds and covering a wide range of m/e values may be required. Similarly, sampling of the fid signal in FT.nmr (p. 411), where information over the complete spectral range is gathered in less than one second, needs to be at a frequency of several megahertz. Thus, the *conversion speed* of an ADC, which is defined as the time taken to sample the analogue signal once and output the digital equivalent, is as important a consideration as resolution in analytical applications. Typical conversion speeds are of the order 5 to 30 μsec.

If several instruments in a laboratory are to be interfaced with a computer, individual ADC units can be fitted to each one and the digitized signals from each transmitted to the computer. This has the advantage of minimizing the effect of electrical noise from outside sources, especially where the instruments and computer are not close together. An alternative arrangement is to use an *analogue multiplexer* which consists of a set of analogue input channels, switches and an ADC with a single digital output. The multiplexer can be controlled by the computer software to sample a single channel or to monitor a series of channels sequentially and at selected time intervals.

A further requirement of an instrument-computer interface and its controlling software is the provision of an accurate and stable *time-base*. This is necessary to ensure the synchronization of data collection with instrument scanning, to facilitate the exact matching of data from repetitive scans or runs, and for the computation of real-time experimental parameters such as retention times in chromatography. The generation of *interrupt* and *control* signals is also used in this context. Reference to the time-base allows interrupt signals to be sent at precise intervals under software control and is an essential feature of an analogue multiplexer system. Control signals passed from an instrument to the computer in the form of voltage pulses are also utilized in some instances.

To re-plot stored analytical data such as chromatograms and spectra on a chart-recorder, *digital-to-analogue converters* (DAC) are used. These are similar in construction and performance to an ADC but function in the reverse direction producing an analogue output signal varying between defined limits. A particular advantage of re-plotting data from an earlier run is the ability to produce a permanent record (*hard copy*) in a different or *enhanced* form. For example, a spectrum might be re-plotted after smoothing, scale-expansion or the subtraction of solvent peaks, or the first derivative of a titration curve might be drawn.

The transmission of data between an ADC or DAC and a computer is usually accomplished by sending one byte at a time along a multiway ribbon cable (*external bus*). The IEEE 488 standard bus has been widely used for this purpose, but other buses are also used. As all bits are handled simultaneously this is known as *parallel* I/O transmission, the data entering or leaving the computer via a *parallel port* or by direct connection to the computer's *internal bus*. Alternatively, one bit at a time can be transmitted along a single line (*serial* I/O transmission). Data enter or leave the computer via a *serial port*, usually an RS-232C. Serial I/O ports are used for communications between computers and various I/O devices (*peripherals*) such as electric typewriters,

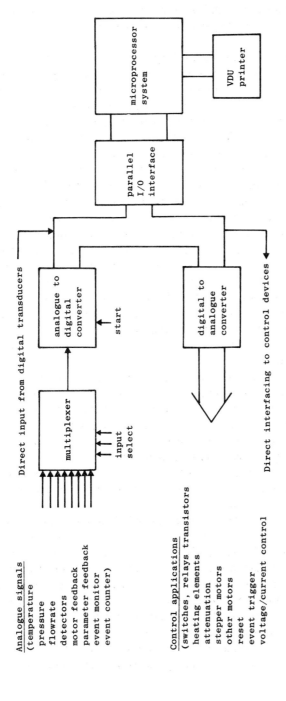

Figure 13.4.   A typical arrangement for interfacing a microprocessor system to laboratory instruments

printers, plotters and VDU's. Serial transmission is relatively slow and fast data processing is better accomplished using parallel transmission, provided that the distance between the various devices does not exceed a metre or so. A typical interfacing arrangement is shown in figure 13.4.

13.4   The Scope of Microprocessor Control and Computers in Analytical Laboratories

Because of the increasing importance of both microprocessors and computers in analytical chemistry, analytical chemists need an appreciation of their scope and limitations.

Most instrumental parameters can now be set and monitored continuously under the control of a microprocessor and employing a limited amount of memory. This facilitates the running of repetitive analyses with improved precision (although not necessarily with improved accuracy) and unattended operation which releases the analyst for other duties. An example of the degree of control available in a modern instrument is shown in figure 13.5.

For laboratories or groups of laboratories where computerized instrumentation is going to make a major contribution, the choice between a relatively large computer linked to many instruments and a series of mini- or microcomputers each *dedicated* to a single instrument or small group of instruments has to be made. Complex analytical instruments such as mass spectrometers and FT nmr spectrometers will continue to require their own dedicated computer to perform specialized data handling routines. The increasing power of microcomputers will enable more sophisticated data reduction and storage to be accomplished by individual instrument-computer combinations. However, there still remains the need for mass data storage such as in libraries of chromatograms or spectra or for quantitative results, for the overall management of large laboratories using LIMs and for communication between groups of laboratories, possibly dispersed throughout the world. These applications require large mini- and mainframe computers and a time-sharing system if individual instruments are to be interfaced. It must be recognized, however, that a central computing facility, whilst attractive in terms of standardization, ease and speed of communication and access to large data banks, can be inconvenient, if not disastrous, each time the computer breaks down. This eventuality is by no means uncommon and strengthens the case for a series of powerful microcomputers with dedicated tasks and independent of each other's transient operating problems. Local area networks (LANs) facilitate communication and data transfer between a group of instruments and microcomputers and allow them to access a mainframe computer when necessary.

(a)

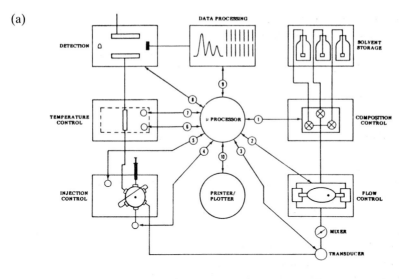

(b)

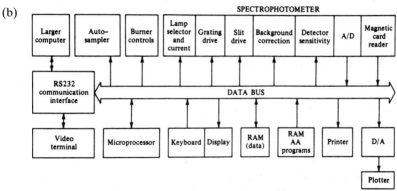

Figure 13.5.   (a) Schematic diagram of a microprocessor-controlled liquid chromatograph. (b) In-line microprocessor-controlled atomic absorption spectrophotometer
(address and control buses omitted for simplicity)

The following examples are intended to illustrate some of the current
capabilities of instruments under microprocessor control or interfaced to a
dedicated   microprocessor.   These   capabilities   are   becoming   more
sophisticated as microcomputers become more versatile and memory size
and speed increases.

1. A MICROPROCESSOR-CONTROLLED POTENTIOMETRIC TITRATOR

A typical instrument (figure 13.6) will have a certain amount of memory for

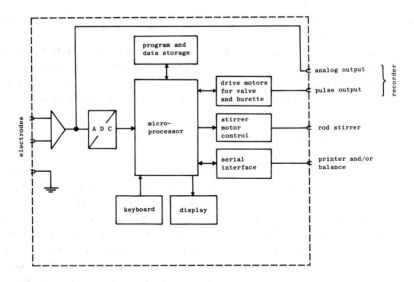

Figure 13.6.   Simplified block diagram of a microprocessor-controlled potentiometric titrator

the storage of methods and data, and will offer a choice of several modes of
titration, e.g.

   incremental (pre-selected increments and time intervals)
   equilibrium (pre-selected increments at variable time intervals)
   end-point (pre-determined end-point in mV or pX)
   recording (continuous, curve drawn on chart recorder)

Sets of parameters such as burette volume, reagent strength, increment size
and time interval, end-point potential, format of results, etc., can be stored
and recalled from memory as standard methods for routine analyses. An
alpha-numeric keyboard is used to enter or change the parameters, to take
individual pX or mV readings and to control the rinsing and the refilling of
the automatic burette. Raw titration data and computed analytical results
can be printed out as a permanent record, and titration curves can be
produced on a chart recorder.

   New methods can be created by automatic optimization of parameters
during a 'trial run' and all methods can be stored permanently in a *non-
volatile* area of memory which is preserved even when the instrument is
switched off. Some instruments provide a means of producing first and second
derivatives of the titration curve (p. 239) which can be advantageous where
the end-point is indistinct or there is more than one end-point to be detected.

Titrators with a substantial amount of RAM incorporate what is in effect a dedicated microcomputer.

Accessories are available for some instruments to enable photometric or voltammetric titrations (e.g. the Karl-Fischer determination of water) to be performed. Others allow the direct transfer of weighings from an electronic balance into RAM where they can be used in the computation of results.

2. AN INFRARED SPECTROMETER INTERFACED TO A DEDICATED MICROCOMPUTER

The microcomputer should incorporate a VDU and have *high density colour graphics* capability (up to $1280(x) \times 1024(y)$). This enables IR spectra to be displayed on the screen of the VDU with excellent definition so that comparisons, the results of scale expansions and other spectral manipulations can immediately be seen. Parameters such as range, scan time and data point interval, etc., are set and monitored under microprocessor control and stored along with the spectrum. During the scanning of a sample, several thousand data points may be collected and stored in RAM which should be able to accommodate and display at least three spectra simultaneously. There is a wide range of manipulations that can be performed by the analyst on stored spectra, e.g.

display of one or more spectra
overlay of two spectra
smoothing of a spectrum with a noisy baseline
flattening of a sloping baseline
conversion of a transmittance spectrum into absorbance and vice versa
averaging repetitive scans of the same sample
enhancing a weak spectrum by signal averaging
addition or subtraction of two spectra
ordinate and abscissa scale expansion
transfer of a spectrum from RAM to disk and vice versa
re-plotting a spectrum in RAM onto a chart recorder
setting up and executing a *command file* consisting of a combination of the above operations

In addition to the above facilities which enable the analyst to save a considerable amount of time and to improve the quality of spectra, there is also the ability to store thousands of spectra on disk in a library of *peak tables*. Each table will consist of the wavenumbers of twenty or thirty of the most significant peaks in the spectrum together with the corresponding peak transmittance values. Several thousand tables can be stored on a single floppy disk and library searches can be conducted in a matter of seconds. After recording the spectrum of an unknown sample, a preliminary search to indicate possible structural features can be initiated. This may be followed by a complete search in which the peak table for the unknown is matched with as many library tables as the analyst has available. The computer then displays a list of ten to fifteen possible compounds in order of closeness of match using a graded scale, e.g. 0 to 9.

3. A COMPUTING INTEGRATOR FOR CHROMATOGRAPHIC ANALYSIS

The automatic recording of retention times, peak heights and areas, the identification of unknowns and quantitative analysis are the main computational requirements of a computing integrator. The sensing of the beginnings and endings of peaks (*peak detection*) and synchronization of the start of a chromatographic run with the computer's own time-base to facilitate the recording of retention times are both essential capabilities. Although a standard means of peak detection makes for a high degree of uniformity in the treatment of successive chromatograms, the Gaussian nature of peaks, the presence of 'tailing' or 'fronting' and the possibility of overlap cause additional problems. The software must be able to distinguish between baseline or baseline drift and noise and the start of a genuine peak. This is generally achieved by continuous monitoring of the slope and/or setting a *threshold* value for the analogue signal from the detector. Slope sensing also allows peak maxima, shoulders and troughs between partially resolved peaks to be detected.

Most computing integrators allow the user to select values for parameters such as threshold levels for peak detection, sampling rate during the chromatographic run and the best geometrical/mathematical procedure to be used for the calculation of peak areas. Sampling rates should be as high as 50 Hz or more for sharp early eluting peaks and especially for those from a capillary gc column; later and broader peaks from packed columns can be sampled at rates below 10 Hz. Sampling rates may be fixed throughout a run or varied automatically with peak widths and under software control. A complete run of twenty minutes will require many thousands of data points which can be stored in RAM for immediate processing or transferred to disk.

Peaks are identified from absolute or relative retention times by comparison with data from previously run standards stored in RAM or in libraries on disk. To take account of the variability of retention times from successive runs, *retention time windows* are used. These are defined as being $t_R \pm x$ per cent for a standard, the unknown being positively identified if its retention time falls within the specified range. The size of the window can be varied by the user to conform with the degree of certainty required. Reference peaks can be selected for the calculation of relative retention times or as internal standards in quantitative analysis (pp. 10, 108).

Raw data and results may be printed out and chromatograms replotted with retention times, peak identities and quantitative information alongside each peak. With the increasing availability of high density graphics, the VDU has become an important component of many computing integrators. Thus data can be displayed in selected formats and complete annotated chromatograms presented as the data is collected. As in the case of computerized ir spectrometry (p. 545) the VDU also enables immediate enhancements and comparisons of two or three chromatograms to be made. Some systems include a multiplexer (p. 540) to allow the sequential sampling of several chromatographs, and *buffered interfaces* which can store complete chromatograms temporarily until such time as the computer can transfer them into RAM or on to disk.

4. A MICROCOMPUTER-BASED X-RAY OR γ-RAY SPECTROMETER

Modern X-ray and γ-ray spectrometers frequently employ semiconductor detectors which produce electrical pulses with sizes proportional to the energies of the incident photons. Using a fast ADC, the signal is digitized and a conventional spectrum generated by a dedicated microcomputer. As many as 8K data points (channels) may be sampled with facilities for perhaps four spectra stored in RAM, together with key information such as calibration data, accumulation time etc.

The very short duration of the pulses comprising the analogue signal from the detector, demands special attention to ADC design. A version which has become widely used is the Wilkinson type linear ramp converter. The popularity derives from very good linearity. Figure 13.7 illustrates the

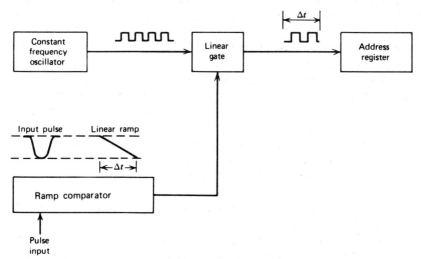

Figure 13.7.   Block diagram of a linear ramp (Wilkinson-type) ADC

essential parts of the ADC circuit. The detector signal is fed to a comparator circuit which responds to the arrival of a pulse by initiating the generation of a linearly increasing ramp voltage and signalling the opening of the linear gate. A continuous comparison of the signal pulse amplitude is made with the ramp voltage until the amplitude of the pulse is matched by the ramp voltage, at which point the linear gate is closed. Thus the linear gate is opened for a time (Δt) which is proportional to the amplitude of the signal pulse. During this time, pulses from a constant frequency clock are admitted to the address register which accumulates them, the number accumulated being proportional to the time the gate is opened and thus to the signal pulse amplitude. To minimize the conversion time, and maintain accurate digitization of the input signal, high clock frequencies are needed. Commercial ADCs are currently available with speeds of up to 400 MHz and are often characterized by quoting the speed. It should be emphasized that this is not the same as the conversion speed (p. 539) which remains in the range of 5–30 μs.

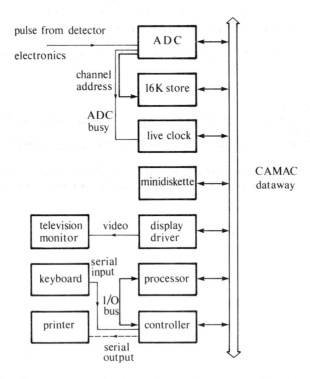

Figure 13.8. Arrangement of hardware for a microcomputer-based X-ray or γ-ray spectrometer

An indication of the flexibility imparted by computer control of such a spectrometer is given in the following list of typical facilities for spectra acquisition, display, storage and analysis.

Display of one or more spectra on a VDU
Overlay of spectra for comparison
Spectrum smoothing
Addition or subtraction of spectra
Peak search
Peak identification by reference to a library
Peak integration
Storage and recall of spectra on magnetic tape or discs
Provision of hard copy by $x-y$ plotter or printer
Energy calibration

Multiscaling which enables the decay of a nominated region of interest in the spectrum to be monitored with respect to time. Thus a half life may be obtained as an aid to isotope identification.
Storage and automatic resetting of instrument parameters
Correction of data for quantitative analysis

The principal hardware required to provide these facilities is shown in figure 13.8.

Further Reading
Analysis-82, Microcomputers in the Laboratory, (Conference), United Trade Press, London, 1982.
BARKER, P., *Computers in Analytical Chemistry*, Pergamon, Elmsford, New York, 1983.
RATZLAFF, K., *Computer-assisted Experimentation*, Wiley, New York, 1987.
STEVENSON, D. and MILLER, K., *Microprocessor Applications*, Wiley, Chichester, 1987.

Index

Absorbance 356
Absorptiometers 354
Absorption
  bands, intensity of 352
  bands, measurements, precision of 359
  bands, photoelectric 459
  broad band 320
  edges 336
Absorptivity, molar 356, 362
Accuracy 14
Acid 37
  conjugate 44
  dissociation constant 44
Addition, standard 32, 108
Adsorption systems 78
Agents, secondary complexing 202, 314
Allowed transitions 286
Alpha-particle spectrometry 463
Analogue to digital converter (ADC) 464, 539
Analyser 275, 427
  auto 521
  multi-channel pulse height 464
  quadruple 427
  single channel pulse height 464
Analysis
  flow injection 522
  frontal 87
  leadspace 103
  of a competitive product 513
  of hydrocarbon products in a catalytic reforming study 516
  pulse height 455, 463
  qualitative 104, 153, 287, 369, 384, 419
  quantitative 106, 153, 248, 287, 370, 391, 424
  repetitive, automation of 520
  sensitivity of neutron activation 471
Analytical
  problems and procedures, examples of 509
  procedures and their automation 504, 519
  scheme for a carburettor and combustion chamber cleaner 512
  standardization of analytical methods 31
Anode 223

Antigens 469
Antiserum 469
Application of statistical tests 21
Arc, AC 289
  DC 289
Assessment of analytical data 14
Association 62
Atmospheric pressure chemical ionization (APCI) interface 131
Atomic
  energy 270
  spectrometry 282
  structure and spectra 282
Atomic absorption
  analysis, graphite furnace for 325
  detection limits for 315
  spectrometry 315
Atomic emission
  sample preparation for 289
  spectroscopy detection limits 302
Atomic spectrometry
  spectrometer 288
  techniques based on 283
Attenuated total reflectance (ATR) 392
Auger effect 313
Autoradiography 455, 464
Auxochromes 365
  effect of 366

Bands, charge-transfer 369
Base 37
  conjugate 44
Bathochromic or red shift 365
Beer-Lambert Law 355
  derivation of 355
  deviations from the 357
  use of the 357
Bioluminescence 459
Boersma 486
Bombardment, fast atom (FAB) 426
  reactions and the growth of radioactivity 453
Bonded-phase chromatography (BPC) 138
Broadening 317
Buffer
  and pH control 47
  solution 47

Calibration 8, 480
  curve 32
  curve-point method of temperature 480
  procedures 31
Calorimetric 486
Calorimetry
  and DTA 493
  applications of 493
  comparison of DSC and DTA for
    $CuSO_4 \cdot 5H_2O$ 492
  differential scanning (DSC) 489
  schematic curve 491
Capacity factor 81
Capillary gel electrophoresis (cge) 177
  electrophores 126
  isoelectric focusing (cief) 177
  zone electrophoresis (cze) 173
Carbon, graphitized 119
Carbon-13 nmr 408
Carrier-gas 88
Carriers
  impurity 462
  stream 522
Cathode 223
  hollow-cathode lamp 319, 321
Cell
  electrolysis 223
  electrolytic 223
  galvanic 223
  photoelectric 280
  photovoltaic or barrier-layer 280
  voltaic 223
Characteristics 246
Charts, quality control 29
Charts
  cusum 31
  Shewhart 30
Chelate
  effect 52
  extraction systems 66
  formation and stability of metal—
    EDTA 201
Chemical effects 327
  pathway studies 465
  shift 399
Chemical reactions in solution 38
Chemical shift reagents 407
Chemiluminescence 459
Chemometrics 6, 33, 529, 530
Chiral stationary phases 119
Chromatograph
  radio-gas 457
  schematic diagram of a gas 88
Chromatographic analysis, computing
  integrator for 546
  mechanisms 75
  optimization function (COF) see CRF
  response function (CRF) 132

Chromatography 75
  exclusion 160
  gas 87
  gas–liquid (glc) 88
  gas–solid (gsc) 88
  high-performance liquid 112
  ion 128, 139
  ion-exchange 154
  pyrolysis-gas 497
  supercritical fluid 143
  thin-layer 146
Chromophores 365
Cluster analysis 34
Coefficient
  distribution or partition 57
  mass absorption 336
Coincidence
  anti-coin circuits 463
  circuits 463
Colorimeters 354
Colorimetry 359
Columns 91, 118
  capillary (open tubular) 94
  guard 118
  microbore 118
  packed 91
  scavenger 118
  wall-coated open tubular or WCOT 94
Compact disc ROM (CD-ROM) 536
Complex 36
  chelated 67
  formation 51, 61
  metal 367
  non-chelated 67
Compton scattering 459
Computers 534
  microcomputer-based X-ray or γ-ray
    spectrometer 547
  mini and micro 536
Concentration polarization 225
Condensed spark 289
Cone
  inner 309
  outer 309
Confidence interval 19
Confidence level 18
Coning and quartering 505
Constants
  coupling 403
  dissociation 39
  force 379
  formation 39, 50
  monitoring and on-line analysis 524
  self-ionization 42
Continuous wave (CW) or scanning nmr
  spectrometer 395
Contour maps 126
  plots 125

Conversion, internal 452
Correlated Spectroscopy (COSY) 416
Coulometry 257, 258
  at constant potential 259
CPU 535, 536
Crystal, lithium drifted 462
Curie-point 480
  comparison of times 480
  pyrolysers 480
Current 244
  background 462
  constant 258
  diffusion 244, 245
  residual 244
  ring 401
Cutting-out and weighing 107 .
CW instrument 398

Data processing and data analysis
  (chemometrics) 529
Data recording and storage 529
Data rejection 18
  analysis of 18
  processing and analysis 529
  recording and storage 529
Decay
  alpha 449
  constant 452
  free induction 412
  gamma (isomeric transition) 482
  kinetics of decay reactions 482
  negatron 439
  reactions 439
De-coupling
  noise 410
  off-resonance 410
Definitions and basic concepts 14
Degree of disorder 40
Degree of freedom 14
Demasking 203
Densitometry, transmission 465
Derivatization 101
Desorption, field (FD) 426
Detection
  end point 186
  limits of 27
  of separated components 149, 168
  practical limit 28
  theoretical limits 28
Detectors 95, 123, 279
  amperometric 128
  capillary electrophoresis 171
  diode array 280
  electrochemical 128
  electron capture (ecd) 99
  flame ionization detector (fid) 97, 98
  flame photometric (fpd) 100
  gas ionization 455, 457

GC detector characteristics 96
  guard 464
  ionization 97
  nitrogen/phosphorus (NP) 98
  semi-conductor 461
  solid state 354
  thermal 280
  thermal conductivity 96, 97
Determinate 16
Development procedures 149
Deviation 15
  relative standard or coefficient of
    variation 16
  standard 15
Diamagnetic shielding 399
  anisotropy 401
Diffusion 82, 83
  molecular 84
Digestion 217
Digital to analogue converters (DAC)
  540
  computers, application of 520
Diode array uv/visible
  spectrophotometers 353
  detectors 280, 301
  spectrophotometers 126
Dipole moment 377
Direct-current plasma (DCP) source 298
Directed search 132
Disc 392
Discharge tubes, electrodeless 322
  of spark 426
Discrimination 91
Discrimination effect 89
Disk operating system 536
Dispersion, prism 276
  coefficient D 523
  diffraction grating 276
Displacement development 87
Distortionless Enhancement by
  Polarization Transfer (DEPT) 415
Distribution ratio 57
  Craig counter-current 71, 75
  discontinuous counter-current 71
Doppler effect 317
DOS 536
Double beam 275
DTA (differential thermal analysis) 452
  applications of 486
  schematic trace 484
  study of calcium oxalate
    $CaC_2O_4.H_2O$ 485, 486

Effects
  conjugation 365
  force constant 382
  inductive 383
  interelement 328

Effects (*contd*)
  isotope 438
  mass 381
  mesomeric (resonance) 366, 383
  solvent 367
Efficiency 82
Electrically heated 500
Electrochemical
  cell 223
  techniques 223
Electrode
  calomel 226
  counter 258
  dissolved oxygen 251, 252
  fluoride 234
  gas sensing 237
  glass 230–233
  J-type mercury pool 206
  liquid membrane 235
  membrane or ion-selective 229
  metallic indicator 228
  reference 226
  response and selectivity 229
  silver–silver chloride 227
  solid state 233
  standard hydrogen (SHE) 223
  systems 228
  working 258
Electrogravimetry 257
Electrokinetic injection 172
Electrolyte
  strong 37
  weak 37
Electromagnetic radiation 267
  adsorption and emission of 271
  spectrum 269
Electron
  impact (EI) 426
  primary 456
Electron beam, interaction of with a solid
  specimen 332
Electron probe microanalysis 333
Electroosmosis 167, 169
Eluotropic series 79, 80, 121
Elution 86
  gradient 87
  profile 86
  stepwise 87
Emission spectra 311
End point 185
Energy
  level diagram 284
  solvation 53
Enthalpy, ΔH 39
Entropy, ΔS 39
Environmental monitoring 473
Enzyme sensors
  applications of 253

  biochemical 251
  principles of 253
Equilibrium
  acid-base 43
  analytical reactions 39
  complexation 50
  conditional constants 40
  constants 38
  kinetic factors in 41
  solubility 53
  temperature effects on equilibrium
    constants 41
  weak acid and weak base 46
Error 14
  determinate 16
  indeterminate 16
  nature and origin of 16
  normal 73
  normal curves 17
Ethanol, determination of in beverages
  533
Ethylenediaminetetraacetic acid (EDTA)
  199
  titrations
    applications of EDTA titrations 227
    end point detection for EDTA
      titrations 224
    metal ion indicators for EDTA
      titrations 225
    potentiometric EDTA titrations
      226
Evaluation of results and methods 18
Exchangeable protons 407
Exchangers
  anion 155
  cation 156
  liquid-ion 157
Excitation 339
  electric charge 289
  laser 289
Excitation process, interferences and errors
  associated with 290
Exclusion 76
Expert systems 533, 534
Extraction
  batch 71
  continuous 71
  methods of 71
  of covalent, neutral molecules 60
  of ion-association complexes 66
  of unchanged metal chelates 62
  selectivity of 58
  systems 59

F-test 20
F, critical values for 20
Filter photometry 275
Fission, spontaneous 449

Flame
    characteristics 308
    degree of ionization of elements in 316
    emission 315
    photometry, applications of 314
    processes 310
Flash vaporizer 89
Floptical disk 536
Flow control 88
Fluorescence 271, 315, 373
    detection 171
    detectors 126
    secondary 336
    yield factor 335
Fluoride, determination in water samples
  510
Fluorimetry 126, 362
Formula, empirical 430, 432
Fourier transform (FT) infrared
  spectrometer 111
Fragmentation, general rules 431
Franck-Condon principle 362
Fronting 77, 93
Fundamental 379
Furnace, micro 500

Gaussian
    concentration profile 75, 76
    curve 73
    distribution 17
    profile 85
    shaped peak 82
GC-infrared spectrometry 111
GC-mass spectrometry 108
Geiger–Muller detector 456
Geiger region 457
Gels, structure and properties of 161
    hydrophilic 161
    hydrophobic 162
Geometric methods 107
Gibbs free energy, ΔG 39
Globulins 469
Glossary of terms 7
Graphitized carbon 119
Gravimetry 184, 211
    applications of 221
    practical procedures 218
    thermo 477
Grob 91
Group frequencies 382
Groups, adjacent 382

Half-cell reactions 223
Half-life, $t_{\frac{1}{2}}$ 453
Haptens 469
Headspace 103
Heat flux DSC 492
Heavy metal pollution 514

Heteronuclear Chemical Shift Correlation
  (HETCOR) 417
Heteronuclear Correlated Spectroscopy
  (HETEROCOSY) 417
HETP 82
High performance capillary electrophoresis
  (hpce) 168
Holes, positive 461
Homologous pair 292
Homonuclear Correlation Spectroscopy
  (HOMOCOR-2D) 416
Hplc-ms 130
Hydrodynamic injection 172
Hydrogen bonding 382
Hydrolysis 48
'Hyphenated' techniques 108, 129
Hypsochromic 366

ICP-MS 472
Immunogens 469
Incredible Natural Abundance Double
  Quantum Transfer Experiment (2-D
  INADEQUATE) 418
Indicator 135
    redox 195
    selection of visual ind. for a redox
      titration 196
    specific and self-indicating reagents 197
    visual 186, 188
Infrared
    microscopy 393
    near region (NIR) 391
    spectrometer 545
Injection
    on-column 89, 91
    sample system 89, 117
    split 89
    splitless 91
    system 103
Injectors
    automatic 91
    valve 117
Instrument optimization 528
    computer interfaces 537
Instrumentation 274, 308, 319, 338, 362,
  395, 425, 484, 491, 495, 497
Integration, automatic 107
Interferometers 278, 279
Internal conversion 373
Internal normalization 107
    standardization 108
Intersystem crossing 373
Ion
    common ion effect 214
    detection and recording system 427
    total ion current 110
    tropylium 434
Ion chromatography 139

Ion-exchange 76
  non-resinous ion-exchange materials 157
  properties of ion-exchange resins 153
  structure of ion-exchange resins 155
Ionic migration 165
  conductances 262
  effect of temperature, pH and ionic
    strength 166
Ionization
  chemical (CI) 426
  field (FI) 426
  gas 455
  secondary 457
Isoeluotropic 136
Isosbestic point 358
Isothermal operation 92
Isotope
  radio-dilution methods 466
  reverse dilution 466

Junction, p-n type 462, 463

Katharometer 96
K-electron capture 451
Kováts 105
Kováts retention index 105

Labelling 447
Laboratory Information and Management
  Systems (LIMS) 6, 533
Laboratory management 530
LANS 536
Laser microprobe 289
Ligand 37
  aminocarboxylic acid 209
  excitations within an organic 367
  multidentate 52
Light pipe 111, 393
Local area network (LAN) 531, 533
L'Vov shelf 324, 325

Masking 41, 203
Mass
  reduced 379
  transfer 83, 84, 92
Matrix
  effects 32
  matching 32
Maxwell–Boltzmann equation 272
Mean 14
Median 14
Method valuation 5
Micellar electrokinetic capillary
  chromatography (mecc or mekc) Mini-
  and microcomputers 536
Microprocessors 534, 536, 538
  -controlled atomic absorption
    spectrophotometer 543

  -controlled liquid chromatograph 543
  -controlled potentiometric titrator 544
Molecular energy 270
Molecules
  diatomic 377
  infrared spectrum of in solution 379
  polyatomic 379
Momentum, angular 395, 397, 398
Monitors, conductance 128
Mull 392
Multichannel pulse height analysis
  system 341
Multiple path term 83, 92
Multiplication, gas phase 457
Multipulse techniques
  1-D NMR 412
  2-D NMR 412

Negatrons 451
Nernst equation 225
Nernst partition law 57
Nitrogen rule 433
Nmr process 397
  carbon-13 408
  pulsed Fourier transform 411
Noise decoupling 410
Nuclear spin energy levels 398
  energy levels 448
  reactions 448
  structure 448
Nucleation
  heterogenous 216
  homogenous 216
Nucleides, stable and the stability line
  450

Off-resonance decoupling 410
Ohmic drop 224
Optimizing separation 132
Orbitals
  atomic 363
  molecular 363
Organic material, analysis of 506
Overall line width 318
Overlapping resolution map (ORM)
  136
Overloading 77
Overtone
  first 378
  second 378
Oxonium systems 69

Particle, size growth 92, 216
Particle-beam interface 130
Partition 76
  systems 80
Pauli exclusion principle 282
Peak slicing 111, 112

Peaks
  base 430
  endothermic (endotherms) 484
  exothermic (exotherms) 484
  isotope 431, 434
  molecular ion 431
  multiplet 402
Peptization 217
Peristaltic proportioning pump 521
pH
  complexation and solubility equilibria
    37
  effect 60
  of salt solutions 48
Phase
  bonded-phase packings 119
  chiral stationary phases 119
  mobile 79, 80, 88, 121, 152
  stationary 78, 80, 93, 118, 151
  vapour in spectra 111
Phosphorescence 373
Photoluminescence 373
Photometers
  filter 354
  UV/visible 123
Photomultiplier tube 280
Physical interference 327
Pirkle phases 119
pKa values 44
Plasma
  DC plasma jet 302
  inductively coupled plasma torch 304
  sample introduction for plasma sources
    305
Plasma mass spectrometry 283
Plate
  height 82
  number 82
PLOT 95
'p'-notation 38
Polarogram 244
Polarographic wave 244
Polarography 243
  applications of 256
  differential pulse 249
  linear sweep oscillographic
    polarography 248
  modes of operation used in 248
Polyatomic organic molecules 363
Potassium bromide 392
Potential
  activation overpotential 225
  activity dependence 225
  asymmetry 232
  concentration overpotential 225
  constant 258
  decomposition 244
  half-wave 245

liquid-junction 224
standard electrode 224
theoretical cell 224
Potentiometric, direct measurements 238
Potentiometry 227
  applications of 242
  direct measurements 238
  null-point 242
Potentiostat 247
Precipitants, inorganic 212
Precipitates
  purity of 217, 221
  rates of formation and particle growth
    215
  solubility of 214
  stoichiometry 220
Precipitation
  co- 217
  methods 212
  organic agents 213
  post 217
  reactions 212
Precision 15, 24
Pretreatment 509
Principal modes of hpce 173
  component analysis 34
Principle 429
Procedures, colorimetric and spectro-
  photometric 370
Proportional gas ionization detector
  456
Proteins, specific binding 459
Protons, exchangeable 407
Pulse damper 114
Pulsed Fourier transform nmr (FT-
  NMR) 411
  (FT) spectrometer 396
Pumps, constant flow reciprocating 114
Pyrolyser, controlled furnace-type 500
Pyrolysis GC 101
  wire 502

Quantitative measurements and
  interferences 326
Quantum efficiency factor (or yield) 374
Quantum numbers 282
Quenching 374, 457
  chemical 459
  colour 459

Radiation
  absorption of characteristic radiation
    317
  detection of emitted 292
  deuterium continuum radiation source
    320
  resonance 322
  sharpline radiation source 318

Radioactivity
  analysis 471
  instrumentation and measurement of
    455
  statistics of radioactive measurements
    465
Radiochemical, methods in analysis 447
Radioimmunoassay 466
  applications of 470
Radionucleides
  analytical uses of 465
  growth and decay of 454
RAM 535
Rapid scanning 125
Ratio
  isotope 432
  magnetogyric 397
  neutron to proton 449
Reagent
  chemical shift 407
  Zimmerman–Rheinhardt 198
Reductor
  Jones 197
  silver 197
Refractive index (RI) monitors 127
Region
  n-type 463
  p-type 463
Rejection quotient, Q 18
  critical values of 19
Relaxation 398
Releasing 314
Reliability of measurements 18
Representative samples and sample
    storage 504
Resolution, R 82, 84
  overlapping resolution map (ORM)
    136
Result, true 14
Retardation factor, R 81
Retention time, t_R 81
Reversible cell 223
R_f values 148
Ring current 401
Robotics, laboratory 526
Robots
  arm systems 526
  Cartesian 526
  cylindrical 526
ROM 536

Sample application 149
Sample concentration and clean-up
  solid phase extraction 506
Samples, biological 509
Sampling and sample pretreatment
    504
Sampling procedures 392

Scintillation 455
  counters 458
Scintillators 459
  liquid 458, 459
  secondary 459
  solid 458
Secondary fluorescence 336
Selection role 286
Selectivity 158
  ratio 233
Self ionization 42
Self-absorption 294, 313
Self-reversal of the line 294
SEM-EDAX 342
Separation
  factor 84
  optimizing a 132
  process 163
  techniques 55
Separator, jet-orifice 108, 109
Sequential basis 305
Sharp line sources 321
Significant figures 26
Silanizing 92
Silanol groups 79
Simplex lattice design 136
Simultaneous instruments 301
Smith–Hieftje technique 320
Solid phase extraction (spe) 506
Solid state 455
Solubility 220
  products 39, 54
Solutes, characterization of 80
Solvent effects 41
Solvents
  acidic 192
  basic 193
  delivery systems 113
  extraction 56
  in analytical chemistry 42
  ionizing 42
  non-ionizing 42, 43
  non-protonic 42
  protogenic 43
  protonic 42
  protophilic 43
  selectivity triangle 136
Sorption 76
  isotherms 76, 77, 93
  mechanisms 76
Spectra 429
Spectral lines, intensity of 272, 286
Spectrometers
  direct reading 289, 294
  quadrupole mass 427
Spectrometry
  analytical techniques 273
  applications of arc/spark emission 296

applications of mass 436
applications of plasma emission 306
applications of X-ray emission 343
arc/spark optical emission 287
atomic fluorescence 329
flame atomic emission 314
flame emission 311
γ-ray 459
glow discharge atomic emission 298
identification of organic compounds 436
inductively coupled plasma-mass (ICP-MS) 309
infrared 377
mass 425
nuclear magnetic resonance (nmr) 393
plasma emission 301
visible and ultraviolet 361
X-ray emission 330
X-ray for environmental monitoring 473
Spectrophotometers 123
diode array 126
diode array UV/visible 353
dispersive 123
double-beam recording IR 353
double-beam recording UV/visible 353
recording 352
Spectrophotometry 359
differential or precision 371
Spectrum
first-order 404
γ-spectrum 460
Spin quantum number 394
Spin-spin coupling 399, 401, 402
Splitting, second-order 404
Spread 15
Sputtering 322
Standards. recognized 32
Stream-splitter 110
Stripping voltammetry 243, 250
Structure, vibrational and rotational fine 362
Substoichiometric yield determination 466
Supercritical fluid extraction 509
Supersaturation, relative 216
Support, solid 92
Supporting medium 167
Surface adsorption 76
System, sample inlet 425

t, values of 19
t-factor 19
t-test 21
Tailing 77, 93
Take off angle 334

Temperature effects 85
programming 92
Tertiary fluorescence 336
Tetramethylsilane (TMS) 396, 401
Thermal desorption 103
Thermal randomization 462
analysis curve 475
analysis techniques 476
differential analysis (DTA) 482
events 475, 476
techniques 475
Thermobalances 479
Thermogram 475
Thermogravimetry (TG) 477
Thermomechanical analysis (TMA) 493
applications of 497
expansion study 495
Thermospray interface 130
Time, dead or paralysis 457
Titrand 185
Titrant 185
Titration 185
acid-base 187, 240
acid-base curves 190
acid-base in non-aqueous solvents 192
amperometric 243, 253, 256
applications of acid-base 192
applications of coulometric 261
applications of redox 197
bi-amperometric 254
complexometric 199
conductometric 261
coulometric 259
EDTA 208
photometric 371
potentiometric 239
precipitation 210, 240
redox 193, 241
visual indicators for redox 195
with complexing agents other than EDTA 207
Titrimetric reactions 185
Titrimetry 184
Trace metal techniques, characteristics of 515
Tracer studies 307
Transition probability 296
Transitions, d-d 367, 368
charge-transfer 367
Transmittance 356
Transport
convective 523
diffusional 523

van Deemter 83
Vaporization
by reduction and hydride generation 326

Vaporization (*contd*)
  flame 323
  flameless 324
  sample 323
Velocity, burning 310
Vibration
  characteristic frequencies 380
  coupled 384
  normal modes of 380
Vibrational relaxation 373
Volume
  adjusted retention volume 104
  comparison of retention volumes 104
  dead 81, 100
  relative adjusted retention volume, α_x 105
  retention volume, V_R 81
  void 81

WCOT 94

X-rays
  absorption and emission of 334
  analysis crystals used in dispersive 341
  emission of primary 331
  energy dispersive analysis of (EDAX) 342
  fluorescence spectrometry 335
  processes 330
  transitions in the molybdenum atom 334

Zeeman effect 320
Zone
  electrophoresis 164
  primary reaction 309
  secondary reaction 309